Study Guide

to accompany

GENERAL CHEMISTRY

SIXTH EDITION

Whitten • Davis • Peck

Raymond E. Davis

Professor of Chemistry
University of Texas, Austin

Saunders College Publishing
A Division of Harcourt College Publishers

Fort Worth Philadelphia San Diego New York Orlando Austin
San Antonio Toronto Montreal London Sydney Tokyo

Printed in the United States of America

ISBN 0-03-021232-4

012 202 7654321

PREFACE

The science of chemistry is a tremendously broad subject of study, with implications ranging from applications in the research laboratory, industry, medicine, and agriculture, through innumerable uses in our daily lives, to the challenging intellectual exercise of making sense of such a vast subject. The only way to make all of this a manageable study is to organize it. Therefore we systematize our observations, trying to see common features in many different experimental results; we use the resulting summaries of observed behavior, called scientific laws, to help us predict chemical and physical behavior in unknown or untested cases; and we try to understand our observations and their summaries in terms of broad concepts such as the atomic theory. The merging of literally millions of chemical and physical observations, along with theories to explain them, into the ever-growing science of chemistry is one of the grandest intellectual achievements of the human mind.

The biggest challenge for many students undertaking their first college-level study of chemistry is usually determining what, in the wealth of detailed information in a chapter, comprises the key central themes of that chapter, and what other material merely supports, explains, or exemplifies these main ideas. This Study Guide was prepared to assist students in their study of chemistry from the text *General Chemistry, Sixth Edition*, by Whitten, Davis, and Peck or *General Chemistry with Qualitative Analysis, Sixth Edition*, by the same authors. It is intended to serve as a supplement to lectures and text reading. In preparing this Study Guide, I have been guided by the belief that the primary functions of such a guide are similar to those of an effective teacher—to summarize, to focus study toward particular goals, to stimulate practice at applying concepts and sharpening skills, and to provide an assessment of how the study is progressing.

Each chapter contains five parts:

1. A **Chapter Summary**, which summarizes the main themes of the chapter, tying together the various main ideas of the chapters, and relating it to previous study and to topics to be encountered later.
2. A list of explicit **Study Goals**. Each includes references to appropriate sections in the text and to text exercises related to that goal.
3. A list of some **Important Terms** from the chapter. Many of these are in the Key Terms list in the text chapter, while others are important terms that first appeared in an earlier chapter.
4. A **Preliminary Test**, consisting of many questions (averaging more than 80 per chapter) of easy to moderate difficulty. These supply extensive practice at applying the terminology, basic concepts, and fundamental calculations of the chapter. They lay the necessary groundwork for practice with more difficult exercises from the textbook. Many students find this the most useful aspect of the Study Guide.
5. The **Answers to the Preliminary Test**, containing answers to *all* Preliminary Test questions, many with additional comments, reasoning, or stepwise solutions presented.

The following section, **TO THE STUDENT**, present suggestions to guide students in developing systematic, productive study habits, coordinating material from classroom, text reading, Study Guide, and homework.

My gratitude and my great respect are due to John Vondeling of Saunders College Publishing, who suggested to me the writing of the first edition of this Study Guide, and who has encouraged and guided me in all aspects of my involvement in the project. I am especially indebted to my friends, Professor Ken Whitten, Professor Larry Peck, and the late Professor Ken Gailey, for their cooperation, their enthusiasm, and their many helpful suggestions at all stages. I also thank the students in my many introductory chemistry classes; our study of chemistry together, their difficulties and successes, and their many questions and discussions have helped to provide the point of view from which the guide was written; their welcome comments have aided in developing the guide into its present form. I would appreciate further comments and suggestions from readers of this guide. I thank my wife, Sharon, for cheerfully typing major portions of this guide. Most of all, I again express my deepest appreciation to my family—Angela, Laura, and Brian, and especially Sharon—for the continued love, understanding, encouragement, and patience they have given me through yet another writing project.

Dedicated with love to the new generation

Ryan Hauser   *Kendall Hauser*

Stuart Davis

Olivia Kane   *Hannah Kane*   *Kevin Kane*

Grace Loera

TO THE STUDENT

The true understanding of chemistry is not just memorizing material, and you will not perform adequately in your chemistry course if you try just to *remember* everything that you hear or read. To be successful in your study of chemistry you must (1) *organize* the material as you study and (2) *practice applying* the concepts and skills you learn to particular problems or experiments. You should always keep these two goals in mind in any study session. You will find that your performance on examinations will be directly related to the amount of real practice that *you* have devoted to thinking about and applying the concepts, and to solving problems and answering questions. You will find that almost every stage of your study of chemistry will depend on a firm working knowledge of material that you have already covered. Do not fall behind in your study, and do not neglect a topic just because you have already taken the exam that covers it!

The most helpful way to study is to take the most active role you can. Just reading (your class notes, the text, examples, ...) rarely does as much good as getting involved—for instance, working the examples out yourself, or outlining a text chapter. Students often underestimate the great help that the act of *writing* something down can be in the learning process. This forces you to pay more attention than if you were just reading, and will help you to remember what you have written down.

The following study suggestions may be helpful to you:

1. You should always read a little ahead of the material that will be covered in class. Your instructor will probably not assume that you have completely understood the text material before class, but it is always helpful if you have read ahead at least lightly so you have some idea what is coming. This Study Guide contains a **Chapter Summary** of the main ideas and a list of definite **Study Goals** for each chapter. Even though you may not understand these in much detail before you hear lectures over the chapter, read them before class anyway. Then some of the words begin to be familiar to you, and you can recognize key ideas when they come up in class. For the same reasons, look in advance at the list of **Important Terms** in this guide so you can pay special attention when these come up in class.

2. Try to take class notes that are sufficient to remind you of the general development being presented in class. You will not be able, nor should you try, to write down everything that is said, or even everything that appears on the board or the screen. It is much more important to pay attention and think about the reasoning being presented.

3. Very soon after each class (the *same day* if possible) rewrite your class notes. This should be done not just to improve their legibility, but to expand them by adding material that was mentioned in class but that you did not have time to write down fully. As you rework your class notes, read the appropriate sections of the text (perhaps your instructor has suggested special pages, sections, or examples to supplement the class discussion). But do not *just* read—incorporate this material to improve your notes. As you do this, you should think carefully about what you are writing—do not just copy. Thus, you will be organizing the material in your mind, relating the various ways the same topics are explained in the lecture, in

the text, and in the Study Guide, and incorporating more examples than in class. Note two important points about this reworking of your notes: (a) it should be done as *soon as possible* after each class, while the material is still fresh in your mind, and (b) it involves *writing* about the material, which forces you to concentrate more effectively than if you just read.

At this stage, too, you should reread the **Chapter Summary** in the guide. The list of **Study Goals** that this guide contains will serve as a framework for organizing your study of the text and lecture notes, and will aid in pointing your study in the right direction. These goals emphasize both what are the central ideas of the chapter and what you should be accomplish with these ideas. So some of the Study Goals will say that you should "Know what is meant by ...," "Relate ... to ...," or "Understand ...," while others emphasize that you should "Be able to calculate ...," or "Know how to ... ." Each goal is accompanied by lists of related text sections and suggested Exercises from the end of the text chapter.

A good technique for summarizing the material is to pretend that you have to teach it to someone else. Write out your own notes, and go over what you would way to help someone else understand the concept or the problem. "Explain" it to yourself in your own words.

4. Refer to the *Saunders Interactive General Chemistry CD-ROM* as suggested in the margins of the textbook chapter. It contains many helpful illustrations, video clips, and practical applications.

5. As early as possible in your study of each chapter, you must get a working knowledge of the important new terms as they arise, and review terms from earlier chapter. Each chapter of this Study Guide contains a brief section entitled **Some Important Terms in This Chapter**. You should try to write down, in your own words, what these terms mean to you. Then look them up in the textbook. Many of them appear in the Key Terms list in the text, while you will have to find others in the chapter (or in preceding chapters). Try to improve your own definitions. Again *do not* just read the definitions or copy them from the text. Putting them into your own words, in your own writing, makes you think.

6. It is essential that you *use* the material as early and as extensively as possible. The best way to do this is to answer questions and to work problems that the concepts you are learning. One of the most dangerous mistakes you can make when you study is to read the question, then read the answer or solution, and then say to yourself, "Yes, I can do that." You are only fooling yourself—and it is better to find out during your study than on an exam that you really did not know how to work that problem. It is crucial to practice using the material as soon as possible after each class—do not wait until a day or two before the examination, or you will be missing something that you might need to know in order to understand the next class meeting! This important phase of your study should follow this general order:

(a) The **Preliminary Test** questions in this Study Guide. These short questions, averaging more than 80 per chapter, are designed to help you master the fundamental terminology, the basic concepts, and the main types of calculations from the chapter. At the end of the guide chapter, you will find **Answers to Preliminary Test Questions**, containing answers to *all* these questions. Many of these answers include comments, reasoning, stepwise solutions, or text study suggestions. These questions are usually easier than homework or questions in the main text, and they help you learn to use terms and concepts a little at a time. And *do not* just read the questions—write out the answers. On multiple choice questions, do not just find the correct answer and go on. Rather, try to understand *why* each wrong answer is wrong; this helps you to think about the underlying concepts from different viewpoints, and make study of the question four or five time as useful. Do not cut this preliminary portion of the study short—it is usually best to "overlearn" these basic terms and operations, because you will be using them in many ways in more difficult questions. Be sure you can walk before you try to run!

(b) Examples from the textbook. Be sure to work these out yourself. It might help to cover up the solution to the example and then uncover it a line at a time, after you have already tried to figure out the next step of the reasoning or calculations. Write down as fully as you can the reasoning for each step.

(c) Homework that your instructor may assign. These may be as simple as the Study Guide questions, or they may involve using several ideas in the same question. In any case, be systematic in answering the questions.

(d) Exercises at the end of the chapter in the Whitten, Gailey & Davis text. Some of these are quite simple, similar to the Study Guide or the homework; others may be moderately or very difficult. Many of these questions require you to combine and use several concepts in your reasoning, to see whether you really understand the material. Again, the **Study Goals** in this guide will point out related exercises from the text, helping you to focus your study. The answers to all even-numbered numerical problems are in the back of the text, and complete solutions to these are available in the *Student's Solutions Manual*, by Tang and Keeney-Kennicutt, which your instructor may wish to make available to you. Work the even-numbered exercises before the other numerical problems, so that you know when you have gotten the correct answer (but do not just start with the known answer and then work backwards). Then work other problems that are similar to these. A most productive technique for working problems is to write out comments for each stage of the reasoning, "explaining" to yourself how and why you did that step. This helps you to focus on how the concepts of the chapter are being applied to the problem.

7. Finally, when you are doing your final preparation for examinations or quizzes, the **Study Goals** and the **Preliminary Test** questions will point out topics that require further work.

Many students find it helpful to set aside a separate portion of the notebook for working problems and answering questions. At every stage of this problem working and question answering, be as thorough and systematic as possible. Write down *why* you are doing each step in solving a problem. This forces you into the discipline of thinking about what you are doing, so it will be easier to remember the next time even if the problem is worded differently. It this way, it will also be easier for you to review this material at exam time. As you do this, you will find that a wide variety of problems and questions actually involve only a few *central concepts*, but that these can be combined in many different ways. The more practice you have had at working and applying these concepts to specific situations, the better prepared you will be.

You may wish to modify this suggested study approach to suit your own learning style, and your instructor may have additional suggestions. Whatever you do, be systematic in your study, and take an active part in working problems—writing is always better than just reading. In this way, you can share in the excitement and enjoyment of the complex, useful, and fascinating subject of chemistry.

SUMMARY OF SUGGESTED STUDY STRATEGY

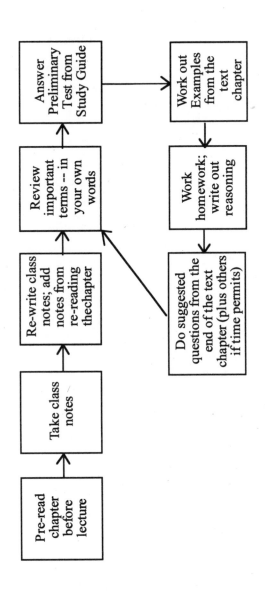

Pre-read chapter before lecture → Take class notes → Re-write class notes; add notes from re-reading the chapter → Review important terms -- in your own words → Answer Preliminary Test from Study Guide

Do suggested questions from the end of the text chapter (plus others if time permits) → Work homework; write out reasoning → Work out Examples from the text chapter

WRITE AT EVERY STAGE OF YOUR STUDY

CONTENTS

1 The Foundations of Chemistry

Chapter Summary

In our study of chemistry, we must manipulate numbers, quantitative ideas, words, and concepts. The main goals of this chapter are (1) to begin to learn what chemistry is about and the ways in which the science of chemistry views the world, and (2) to acquire some skill in manipulations that are useful and necessary in the understanding of chemistry.

We say "begin", because our progress in understanding chemistry, like any other complex subject, does not go in a single straight line. Rather we start with a few main ideas, perhaps oversimplified or understated at this stage. As we develop these ideas and expand them into others, we find that we must keep coming back to rethink these ideas. This sharpens our thinking and our understanding of what the ideas mean.

For example, very early in this chapter we encounter the idea of **energy**. This concept has a central role in our understanding of chemistry. Yet our first introduction to it (Section 1-1) seems rather formal and perhaps not too useful. If the authors told us now everything we will need to know about energy for our study of chemistry, it would be a very long and confusing chapter. Instead, we are told enough to get started. Our ideas of energy are then developed further and used at many stages throughout the book: later in Chapter 1, in the first introduction to heat as a form of energy; in Chapter 5, where we learn about light as a form of energy as we study atomic structure; in Chapters 12 and 13, to help explain the properties of gases, liquids, and solids; in Chapter 15, where study of energy changes helps us to predict whether a reaction can occur; and many other places. At each stage, we learn more about energy, and more about chemistry.

You should be willing to have your interest aroused in many of the topics that appear early in this first chapter—matter, energy, physical changes, chemical changes, and so on. You should not insist on a complete definition or understanding of all of these new concepts at the first encounter.

In the later sections of the chapter, 1-7 through 1-13, you will learn some of the very important quantitative skills that you will use throughout your study of chemistry. Your progress in studying chemistry will depend on how well you can perform the calculations you learn here. Developing these skills takes work and personal practice. It is easy to watch your instructor or a study partner work some problems and then to nod and say, "Yes, I can work that kind of problem." You probably cannot, until after *you* have worked such problems, and many of them, with less and less reliance on text, instructor, or study partner. Only then will you have the skill at problem solving that you will need in your further study of chemistry. And then you need to keep that skill sharp by frequent review and practice, just as you could keep a hard-earned ability to play a musical instrument or to ride a unicycle only by further practice.

The chapter opens with some questions that indicate the vast scope of chemistry, a science that touches every aspect of our lives. Section 1-1 then introduces the two central concerns of chemistry—matter and energy. Early in Chapter 1 we read that chemistry is the science that describes matter—its chemical and physical properties, the chemical and physical changes that it undergoes, and the energy changes that accompany those processes. We need to learn what such words mean. We summarize our experience in various statements known as **laws**. It is important to realize that in science a law is not an arbitrary rule—rather it is a statement that summarizes what is common among a large number of observations. Then we use that generalization to predict the outcome of further experiments. We learn here some important ideas that we will use often throughout our study of chemistry—the Law of Conservation of Matter (to be used, for example, in Chapters 2 and 3) and the Law of Conservation of Energy (a central idea in Chapter 15).

Matter exists in one of three **states**—**gas, liquid,** and **solid** (Section 1-2). We shall study this topic in much more detail in Chapters 12 and 13. We describe any sample of matter in terms of its **properties** (Sections 1-3 and 1-4). These may be **chemical properties**, involving the change of the substance into a different substance (a **chemical reaction**), or **physical properties**, not involving such a reaction. We learn (Section 1-5) that any sample of matter may be either a **substance** or a **mixture**. Any specimen of a substance has identical properties to those of any other sample of the same substance. A mixture has properties that can vary with gradually differing composition. (In Chapter 14, we shall study the composition and properties of a **solution**, which is a particular kind of mixture.) Further, a substance may be either an **element** or a **compound**. The elements are the approximately 100 substances that cannot be decomposed into simpler stable substances. A compound is made up of two or more different elements in a **fixed ratio**.

The rest of the chapter concerns some basic skills that we must master before further study. Sections 1-6 and 1-7 deal with the systems of units that we use to express the results of measurements—the **metric system** and the related **SI system** of units. Each system of units arbitrarily defines a certain amount of the quantity we wish to describe (distance, mass, volume, time, energy, etc.) to correspond to a particular unit. Once we accept the meaning of each of these units, we still may wish to alter the size of units so as to keep numbers to a manageable size. In our English system of measurements, we prefer to describe some lengths or distances in inches whereas others are more conveniently (or more conventionally) described in feet, yards, miles, or light-years. In the metric or SI systems we also related units of different magnitudes, but these related units can be described by the same set of prefixes applied to any basic unit. Each prefix alters the unit by multiplying or dividing it by some power of ten. This makes conversion within the system very easy—just a matter of shifting the decimal point.

Two aspects of dealing with numbers are introduced in Section 1-8. **Scientific** or **exponential notation** helps us to represent large or small numbers in a convenient way. **Significant figures** indicate how well we know a quantity we have measured or one we have calculated from a measured quantity. The **unit factor method** (Section 1-9) is a simple method for converting from any unit to any other related one. This method is explained, with many examples. Practice this method until it is an easily used and reliable tool. Section 1-10 presents percentage as a unit factor calculation.

The remainder of the chapter (Sections 1-11 through 1-13) introduces you to other important quantities such as density, specific gravity, heat, and temperature. Here you will learn some calculations related to these quantities.

IMPORTANT. In your study of Sections 1-7 through 1-13, you develop and sharpen some tools and skills that you will need for your further understanding of chemistry. Do not just learn to go through the motions in a prescribed way to work or answer a certain kind of question. Instead, keep in mind *why* you (or the textbook or your instructor) are approaching a problem in a particular way. In each example throughout the text, the reasoning or strategy is given first -- the **Plan**; then the detailed **Solution** is shown. Study the plan carefully so you understand how the general concepts are being applied to the specific question or problem to be answered.

Look at each question first to see what is *given* and what is *asked for*. This will often help you start on a question even if it is not a numerical problem. Writing down a list of these quantities, with units, is often helpful. Then try to recall a relationship you can use to relate the given to the unknown quantities -- that is, develop *your own plan* for answering the question. Sometimes this will require using two or more relationships in sequence. Several examples in Sections 1-7 through 1-13 illustrate this. Once you have found the right relationship, you might need to manipulate it before you carry out the arithmetic required. Sometimes you need to rearrange the relationship algebraically, to isolate the unknown quantity in terms of the known ones. It is always helpful to carry through the units on all quantities. Pay special attention to the *Problem Solving Tip* boxes that you find frequently in the text. They will not only alert you to common errors in problem solving, but they will often give you a new insight into some aspect of problem solving.

Finally, it is always a good idea to *think* about whether the answer you obtained makes sense. Suppose we had used conversion factors to calculate that 1 ft = 4.72 cm. If we recall that 1 cm is about the width of a little fingernail, it would be clear that the numerical result *must* be wrong—it obviously takes more than 4.72 human-fingernail widths to span 1 ft. (When we go back over our calculation, we find that we divided when we should have multiplied. We could have easily avoided this error if we had written units throughout the calculation.) Checking to see whether the answer is reasonable is usually not a guarantee that you have the correct answer. But frequently it can tell you that your answer is certainly an *incorrect* one. Then you can begin to search for your error.

Above all, you should discipline yourself to think about and work problems systematically. Do not just switch on your calculator when you see that a problem involves numbers. Think your way through the problem, at least in broad terms, to develop a plan before you begin solving the problem. Then write down the steps by which you will arrive at your answer before you start the actual computations, being careful to include units for each term. In this way, if you arrive at an incorrect answer, you can review your solution to the problem and perhaps find the mistake. If you get the correct answer, it will be easier for you to review and check the result, and then to review in a later study session. And when you are finished, think about the answer.

> The section entitled "To the Student" at the beginning of this Study Guide suggests how you might organize your study of chemistry. Begin that systematic approach with this important chapter.

Study Goals

These study goals will help you determine specific directions your study should take. Use them to help organize your class notes and text study. Each study goal refers you to related sections in *General Chemistry, Sixth Edition*, by Whitten, Davis, and Peck. When appropriate, some related exercises at the end of the main text chapter are suggested. In addition, you should always work some of the "Mixed Exercises" for each chapter, so you learn to recognize different types of questions.

1. State and use (a) the Law of Conservation of Matter, (b) the Law of Conservation of Energy, and (c) the Law of Conservation of Matter and Energy. Be able to express these in words other than those used in the text. Give examples of each. (Section 1-1; Exercises 3, 5, and 6)
2. Describe and distinguish among the general properties of gases, liquids, and solids. (Section 1-2; Exercise 4)

3. Define, distinguish among, and give examples of (a) matter, (b) energy, (c) mass, (d) a chemical change, (e) a physical change, (f) a chemical property, (g) a physical property, (h) an intensive property, and (i) an extensive property. (Sections 1-3 and 1-4; Exercises 2, 13 through 22)

4. Define, distinguish among, and give examples of (a) substances, (b) heterogeneous mixtures, (c) homogeneous mixtures, (d) elements, and (e) compounds. (Section 1-5; Exercises 7 through 12)

5. Know the names and symbols of the elements in Table 1-2. (Section 1-5)

6. Know the fundamental SI units in Table 1-4 and the meanings of the prefixes listed in Table 1-5. Know at least one conversion factor (Table 1-7) relating metric and English units of (a) mass, (b) length, and (c) volume. (Sections 1-6 and 1-7; Exercises 29 and 30)

7. Be familiar with conventions regarding exponential notation and significant figures. Apply them properly when doing mathematical operations. (Section 1-8; Exercises 23 through 28)

8. Know how to construct unit factors from equalities. Use these in calculations that involve conversions from one set of units to another (dimensional analysis). (Section 1-9 and 1-10; Exercises 30 through 35)

9. Distinguish between (a) mass and weight; (b) accuracy and precision; (c) density and specific gravity; and (d) heat and temperature. (Sections 1-7, 1-11 and 1-12; Exercises 1 and 2)

10. Carry out calculations relating density, specific gravity, mass, and volume. (Section 1-11; Exercises 36 through 43)

11. Relate the Celsius, Fahrenheit, and Kelvin temperature scales. Convert a specified temperature on one scale to the corresponding temperature on the other two scales. (Section 1-12; Exercises 44 through 52)

12. Distinguish between endothermic and exothermic changes. Be able to carry out calculations that relate heat capacity or specific heat to the heat transfer that accompanies temperature changes. (Section 1-13; Exercises 10, 53 through 58)

13. Use your understanding of this chapter to recognize and solve a variety of types of questions. (Mixed Exercises, 59 through 68)

14. Use your understanding of this chapter to answer conceptual questions, which often do not involve calculations. (Conceptual Exercises, 69 through 79)

Some Important Terms in This Chapter

IMPORTANT. Each new topic that you encounter in your study of chemistry will include some new terms. You must understand clearly what these new terms mean in order to understand explanations that use them. You will need these terms to communicate clearly when you discuss the subject matter, when you ask your instructor questions and when your give your answers on examinations. Here are some important new terms from Chapter 1. In the space provided, write *in your own words*, as carefully as possible, your understanding of what these words mean. *Do not just look them up and copy them.* After you have written your definitions, check the Key Terms list at the end of the text chapter for brief definitions. Also read the more detailed explanations and examples in the chapter and then try to write a better definition in your own words. These are only some of the most important terms in the chapter—pay attention to others you encounter in the Key Terms list or in your reading.

matter

mass

energy

Law of Conservation of Energy

Law of Conservation of Matter

Law of Conservation of Matter and Energy

kinetic energy

potential energy

physical change

chemical change

properties

physical property

chemical property

compound

element

heterogeneous mixture

homogeneous mixture

unit factor

density

specific gravity

heat

temperature

specific heat

heat capacity

Preliminary Test

Each chapter of this study guide contains questions in the form of a preliminary test. The questions in these tests involve the fundamental terminology and basic concepts. Many questions involve only one idea or one skill at a time. Use the questions of these preliminary tests as an initial check of your understanding of fundamental terms, basic concepts, and main types of problems in the chapter. When you do well on the preliminary test for a chapter, you are ready to proceed with further detailed study, assigned homework or problem solving from the text. You should also work many problems and answer many questions at the end of the chapter. This practice will give you the skills and confidence that are necessary to understand and apply the material of the chapter. Refer to the textbook for additional data. Table 1-7 will be a useful source of conversion factors.

True-False

Mark each statement as true (T) or false (F).

F 1. The Law of Conservation of Matter probably does not apply outside our solar system.

_____ 2. Although a sample of matter can change its kinetic energy, its potential energy is always fixed unless it undergoes a chemical reaction.

T 3. If we have two samples of matter, the one that is traveling at the higher speed has the higher kinetic energy.

_____ 4. For most substances, the gaseous state is less dense than is the liquid state.

_____ 5. For most substances, the solid state is more dense than is the liquid state.

_____ 6. When we do not stir a liquid its molecules are motionless.

_____ 7. To observe the chemical properties of a substance we must convert at least some of the substance into other substances.

_____ 8. A substance whose melting point is –7.1°C and whose boiling point is 58.8°C is a liquid at room temperature (about 21°C).

_____ 9. The substance referred to in Question 8 must be bromine.

_____ 10. When a liquid on a surface evaporates it removes some heat from the surface.

_____ 11. Most of the known elements actually occur in very small amounts on the earth.

_____ 12. Most of the naturally occurring elements occur in combination with other elements, rather than as free elements.

_____ 13. One of the elements that occurs in our atmosphere in considerable quantity is present in nature both as the free element and in compounds.

_____ 14. Different samples of a compound can have compositions that are different but only slightly so.

_____ 15. A mixture has properties that are similar to those of its component substances.

_____ 16. A compound has properties that are similar to those of its constituent elements.

_____ 17. Different parts of a solution have different properties.

_____ 18. A T-bone steak is an example of a heterogeneous mixture.

_____ 19. A cup of sweetened hot tea is an example of a heterogeneous mixture.

_____ 20. A glass of iced tea is an example of heterogeneous mixture.

_____ 21. To convert the mass of an object from grams to milligrams, we divide by 1000.

_____ 22. The weight of a body depends on where it is; its mass does not.

_____ 23. In the SI system of units, the meter (m) is more important than the kilogram (kg).

_____ 24. One gram is about the same mass as one ounce.
_____ 25. It is easy to tell, looking from across the room, the difference between a 2-L bottle and a 2-qt bottle.
_____ 26. Because 1 m = 39.37 in, the number of inches in 13.83 m is correctly described as 13.83 x 39.37 = 544.4871 in.
_____ 27. Because one dozen is 12 of the objects in question, the number of eggs in 16 dozen is written as 16 x 12 = 192 eggs.
_____ 28. Density and specific gravity are the same thing.
_____ 29. The mass of a 1.25-L sample of saturated brine solution, specific gravity 1.45, is 1.81 kg at 25°C.
_____ 30. When we add heat to an object, we always raise its temperature.
_____ 31. No matter what temperature scale we use, one degree is always the same increment in temperature.
_____ 32. In order to tell the specific heat of an object, it is sufficient to know of what substance the object is composed; in order to tell its heat capacity, it is necessary to know also how much of the substance we have.
_____ 33. When a piece of iron is allowed to rust, the resulting rust is observed to weigh more than the original piece of iron; this is an example of an exception to the Law of Conservation of Matter.

Short Answer

Answer with a word, a phrase, a formula, or a number with units as necessary.

1. An example of an exothermic process is _____.
2. Three examples of mixtures are _____, _____, and _____.
3. Three examples of elements are _____, _____, and _____
4. Three examples of compounds are _____, _____, and _____.
5. Arrange the following in order from smallest to largest: meter, kilometer, millimeter, centimeter, micrometer.
6. Of the units in Question 5, the one most convenient for measuring the distance from Philadelphia to Paris would be the _____, whereas the distance from home plate to first base on a baseball field would be more conveniently given in terms of the _____.
7. The distance from home plate to first base (90 ft), expressed in metric units, is _____. (Refer to Table 1-7 for needed conversion factors.)
8. The weight of my 2.8-kg dog, expressed in pounds, is _____.
9. The distance 221,463 miles, expressed in kilometers, is _____.
10. The mileage rating of an automobile that is capable of 27.5 miles per gallon could also be expressed as _____ km/L.
11. The number 221,463, expressed in scientific notation, is _____.
12. The number 0.000473, expressed in scientific notation, is _____.
13. Another name for scientific notation is _____.
14. In order to comply with a request to keep a room at 68.0°F, we would have to set a thermostat calibrated on the Celsius scale at _____.
15. The heat capacity of a 250-g block of metal whose specific heat is 0.220 J/g·°C is _____. (Be sure to supply units.)

8

16. The amount of heat necessary to raise the temperature of N grams of water by M degrees Celsius is _____. (Be sure to supply units.)

17. In trying to identify a sample of a pure substance, we make the following observations:
    (a) Its mass is 142.6 g.
    (b) It is a shiny solid at room temperature.
    (c) It is easily etched by hydrochloric acid.
    (d) It melts when heated to 540°C.
    (e) It is 24.4 cm long.
    (f) It is a good conductor of electricity.
    (g) It burns in air.
    Of these observations, the ones that would be helpful in identifying the substance of which the sample is composed are _____.

18. Of the properties listed in Question 17, _____ are chemical properties, whereas _____ are physical properties.

19. Milk of Magnesia contains magnesium hydroxide. Analysis of any pure sample of magnesium hydroxide shows that it consists of 41.7% Mg, 54.9% O and 3.4% H by mass. This information represents an example of the Law of _____.

20. The information given in Question 19 shows that magnesium hydroxide is a (an) _____.

21. The substance(s) listed in Table 1-1 of the text that is (are) gaseous at room temperature (about 21°C) is (are) _____.

22. Fill in the missing entries in the following table of elements.

Element Name	Element Symbol	Element Name	Element Symbol
hydrogen	H	phosphorus	P
Carbon	C	Potassium	K
Calcium	Ca	Sulfur	S
fluorine	Fl	zinc	Z
oxygen	O		Hg

Multiple Choice

Each multiple choice question in this guide has only one best answer. Select the best answer. If you use them properly, multiple choice questions that are not calculations can be especially valuable for study. Do not just look for the correct answer and then go on. Try to see what is wrong with each of the other responses or how the correct response is better than the others. This enables you to look at the ideas of the question from several different viewpoints. Experience with this viewpoint will be especially useful if you encounter multiple choice questions on an exam.

_____ 1. Based on all of the evidence available, we believe that the Law of Conservation of Matter and Energy is
    (a) always true.   (b) usually true.   (c) always false.   (d) usually false.

_____ 2. Which of the following is *not* an example of a compound?
    (a) table salt   (b) baking soda   (c) nitrogen   (d) carbon monoxide

_____ 3. Which of the following SI prefixes that corresponds to a multiplication factor of 1000?
    (a) milli-   (b) kilo-   (c) micro-   (d) centi-   (e) nano-

_____ 4. Which of the following units of volume is largest?
    (a) milliliter   (b) liter   (c) centiliter   (d) deciliter   (e) quart

_____ 5. Refer to Table 1-7 of the text. The conversion factor by which we would multiply a volume in quarts to convert it into milliliters is

$$\text{(a)} \ \frac{1.057 \ qt}{1 \ L} \times \frac{1 \ L}{1000 \ mL} \qquad \text{(b)} \ \frac{1.057 \ qt}{1 \ L} \times \frac{1000 \ mL}{1 \ L} \qquad \text{(c)} \ \frac{1 \ L}{1.057 \ qt} \times \frac{1 \ L}{1000 \ mL}$$

$$\text{(d)} \ \frac{1 \ L}{1.057 \ qt} \times \frac{1000 \ mL}{1 \ L}$$

_____ 6. Refer to Table 1-7 of the text. The conversion factor by which we would multiply an area in square miles (mi^2) to convert it to square kilometers (km^2) is

$$\text{(a)} \ \frac{2.588881 \ km^2}{1 \ mi^2} \qquad \text{(b)} \ \frac{2.589 \ km^2}{1 \ mi^2} \qquad \text{(c)} \ \frac{1 \ mi^2}{2.588881 \ km^2} \qquad \text{(d)} \ \frac{1 \ mi^2}{2.589 \ km^2}$$

_____ 7. The prefix "centi-" means
    (a) the same as the unit to which it is attached.
    (b) 1/1000 of the unit to which it is attached.
    (c) 1/100 of the unit to which it is attached.
    (d) 100 of the unit to which it is attached.
    (e) 1000 of the unit to which it is attached.

_____ 8. Two samples, each of them a pure substance, are found to have different melting points. Which of the following statements about the substances is true?
    (a) The two samples are certainly different pure substances.
    (b) The two samples are probably different pure substances, but we need more information to tell for sure.
    (c) The two substances are certain to have identical chemical formulas.
    (d) Both substances are certain to be compounds and not elements.
    (e) The two substances are certain to have different densities.

_____ 9. Which of the following processes is an example of a chemical change?
    (a) Fermentation of wine
    (b) Formation of salt by evaporation of sea water
    (c) Separation of a solid from a liquid by filtration
    (d) Condensation of moisture on a cold surface in a humid room
    (e) Writing on a piece of paper with a pencil

_____ 10. Which of the following is greatest?
    (a) 100.0 cm   (b) 0.0010 km   (c) 1.000 m   (d) 1000 mm
    (e) All of the preceding quantities are equivalent

_____ 11. The melting point (freezing point) of mercury is $-35°C$. What is the temperature, in degrees Fahrenheit, below which a mercury thermometer would not be usable because the mercury in it would be frozen?
    (a) 35°F   (b) −63°F   (c) −31°F   (d) −5.4°F

_____ 12. We know that air is a mixture and not a compound because
      (a) it can be heated.   (b) it can be compressed.   (c) it is colorless.
      (d) its composition can vary.

_____ 13. Which of the following is an extensive property?
      (a) density (b) temperature (c) volume (d) color (e) reactivity with hydrochloric acid

Answers to Preliminary Test

Do not just look up answers. Think about why the answers that appear here are correct. Reread sections of the text, if necessary. Practice many of the suggested exercises in the textbook. Also work several of the Mixed Exercises in the text, to gain practice at recognizing different types of questions.

True-False

1. False. We have never observed any exceptions to this law, and we believe that exceptions do not occur.
2. False. Its gravitational potential energy, for instance, depends on its vertical position.
3. False. Kinetic energy depends on both mass and speed. Remember that energy is the capacity for doing work. It might be easier to break a window with a very heavy rock thrown at low speed than with a very lightweight rock thrown at a higher speed.
4. True. (In fact, this is true for all substances.)
5. True. Water is one of the very few exceptions to this generalization.
6. False. Think about the diffusion of a drop of red dye throughout an unstirred glass of water. Try to relate what you study to your own experience.
7. True
8. True. Suppose we start with a sample of the substance at such a low temperature that it is a solid, and then warm it. When it reaches –7.1°C, it melts—i.e., the solid changes to liquid. Then we would need to heat the liquid to 58.8°C for it to boil. Thus, it is a liquid over the temperature range –7.1 to 58.8°C.
9. False. It may be (see Table 1-1), but we cannot be sure just from these two data. Perhaps there is another substance that coincidentally has the same melting and boiling points as bromine but with other very different properties.
10. True. For example, recall the cooling effect of the evaporation of perspiration.
11. True. See Table 1-3 and the accompanying discussion in Section 1-5.
12. True. Recall (Section 1-5) that only about one-fourth of the elements occur as free elements.
13. True. Oxygen occurs both as a free element and in compounds such as carbon dioxide.
14. False. A given compound always has the same composition of its elements (see Section 1-5).
15. True
16. False
17. False. A solution is a homogeneous mixture. Be sure you understand the difference between heterogeneous and homogeneous mixtures (Section 1-5).
18. True. Fat, bone, and lean meat are parts of a steak that obviously have different properties. See Section 1-5.
19. False. The sweetener and tea are completely dissolved in the hot water; this is a homogeneous mixture.

20. False. The solid ice and the liquid tea have different properties.
21. False. A milligram (1/1000 of a gram) is *smaller* than a gram; even if we had not learned about conversion factors, common sense tells us that the number of milligrams in an object must be *greater* than the number of grams, so we must multiply by a conversion factor *greater* than one.
22. True. See the discussion in Section 1-7. But we should keep in mind that a chemical reaction usually takes place at constant gravity. Therefore, weight relationships are just as useful as mass relationships for most applications in chemistry.
23. False. One (the meter) is the fundamental unit for the measurement of length, whereas the other (the kilogram) is the base unit for the measurement of mass.
24. False. Perhaps you have read in the grocery store that a 1-lb (16-oz) can of food contains 454 grams.
25. False. Many soft drinks are now being packaged in 2-liter bottles, which contain only about 6% more than 2 quarts.
26. False. Significant figures! The answer would be correctly written as 544.5 inches. See Section 1-8 in the text and pay attention to Exercises such as 23 through 26 at the end of the chapter.
27. True. These are all exact numbers, not obtained from measurement. See Section 1-9.
28. False. Density is the ratio of the mass of a sample of a substance to its volume Thus it has units expressed as mass/volume. It could have different values for the same object, depending on the units used to describe mass and volume. Specific gravity is the ratio of the density of a substance to that of water at the same temperature and is thus dimensionless. See Section 1-11.
29. True. Remember that the density of water is 1.00 g/mL at 25°C, so at that temperature density and specific gravity are numerically equal. Thus the density of the brine solution is 1.45 g/mL. Constructing the necessary unit factors,

$$\underline{?} \text{ g solution} = 1.25 \text{ L} \times \frac{1000 \text{ mL}}{1 \text{ L}} \times \frac{1.45 \text{ g solution}}{1 \text{ mL solution}} \times \frac{1 \text{ kg}}{1000 \text{ g}} = 1.81 \text{ kg}$$

The result of carrying out the computation on a calculator is 1.8125. Rounding to the correct number of significant figures gives the answer 1.81.
30. False. Consider, for example, that when we add some heat to ice at 0°C, we melt some of it to form water, but the temperature does not change. Read Section 1-4 and then study Section 1-12 carefully.
31. False. For instance, on the Celsius scale the difference between the freezing and boiling points of water is $100 - 0 = 100$ degrees, whereas measured on the Fahrenheit scale this *same difference* in temperature is described as $212 - 32 = 180$ degrees.
32. True. Specific heat is an intensive property, whereas heat capacity is an extensive property. Refer to Section 1-3 for a reminder of the terms "intensive" and "extensive", and then study Section 1-13 carefully.
33. False. The iron reacts with the moist air, and the increase in weight is due to something (oxygen) being added to the iron. There are no known violations of the Law of Conservation of Matter in ordinary chemical reactions. See Section 1-1.

Short Answer

1. Many valid examples come to mind. Any process in which energy (often heat) is given off to the surroundings (Section 1-13) could be cited. Burning a piece of metallic magnesium in air

(Section 1-1), combustion of a fuel, conversion of matter into energy as in a nuclear reaction (Section 1-1), and freezing of a liquid (Section 1-4) are among the many adequate answers.

2. Again, many answers would suffice. Among the ones mentioned in Chapter 1 are a solution of copper(II) sulfate in water (Figure 1-4c), a mixture of salt and sugar, a mixture of sand and salt, a mixture of iron powder and powdered sulfur (Figure 1-6), air, and vegetable soup.

3. There are presently more than 100 known answers to this question. Some of the elements mentioned in this chapter are hydrogen, oxygen, fluorine (note the spelling!), krypton, silicon, aluminum, iron, sulfur, and chlorine. How many others can you find mentioned in the chapter? Of how many elements have you seen a sample? You should memorize the list of elements and symbols in Table 1-2.

4. Now there are several million correct answers! Among those mentioned in the chapter are copper(II) sulfate, sodium chloride, water, and carbon dioxide.

5. micrometer, millimeter, centimeter, meter, kilometer. You should know the prefixes given in Table 1-5. Be sure that you know that list.

6. kilometer, meter

7. 27.43 m. The conversion is

$$? \text{ m} = 90 \text{ ft } \times \frac{12 \text{ in}}{1 \text{ ft}} \times \frac{1 \text{ m}}{39.37 \text{ in}} = 27.43 \text{ m}$$

Notice that the limitation on the number of significant figures is in the relationship 1 m = 39.37 in. One foot is exactly 12 inches, and the distance required is, by definition, exactly 90 ft.

8. 6.2 lb. $? \text{ lb} = 2.8 \text{ kg } \times \dfrac{1000 \text{ g}}{1 \text{ kg}} \times \dfrac{1 \text{ lb}}{453.6 \text{ g}} = 6.2 \text{ lb}$. Watch the significant figures. The number on the calculator was 6.1728395, but I certainly do not know the weight of my dog that precisely.

9. 3.563×10^5 km. $? \text{ km} = 221{,}463 \text{ mi } \times \dfrac{1.609 \text{ km}}{1 \text{ mi}} = 6.2 \text{ lb}$

10. 11.5 km/L. $? \text{ km/L} = 27.5 \dfrac{\text{mi}}{\text{gal}} \times \dfrac{1.609 \text{ km}}{1 \text{ mi}} \times \dfrac{1 \text{ gal}}{4 \text{ qt}} \times \dfrac{1.057 \text{ qt}}{1 \text{ L}} = 11.5 \text{ km/L}.$

11. 2.21463×10^5

12. 4.73×10^{-4}

13. exponential notation

14. 20.0°C. See how the rules for significant figures are applied in this case. See Section 1-12 and especially Example 1-16 for conversions of this type. Be sure to pay attention to problems such as Example 1-17, where two stages of conversion are necessary. Exercises 45 and 46 will provide useful practice.

15. 55.0 J/°C. See the definition of heat capacity of an object, Section 1-13.

$$? \text{ cal/°C} = 0.220 \frac{\text{J}}{\text{g·°C}} \times 250 \text{ g} = 55.0 \text{ J/°C}$$

16. $N \times M$ degrees Celsius. Exercises 53 through 58 the text give useful practice in calculations involving heat transfer, specific heat, and heat capacity.

17. (b), (c), (d), (f), and (g). These are properties that are characteristic of the *substance* under study. The other properties listed describe *how much* of the substance is present, but not the substance itself.

18. (c) and (g) are chemical properties; the others are physical properties.

19. Definite Proportions. This is also called the Law of Constant Composition (Section 1-5).
20. compound. A compound is a pure substance consisting of two or more different elements in a fixed ratio. See Section 1-5 for an introduction to compounds.
21. oxygen and methane. From the boiling points, we can see that only methane and oxygen are above their boiling points (and hence are gases) at 21°C. Oxygen is an element and methane is a compound composed of carbon and hydrogen.
22.

Element Name	Element Symbol	Element Name	Element Symbol
hydrogen	H	phosphorus	P
carbon	C	potassium	K
calcium	Ca	sulfur	S
fluorine	F	zinc	Zn
oxygen	O	mercury	Hg

Learn all of the element names and symbols in Table 1-2. Remember that single-letter element symbols are always capitalized; if the symbol consists of two or three letters, the first is capitalized and subsequent letters are always lower case.

Multiple Choice

1. (a)
2. (c)
3. (b). You should know the names and meanings of the prefixes in Table 1-5
4. (b). The action of each prefix is given in Table 1-5. See also Table 1-7.
5. (d)
6. (b). Be careful of significant figures.
7. (c)
8. (a)
9. (a)
10. (e)
11. (c). Carry out this conversion as shown in Section 1-12.
12. (d)
13. (c). See Section 1-3.

2 Chemical Formulas and Composition Stoichiometry

Chapter Summary

In this chapter we learn about the language that chemists use to describe substances. This language is based on the idea that all matter is composed of atoms. These atoms combine in various ways to form substances.

A chemical formula tells us (1) which elements are present in the compound and (2) the relative numbers of atoms of the elements involved. But we cannot handle or count individual atoms. How can we interpret chemical formulas in terms of some quantity we *can* measure, such as the masses of the elements present in the compound? What is the relationship between the mass of an element or compound and the number of atoms or molecules it contains? How can we use the chemical formula of a substance to tell us the masses of elements contained in that substance? How are chemical formulas determined from measurements of the amounts of the elements present in compounds? How do we describe samples that are not pure?

These are some of the questions that we address in this chapter. Many of the ideas mentioned here are probably new to you and might sound somewhat complicated. Do not worry about that—in your study of this chapter, try to progress steadily through the material. New material often uses the skills you should have learned earlier. Be sure to master the ideas and methods associated with one concept or one type of problem before you try to proceed. Calculations and principles of the types you will learn in this chapter are used throughout science. Nearly every chapter in this text, or in any course in chemistry, will use the ideas and skills you will develop in your study of this chapter.

Do not be alarmed by the obviously quantitative or mathematical nature of the chapter material. The math skills needed here are quite minimal—but you must develop your problem-solving ability. As you study, keep in mind the approaches in Chapter 1 and the study suggestions in "To the Student" at the beginning of this guide. Strive to keep aware of the reasons for approaching the problems as we do. Do not just try to learn to work "that kind of problem" from memory, but be aware of the concepts being used. And *think* about your answers.

The chapter opens with a description of the **Dalton atomic theory**, Section 2-1. According to this idea, matter is composed of extremely small particles called **atoms**. **Compounds** represent the combination of these atoms in definite ratios. A **molecule** is the smallest particle of an element or compound that can have a stable independent existence. We indicate the chemical composition of a substance by its **chemical formula** (Section 2-2). In this formula, we use subscripts to tell the number of atoms of each element in a molecule. The **structural formula** shows the order in which these atoms are connected together. Two other representations of molecules, the **ball-and-stick model** and the **space-filling model**, are also introduced in this section. The concept of chemical formulas also helps us understand such important chemical laws as the **Law of Definite Proportions (Constant Composition)**.

Many compounds consist of collections of **ions**—atoms or groups of atoms that possess an electrical charge (Section 2-3). **Cations** have positive charge and **anions** have negative charge. The formula for an **ionic compound** indicates the ratio of atoms (ions) that are present in the compound.

Ionic compounds do not contain molecules. They consist of extended arrays of the cations and anions in stable arrangements with each type of ion closest to ions of opposite charge.

Later we will learn systematic ways to name compounds. In this chapter, we begin that study (Section 2-4). Because there are so many compounds to deal with, you cannot learn the names and formulas for all the ones you will need. It is a good idea to memorize those that are shown in Table 2-1. You will gradually learn the names and formulas for many other compounds as you encounter them. We write the name or formula of an ionic compound by combining the names or the formulas of the individual ions. You should memorize the names, formulas, and charges of the common ions given in Table 2-2. Be sure you include the charge when you write the formula of any ion. We use these formulas and charges to write the formulas and names of many ionic substances. Any compound is electrically neutral, so the formula of an ionic compound can be written by finding the smallest ratio of ions that would balance each other's charge. Though the formula for an ionic compound usually does not include the charges on the ions, we must always keep them in mind.

Further study (Chapters 6 and 7) will enable you to predict which substances consist of molecules and which contain ions. Likewise, you will gradually acquire some knowledge as to whether a substance is likely to be a solid, a liquid, or a gas. Don't let this worry you at this stage. The main goal of this chapter is to learn how to obtain quantitative information from chemical formulas.

Section 2-5 introduces a topic that we use throughout chemistry. Experiments can measure relative atomic masses (usually called **atomic weights**). Then we put these on a convenient scale by convention. The numbers that we look up (for example, in the tables inside the front cover of the text) give us the relative masses (weights) of atoms of the elements.

One of the most useful ideas in chemistry, the **mole concept**, is introduced in Section 2-6. Just as we describe 12 of anything as a dozen, we use the word "mole" to describe 6.022×10^{23} of anything. This large number, called **Avogadro's number**, is the number of atoms in a mole of atoms, the number of molecules in a mole of molecules, or even the number of eggs in a mole of eggs. A mole of atoms or molecules is a convenient amount of a substance to work with in the laboratory. The important mole concept lets us relate something we *cannot* measure (the number of atoms or molecules in a sample) to something we *can* measure (the mass of the sample). Section 2-6 establishes the relationship between the atomic weight and the number of atoms in a given mass of an element.

Section 2-7 then extends this relationship to apply it to the masses of molecules (or more generally the masses of **formula units**) and the number of molecules or formula units in a given mass of a substance. We add the atomic weights of the atoms indicated by the formula to determine the **formula weight** of any substance. For substances consisting of molecules, the formula weight can also be called the **molecular weight**. One mole of the substance is then an amount of that substance whose mass, measured in grams, is numerically equal to its formula weight.

Putting the mole idea together with the concept of relative atomic weights, we develop a way of measuring out equal (or the desired relative) numbers of different atoms or molecules. As an analogy, suppose that we know that all apples are identical, and that an apple weighs one and one-half times as much as an orange, all of which are also identical. If we wanted to have a very large number of oranges and the same very large number of apples, we could avoid counting them out individually. We could just weigh out apples and oranges in a $\frac{3}{2}$-to-1 ratio by weight. For instance, 300 pounds of apples would contain the same number of individual fruits as would 200 pounds of oranges, or 90 tons of apples the same number as 60 tons of oranges. Thus, we have replaced the tedious (or in the atomic and molecular case, impossible) task of counting individual things by the much more convenient approach of weighing out the amounts needed. The rest of this chapter is based on this application of the mole concept. Be sure you understand the ideas of Section 2-6 and 2-7 and can apply them with ease before you proceed. The time you spend learning the mole concept, understanding and using the ideas of the mole, the formula unit, the formula weight, and the molecular weight, and practicing the associated calculations will be well repaid later. Exercises 22 through 45 at the end of the textbook chapter will provide you with much practice in applying the ideas of these two important sections.

Compounds are composed of more than one element. We sometimes need to know what fraction of the total mass of a sample is due to each type of atom present—questions such as "In 50 g of water, how much of the mass is due to hydrogen and how much is due to oxygen?" In Section 2-7, we learn how to interpret chemical formulas in this way. We commonly represent this information as the **percent composition** of the compound. (Remember that in chemistry, *percent* always is taken to mean *percent by mass*, unless we specify some other basis.) What we are actually doing is applying a generalization that is based on the analysis of thousands of compounds—the Law of Definite Proportions (Constant Composition).

Of course, sometimes we do not know the identity of the compound with which we are dealing. We can approach this problem by carrying out some experiments to find out which elements are present (**qualitative analysis**), and then other experiments to find out the amounts of the elements present (**quantitative analysis**). Using this information, we can determine the **simplest (empirical) formula** of the compound. This approach, which is discussed in Section 2-9, is the reverse of the calculation of percent composition that was dealt with in Section 2-8. Such analysis and deduction of formula is important in characterizing any new or unidentified compound.

Section 2-11 discusses how we determine the **molecular formula** of a compound. Often the amount of each element is found by converting it to a compound and then measuring the mass of the new compound formed. The molecular formula is always an integer multiple of the simplest formula. The **Law of Multiple Proportions** summarizes the relationship between compositions of different compounds composed of the same elements.

In Section 2-10, we learn about some other important interpretations of chemical formulas. As you can see, each of these is based on the concepts your learned in Section 2-6 through 2-10.

The final section of the chapter tells how we can describe samples that are not pure. When dealing with **percent purity**, we always mean percent *by mass*, unless we specify otherwise.

IMPORTANT. The concepts and types of problems in this chapter are basic to your further study of chemistry. Work many problems, and always think carefully about the concepts you are using, so you will build a sound foundation to go ahead.

Study Goals

Use these goals to help you organize your study. As in other chapters, some textbook Exercises related to each study goal are indicated.

1. Understand the postulates of the Dalton atomic theory. Distinguish between atoms and molecules. Know the names and symbols of the elements in Table 1-2. (Review Section 1-5; Section 2-1; Exercises 1 through 5)
2. Be able to interpret a chemical formula in terms of (a) the type and number of atoms present and (b) the relative masses of elements present. (Sections 2-1 and 2-2; Exercises 7 through 10, 46, and 47)
3. Know the names and formulas of the molecular compounds in Table 2-2. (Section 2-2; Exercises 9 and 10)

4. Describe ionic compounds. Know the names, formulas, and charges of the ions listed in Table 2-3. Be able to combine these to write names and formulas for some ionic compounds. (Sections 2-3 and 2-4; Exercises 11 through 21).

5. Use the concepts of mole, atomic weight, formula weight, and molecular weight to relate masses of substances to numbers of atoms, molecules, or ions present. (Sections 2-5 through 2-7; Exercises 22 through 45)

6. Given the formula of a compound, calculate its percentage composition. (Section 2-8; Exercises 46 through 50)

7. Given the elemental composition of a compound, determine the simplest formula of the compound. (Section 2-9; Exercises 51 through 53)

8. Distinguish between the simplest (empirical) formula and the molecular formula. Given information about molecular weight or the number of atoms in a molecule, determine its molecular formula. (Section 2-10; Exercises 54 through 56, 63 through 66)

9. Use data from combustion and similar analysis of a compound to find the amounts of elements present in the compound. (Section 2-10; Examples 57 through 62)

10. Summarize the laws of chemical combination: the Law of Conservation of Matter, the Law of Definite Proportions (Constant Composition), and the Law of Multiple Proportions. Be sure you understand how each of these is based on observation of chemical reactions. Give examples and be able to use these laws in calculations. (Sections 2-2 and 2-8 through 2-10; Exercises 67 through 70)

11. Apply the concepts of Sections 2-5 through 2-10 to carry out a variety of calculations based on chemical formulas. (Section 2-11, Exercises 71 through 78)

12. Describe and use information about the purity of a sample. (Section 2-11; Exercises 79 through 84)

13. Use the concepts of the chapter to recognize and solve a variety of types of questions. (Mixed Exercises, 85 through 88)

14. Answer conceptual questions based on this chapter. (Conceptual Exercises 100 through 105)

15. Apply concepts and skills from earlier chapters to the ideas of this chapter. (Building Your Knowledge exercises, 106 through 110)

Some Important Terms in This Chapter

Remember—try to write the meanings *in your own words*. Check the Key Terms list and the chapter reading. Then rewrite your definitions, still in your own words, to improve them. Do not overlook the other new terms in this chapter.

atom

molecule

ion

cation

anion

polyatomic ion

ionic compound

formula

allotrope

atomic mass unit (amu)

atomic weight

composition stoichiometry

formula weight

molecular weight

mole

Avogadro's number

formula unit

percent composition

simplest formula

molecular formula

% purity

Preliminary Test

As in other chapters, this test will check your understanding of basic concepts and types of calculations. Be sure to practice *many* of the additional textbook exercises, including those indicated with the Study Goals.

True-False

Mark each statement as true (T) or false (F).

_____ 1. In writing a chemical symbol, we can use either capital or lower case letters, as we prefer.
_____ 2. The chemical symbol for mercury is Me.
_____ 3. I have seen a mole of water, but I have never seen a molecule of water.
_____ 4. The term "atom" can apply only to an element, not to a compound.
_____ 5. The term "molecule" can apply only to a compound, not to an element.
_____ 6. Molecules always consist of more than one atom.
_____ 7. Elements that do not exist in stable form as single atoms are always diatomic.
_____ 8. A molecule of a compound must consist of at least two atoms.
_____ 9. When we say that the atomic weight of helium is 4.0, we mean that one atom of helium has a mass of 4.0 grams.

_____ 10. The number of oxygen atoms in 16 grams of oxygen is 6.022×10^{23}.

_____ 11. The number of oxygen molecules in 16.0 grams of oxygen is 6.022×10^{23}.

_____ 12. One mole of boron has a mass of $10.8 \times 6.022 \times 10^{23}$ grams.

_____ 13. If we know only the simplest formula of a compound and are given the required atomic weights, we can calculate the percent by mass of any element in the compound.

_____ 14. If we know only the formula of a molecular compound are given the required atomic weights, we can calculate the percent by mass of any element in the compound.

_____ 15. If we know only the percent composition by mass of a compound and are given the required atomic weights, we can calculate the simplest formula of the compound.

_____ 16. If we know only the percent composition by mass of a compound and are given the required atomic weights, we can calculate the molecular formula of the compound.

Short Answer

Answer with a word, a phrase, a formula, or a number with units as necessary.

1. The symbol for the reactive metallic element magnesium is _____.

2. The element whose symbol is Na is _____.

3. The number of hydrogen atoms in a molecule of tetraethyl lead, $Pb(C_2H_5)_4$, is _____.

4. Carbon and oxygen form two compounds. One is carbon dioxide, whose formula is _____, and the other is carbon monoxide, whose formula is _____.

5. The compound HNO_3 is called _____; the compound CH_4 is called _____.

6. The compound HCl is called _____; when it is dissolved in water, it is called _____.

7. The formula for the sodium ion is _____, the formula for the magnesium ion is _____, and the formula for the copper(II) ion is _____.

8. The name of Ag^+ is _____, the name of Al^{3+} is _____, and the name of Fe^{2+} is _____.

9. The formula for the chloride ion is _____, the formula for the hydroxide ion is _____, and the formula for the sulfate ion is _____.

10. The name of CH_3COO^- is _____, the name of O^{2-} is _____, and the name of PO_4^{3-} is _____.

11. The number of nitrogen molecules in one mole of N_2 is _____; this amount of nitrogen contains _____ N atoms.

12. One mole of hydrogen peroxide, H_2O_2, contains the same number of hydrogen atoms as _____ mole(s) of water.

13. One mole of hydrogen peroxide, H_2O_2, contains the same number of oxygen atoms as _____ mole(s) of water.

14. The number of oxygen atoms in n moles of a substance whose formula is $C_aH_bO_c$ could be expressed as _____.

15. In order to calculate the formula weight of a given compound from its known formula, we _____.

16. According to one theory of the origin of the universe, the age of the universe is about 10 billion years. The number of seconds that have elapsed since the origin of the universe would be _____ times Avogadro's number.

17. Suppose we calculate the masses of carbon that would combine with 1.0 g of oxygen in the two compounds carbon monoxide and carbon dioxide. We could then use these two results as an illustration of the Law of _____.

18. The chemical formula we derive only from information about percent composition of the elements in the compound is referred to as the _____ or the _____ of the compound.

19. In order to tell whether the simplest formula, as derived from percent composition data, corresponds to the molecular formula, we need to make some independent measurement or determination of the _____ of the compound.

20. A copper ore is 18.3% $CuFeS_2$. The mass of $CuFeS_2$ in 12.7 kg of the ore is _____ grams.

21. The mass of aluminum in a sample that is 8.52% Al_2O_3 (and contains no other source of aluminum) is _____ grams.

Multiple Choice

_____ 1. What is the number of oxygen atoms in one molecule of sulfur trioxide?
   (a) 1   (b) 2   (c) 3   (d) 4   (e) 5

_____ 2. Which of the following compounds has the highest formula weight?
   (a) water   (b) propane   (c) ammonia   (d) sulfuric acid
   (e) acetic acid

_____ 3. A sample of matter that contains three kinds of atoms could be
   (a) a solution.
   (b) a heterogeneous mixture.
   (c) a compound.
   (d) a homogeneous mixture.
   (e) Any of the other answers could be correct.

_____ 4. How many grams of hydrogen would combine with 12.0 grams of carbon to form the compound ethylene, C_2H_4?
   (a) 1.0 g   (b) 2.0 g   (c) 4.0 g   (d) 8.0 g   (e) 12.0 g

_____ 5. Which of the following is *not* a correct description of 16.0 g of methane, CH_4?
   (a) one mole of methane
   (b) the amount of methane that contains 12.0 g C
   (c) 16.0 x 6.022 x 10^{23} molecules of methane
   (d) the amount of methane that contains 4 x 6.022 x 10^{23} hydrogen atoms

_____ 6. Suppose we have 100 g of each of the following substances. Which sample contains the largest number of moles?
   (a) H_2O, formula weight 18.0
   (b) HCl, formula weight 36.5
   (c) $AlCl_3$, formula weight 133.3
   (d) $MgCO_3$, formula weight 84.3
   (e) impossible to tell, unless we know what reaction will take place with the substance.

_____ 7. An unknown element, (call it X), combines with oxygen to form a compound known to have formula XO_2. We observe that 24.0 g of element X combines exactly with 16.0 g of O to form this compound. What is the atomic weight of element X?
   (a) 48.0   (b) 32.0   (c) 24.0   (d) 16.0   (e) 12.0

_____ 8. What mass ratio of carbon to oxygen is contained in carbon dioxide, CO_2?
- (a) 0.375 g C/g O
- (b) 1.33 g C/g O
- (c) 2.67 g C/g O
- (d) 0.75 g C/g O
- (e) 0.67 g C/g O

_____ 9. For a molecular substance, which of the following statements is true?
- (a) The molecular formula must be the same as the simplest formula.
- (b) The molecular formula must be proportional to the empirical formula, but with all integer subscripts multiplied by an integer greater than 1.
- (c) The molecular formula must be proportional to the empirical formula, but with all integer subscripts divided by an integer greater than 1.
- (d) The molecular formula may be the same as the simplest formula.
- (e) The molecular formula and the simplest formula are not necessarily related.

_____ 10. The simplest (empirical) formula of a compound is known to be CH_2O. By another experiment, the molecular weight of this compound is found to be 180. How many hydrogen atoms are in one molecule of the compound?
- (a) 2
- (b) 6
- (c) 8
- (d) 12
- (e) 18

_____ 11. When a 1.50-gram sample of a compound that contains only C, H, and O was burned in a combustion analysis, 3.00 g of CO_2 and 1.23 g of H_2O were collected. The simplest formula for the compound is _____.

Answers to Preliminary Test

Do not just look up answers. Think about the reasons for the answers.

True-False

1. False. When a symbol consists of only one letter, it is always a capital letter. When it consists of two letters, the first is always capitalized and the second is always lower case. Even though this convention seems (and is!) arbitrary, like any convention of communication it must be applied in the same way by everyone who hopes to use it to communicate. For example, to a chemist CO cannot mean cobalt—it means a compound consisting of carbon and oxygen atoms in a 1:1 ratio.
2. False. Mercury is Hg. See Table 1-2.
3. True. Individual molecules are too small to see or measure, but 1 mole of water is 18 g, which is about 1.2 tablespoons of water. (Can you still carry out the conversions to show that 1 mole of water is 1.2 tablespoons? The density of water is 1.0 g/mL, there are 4 cups in a quart, there are 16 tablespoons in a cup, and there are more conversion factors in Table 1-7.)
4. True. See Section 2-1.
5. False. See Section 2-1.
6. False. See the descriptions of the noble gases in a marginal note on page 49.
7. False. P_4 and S_8 are two exceptions discussed in Section 2-1. See Figure 2-3.
8. True. A compound is, by definition, composed of two or more elements, and the atoms of each element are different from the atoms of any other element. See Section 2-1.
9. False. What we mean is that it has a mass of 4.0 arbitrary units on a scale relative to which a carbon atom has a mass of 12.0 units. This is a situation where thinking about the size of the number helps to avoid a wrong answer. A newly minted U.S. nickel has a mass of about 5 g—do you think a single helium atom has a mass of 4 g?

10. True. Sixteen grams of oxygen is 1 mole of oxygen atoms. One mole always contains Avogadro's number of atoms.
11. False. Oxygen is a diatomic substance (O_2), so 16.0 grams of oxygen, which contains 6.022×10^{23} oxygen atoms, would contain only half that many oxygen molecules, or 3.011×10^{23} O_2 molecules.
12. False. One mole of an element is the amount of the element whose mass in grams is numerically equal to the atomic weight. Thus, 1 mole of boron has a mass of 10.8 g. Think about the size of the number. We should remember that a mole of atoms of any substance is a convenient amount of material to handle in the laboratory. The number stated in the question, $10.8 \times 6.022 \times 10^{23}$ g, is larger than the mass of the earth!
13. True. See Section 2-8.
14. True. See Section 2-8.
15. True. See Section 2-9.
16. False. Review the distinction between the molecular formula and the simplest formula, Section 2-10.

Short Answer

1. Mg
2. sodium
3. 20
4. CO_2; CO. Know the names and formulas listed in Table 2-2.
5. nitric acid; methane. See Table 2-2.
6. hydrogen chloride; hydrochloric acid. See Table 2-2.
7. Na^+; Mg^{2+}; Cu^{2+}. You should know the names, formulas, and charges of the ions listed in Table 2-3. An ion formula written without the charge is incorrect.
8. silver ion; aluminum ion; iron(II) ion or ferrous ion. For some ion formulas, we indicate the charge by a Roman numeral suffix as part of the name, while for others this is not included. Naming ions will be covered more systematically in later chapters. For now, learn the list in Table 2-3.
9. Cl^-; OH^-; SO_4^{2-}. See Table 2-3.
10. acetate ion; oxide ion; phosphate ion.
11. 6.022×10^{23} N_2 molecules; $2 \times 6.022 \times 10^{23}$ or 1.204×10^{24} N atoms.
12. one
13. two
14. $n \times c \times 6.022 \times 10^{23}$ O atoms. There are 6.022×10^{23} molecules in a mole. Each molecule contains c O atoms.
15. add the atomic weights for the number and type of element given in the formula.
16. 5.2×10^{-7}, or 0.00000052. Put another way, when the universe is *1.9 million times* as old as it is now, Avogadro's number of seconds will have elapsed. This gives some notion of what a large number this is. Recall, however, that Avogadro's number of carbon atoms has a mass of 12 g. That number of water molecules is about 1.2 teaspoon. This gives you an idea how small atoms and molecules are!
17. Multiple Proportions. See Section 2-10.
18. simplest formula; empirical formula
19. molecular weight or formula weight

20. 2.32×10^3 g. Remember that percent can be interpreted as "parts per hundred." The answer can be calculated as $12.7 \text{ kg ore} \times \dfrac{1000 \text{ g ore}}{1 \text{ kg ore}} \times \dfrac{18.3 \text{ g CuFeS}_2}{100 \text{ g ore}}$.

21. 573 g Al. The calculation is set up as follows:

$$12.8 \text{ kg ore} \times \dfrac{1000 \text{ g sample}}{1 \text{ kg ore}} \times \dfrac{8.52 \text{ g Al}_2\text{O}_3}{100 \text{ g sample}} \times \dfrac{1 \text{ mol Al}_2\text{O}_3}{102.0 \text{ g Al}_2\text{O}_3}$$
$$\times \dfrac{2 \text{ mol Al atoms}}{1 \text{ mol Al}_2\text{O}_3} \times \dfrac{27.0 \text{ g Al}}{1 \text{ mol Al atoms}}$$

Multiple Choice

1. (c). You should know the names and formulas in Table 2-2.
2. (d). You need to know the names and formulas in Table 2-2. Then calculate formula weight for each compound. Formula weights are: $H_2O = 18.0$; $C_3H_8 = 44.0$; $NH_3 = 17.0$; $H_2SO_4 = 98.1$; $CH_3COOH = 60.0$.
3. (e)
4. (b). From the formula, there are (4 x 1.0 g H)/(2 x 12.0 g C) or 1 g H/6 g C. Thus, 12 g of carbon would require 2 g of hydrogen. Notice that we do not need the reaction by which the ethylene is formed in order to answer the question. It is immaterial that hydrogen gas is diatomic (H_2), because we are dealing with the mass of hydrogen.
5. (c). Because 16.0 grams is one mole of methane, it contains 6.022×10^{23} molecules of CH_4.
6. (a). We do not need to actually calculate the number of moles. For each compound, the number of moles is equal to the number of grams divided by the formula weight. Thus, for the same mass present, the compound with the lowest formula weight is present in the highest number of moles.
7. (a). According to the formula, 1 mole of X atoms would be combined with 2 moles of O atoms, or 32.0 g of O. We observe that 24.0 g of X combines with 16.0 g of O, so it would take 48.0 g of X to combine with 32.0 g of O at the same ratio. Hence the atomic weight of X is 48.0.
8. (a). If you got answer (c), you probably calculated the inverse of the requested ratio. If you got (d), you forgot that there are two O atoms for every C atom. If you got (b), you made the mistake in (d), and also calculated the inverse ratio.
9. (d). See Section 2-10.
10. (d). The formula weight of the simplest formula is 30, and the weight of the molecular formula is six times that. Thus, the molecular formula must be.
11. (a). Apply the methods of Section 2-10, as illustrated in Examples 2-15 and 2-16.

3 Chemical Equations and Reaction Stoichiometry

Chapter Summary

In Chapter 2, we were concerned with one compound at a time. In Chapter 3 we answer some very important questions about chemical reactions, including the following. How do we use symbols to represent a chemical reaction? How can we interpret chemical equations so that they tell us about the amounts of substances that can react with one another or that can be formed? How do we tell when a chemical reaction reaches completion? Many important reactions are carried out in solutions, so how can we conveniently express the amount of a substance present in a solution? How can we tell how much of a solution should be added for a reaction to occur to completion?

We first represent a reaction in a quantitative form that we call the **chemical equation** (Section 3-1). To be consistent with the Law of Conservation of Matter, we must write each equation in **balanced** form, that is, with equal numbers of atoms of all kinds on the two sides of the equation. (We will see later that when the equation involves ions, the total charges on the two sides of the equation must also be balanced. For the applications in this chapter, all substances are written as neutral elements or compounds.) We then interpret the balanced chemical equation to find out the relative amounts of **products** and **reactants** involved in the reaction.

All calculations about amounts of reactants and products involved in reactions are based on the following idea (Section 3-2) : *The coefficients in the balanced chemical equation tell the relative numbers of moles of those substances that react or are formed in the reaction.* From that starting point, we can "translate" the mole terminology into any other units desired—numbers of molecules, masses of substances, or even gas volumes (as in Chapter 12). In this very important section, you will learn how to construct **unit factors** to represent the chemical equivalence of two substances. This central idea can then be expanded to solve many questions about chemical reactions. One of the most important of these is the question of finding which reactant limits the progress of a reaction, the **limiting reactant concept** (Section 3-3). In Section 3-4, we learn to use the idea of **percent yield** to describe reaction conditions in which less than the theoretically possible amount of a product is isolated. You can apply the same ideas to impure reactant samples. Section 3-5 deals with **sequential reactions**, which take place in a series of steps or individual reactions. As you study these topics, keep in mind the central idea stated above, and try to see how all these apparently different kinds of problems are just various applications of that central idea. In this way, you will avoid the trap of just memorizing how to work certain kinds of problems.

There are many reasons why mixtures are involved in chemical reactions; when the mixtures are homogeneous, we call them **solutions**. We usually do not add pure substances to one another to carry out a reaction. Sometimes the reaction forms products that are mixed with (dissolved in) other substances. Many substances are dangerous or difficult to handle when they are pure, or may not even exist as pure substances, so we supply them in the form of mixtures. The use of solutions makes it very easy to measure the amount of a reactant to add to a reaction mixture. We can use a solution to measure a very small amount of a reactant more precisely than we could with the pure substance. The remainder of this chapter deals with solutions and their use in chemical reactions.

First we learn some methods for expressing concentrations of solutions (Section 3-6). A simple method of telling how much **solute** is contained in a given solution is in terms of **percent by mass**. (Remember that when we describe anything in chemistry in terms of percent, we mean "percent by mass" unless we specify otherwise.) As we have seen in this chapter, the relative amounts of substances involved in chemical reactions are most simply expressed in terms of moles. Further, it is more convenient to measure the volume than the mass of a liquid solution. **Molarity** relates the number of moles (which we need to know) to the volume of the solution (which we can easily measure). The calculations we need to do when we **dilute** a solution are discussed in Section 3-7.

We use the molarity method as the most convenient way of expressing the amount of a solution that contains the required amount of a substance for a chemical reaction. Calculations of this type, discussed in Section 3-8, are quite useful both in the laboratory and in practical applications. They will also be important in our discussion of the properties of solutions (Chapter 14), the rates of chemical reactions (Chapter 16), the extent to which reactions proceed before reaching equilibrium (Chapter 17 through 20), electrochemistry (Chapter 21), and many other topics.

Study Goals

1. Balance simple chemical equations. Know what information is contained in a balanced chemical equation. (Section 3-1; Exercises 1 through 7)
2. Use the concepts of this chapter to do calculations based on balanced chemical equations. (Section 3-2; Exercises 8 through 25)
3. Given the amounts of two or more reactants, determine which is the limiting reactant and which is (are) present in excess. Determine the extent of the reaction. (Section 3-3; Exercises 26 through 35)
4. Given the amount of a limiting reactant present in a chemical reaction and the amount of a product formed, determine the percent yield. Given the percent yield and one of the other two variables, calculate the missing quantity. (Section 3-4; Exercises 36 through 45)
5. Carry out calculations for a series of reactions or reaction steps. (Section 3-5; Exercises 46 through 54)
6. Relate the given amount of solute in a given amount of solvent or solution (a) to percent by mass of the solute and (b) to molarity. (Section 3-6; Exercises 55 through 69)
7. Relate the concentration and volume of a solution prepared by dilution to the concentration and volume of the concentrated solution from which it was prepared. (Section 3-7; Exercises 70 through 75)
8. Relate the amount of substance required or produced in a reaction to the concentration of a solution containing that substance. (Section 3-8; Exercises 87 through 93)
9. Recognize and work various kinds of problems from this chapter. (Mixed Exercises, 91 through 95)
10. Answer conceptual questions based on this chapter. (Conceptual Exercises 94 through 101)
11. Apply concepts and skills from earlier chapters to the ideas of this chapter. (Building Your Knowledge exercises, 102 through 108)

Some Important Terms in This Chapter

Remember—try to write the meanings *in your own words*. Then check the Key Terms list and the chapter reading. Then rewrite your definitions, still in your own words, to improve your understanding of them. Do not overlook the other new terms in the chapter.

balanced chemical equation

reactants

products

reaction stoichiometry

limiting reactant

theoretical yield

actual yield

percent yield

sequential reaction

solution

solvent

solute

percent by mass

molarity

dilution

Preliminary Test

As in other chapters, this test will check your understanding of basic concepts and types of calculations. Be sure to practice *many* of the additional textbook Exercises, including those indicated with the Study Goals.

True-False

Mark each statement as true (T) or false (F).

_____ 1. Before we can use a chemical equation to determine the amounts of different substances involved in the reaction, we must balance the equation.

_____ 2. In a balanced chemical equation, the total number of moles of reactants must equal the total number of moles of products.

_____ 3. The total number of grams of the reactants that react must equal the total the number of grams of products formed.

_____ 4. The total number of atoms in the reactants that react must equal the total number of atoms in the products.

_____ 5. The equation

$$A + 2B \rightarrow 3C + D$$

for the reaction involving the hypothetical substances A, B, C, and D, implies that the products C and D are always produced in a three-to-one mole ratio.

_____ 6. The equation shown in question 5 implies that in any reaction involving A and B as reactants, A must be the limiting reactant.

_____ 7. The equation $2Na + H_2O \rightarrow 2NaOH + H_2$ is balanced.

_____ 8. To balance the chemical equation

$$KClO_3 \rightarrow KCl + O_2$$

we change O_2 to O_3.

_____ 9. The balanced equation

$$Ca + 2H_2O \rightarrow Ca(OH)_2 + H_2$$

tells us that 1 mole of calcium atoms would react to give 1 mole of H_2.

_____ 10. According to the balanced equation in Question 9, 18 grams of water would be required to react with 40 grams of calcium.

_____ 11. When we are told the amounts of two different reactants that are put into a reaction mixture, we should assume that all of both reactants would be used up.

_____ 12. To tell how to prepare a 20.0% solution of a compound in water, we would have to know the formula of the compound.

_____ 13. To tell how to prepare a 0.25 M solution of a compound in water, we would have to know the formula of the compound.

_____ 14. If we wish to make a 1.00 M solution of sodium chloride (NaCl, formula weight 58.4), we would add 58.4 g of NaCl to 1.00 L of water.

Short Answer

Answer with a word, a phrase, a formula, or a number with units as necessary.

1. Balancing of a chemical equation is based on the Law of _____

2. When the equation

$$Ca(OH)_2 + HNO_3 \rightarrow Ca(NO_3)_2 + H_2O$$

is balanced, the coefficient of $Ca(OH)_2$ is _____.

3. When the equation

$$C_3H_8 + O_2 \rightarrow CO_2 + H_2O$$

is balanced, the coefficient of O_2 is _____.

4. When the equation

$$Al + CuSO_4 \rightarrow Al_2(SO_4)_3 + Cu$$

is balanced, the coefficient of Cu is _____.

- -

In questions 5 through 7, balance the equations with the smallest whole-number coefficients. Then find the sum of the coefficients in the balanced equation. Don't forget coefficients of 1.

5. $CS_2 + Cl_2 \rightarrow CCl_4 + S_2Cl_2$ _____

6. $C_7H_{16} + O_2 \rightarrow CO_2 + H_2O$ _____

7. $Ag + H_2S + O_2 \rightarrow Ag_2S + H_2O$ _____

- -

8. Balance the following chemical equation:

$$3Ca(OH)_2 + 2H_3PO_4 \rightarrow Ca_3(PO_4)_2 + 6H_2O$$

The unit factor that relates the number of moles of $Ca_3(PO_4)_2$ formed to the number of moles of $Ca(OH)_2$ reacting is _____.

9. Using the balanced chemical equation in Question 8, the unit factor that relates the mass of $Ca_3(PO_4)_2$ formed to the mass of $Ca(OH)_2$ reacting is _____.

10. Any chemical reaction must stop when one of the reactants is entirely consumed, even if some of the other reactants remain. The reactant that is consumed first is called the _____.

- -

We place 2.0 moles of CS_2 and 7.5 moles of O_2 in a closed container and burn them, according to the balanced equation

$$CS_2 + 3O_2 \rightarrow CO_2 + 2SO_2 .$$

The next four questions, 11 through 14, refer to this reaction.

11. The limiting reactant is _____.
12. When the reaction stops, the reactant left over is _____; there would be _____ moles of this reactant left.
13. The number of moles of CO_2 formed would be _____.
14. The number of moles of SO_2 formed would be _____.

- -

15. In the reaction in Question 8, the maximum mass of $Ca_3(PO_4)_2$ that can be prepared from 19.6 g of $Ca(OH)_2$ and 18.6 g of H_3PO_4 is _____ grams.
16. The maximum amount of a particular product that could be obtained from a reaction is called the _____.
17. The amount of a particular product that is actually isolated from a reaction is called the _____.
18. The percent yield of a chemical reaction is defined as _____.
19. The maximum percent yield that a reaction can give is _____.
20. If the reaction

$$P_4 + 6I_2 \rightarrow 4PI_3$$

gives a 76.3% yield, the mass of PI_3 that could be obtained from 250 g of I_2 and excess phosphorus is _____ grams.
21. In order to dilute a solution, we add more _____.
22. To make a 3.00 M HNO_3 solution, we would have to dilute 500 mL of a 12.0 M HNO_3 solution to a final volume of _____ mL.
23. The molarity of a solution that contains 0.02570 mol of solute in 518 mL of solution is _____.

24. Suppose we know the molarity of an aqueous solution of sulfuric acid (formula weight 98.1) and we wish to express the concentration as mass percent sulfuric acid. The one additional value that would allow us to calculate the conversion is the _____ of the solution.

Multiple Choice

_____ 1. Ethylene, C_2H_4, burns in oxygen to produce carbon dioxide and water. The correct form of the chemical equation that describes this reaction is

(a) $C_2H_4 + O \rightarrow CO_2 + H_2O$ (b) $2C_2H_4 + O_2 \rightarrow 2CO_2 + H_2O$

(c) $C_2H_4 + 2O_2 \rightarrow 2CO + 2H_2O$ (d) $C_2H_4 + 3O_2 \rightarrow 2CO_2 + 2H_2O$

_____ 2. In the reaction

$$CaCO_3 \rightarrow CaO + CO_2$$

100 grams of calcium carbonate, $CaCO_3$, is observed to decompose upon heating into 56 grams of calcium oxide, CaO, and 44 grams of carbon dioxide, CO_2. This illustrates

(a) the Law of Conservation of Matter.

(b) the Law of Conservation of Energy.

(c) the Law of Gravity.

(d) the Law of Multiple Proportions.

(e) the conversion of matter into energy.

_____ 3. If 5.0 grams of H_2 is reacted with 10.0 grams of O_2 in the reaction

$$2H_2 + O_2 \rightarrow 2H_2O$$

which would be the limiting reactant?

(a) H_2 (b) O_2 (c) H_2O

_____ 4. We place 8.0 g of S in a closed container with 14.0 g of O_2. These elements then react as completely as possible, by the equation

$$2S + 3O_2 \rightarrow 2SO_3$$

At the end of the reaction, how many grams of S and O_2, respectively, would be left over unreacted? (Use atomic weight S = 32, O = 16.)

(a) 2.0, 2.0   (b) 0.0, 6.0   (c) 2.0, 0.0   (d) 0.0, 2.0   (e) 0.0, 0.0

_____ 5. Two reactions that occur in the production of iron (Fe) are

$$2C + O_2 \rightarrow 2CO \quad \text{followed by} \quad Fe_2O_3 + 3CO \rightarrow 2Fe + 3CO_2.$$

Note that the carbon monoxide, CO, formed in the first reaction is then used in the second reaction. According to these equations, how many moles of oxygen, O_2, are required to produce 800 moles of Fe?

(a) 300   (b) 600   (c) 900   (d) 1200

(e) None of the preceding answers is correct.

_____ 6. What is the total mass of products formed when 3.2 grams of CH_4 is burned in air?

$$CH_4 + 2O_2 \rightarrow CO_2 + 2H_2O$$

(a) 16 g   (b) 36 g   (c) 44 g   (d) 80 g   (e) 160 g

_____ 7. What is the percent yield of elemental sulfur if 7.54 grams of sulfur are obtained from the reaction of 6.00 grams of SO_2 with an excess of H_2S?

$$2H_2S + SO_2 \rightarrow 2H_2O + 3S$$

(a) 76.1%   (b) 79.4%   (c) 83.7%   (d) 88.4%   (e) 91.4%

_____ 8. A 1.00% solution of NaCl (formula weight 58.4) weighing 200 grams contains how much NaCl?

(a) 1.00 moles (b) 1.00 grams (c) 0.584 moles (d) 0.584 grams (e) 2.00 grams

_____ 9. What is the molarity of a solution containing 3.50 grams of NaCl (formula weight 58.4) in 500 mL of water?

(a) 0.0598 M (b) 0.120 M (c) 0.0769 M (d) 0.0154 M (e) 0.600 M

_____ 10. We add 0.010 moles of $BaCl_2$ to enough water to make a total of 100 mL of solution. What is the molarity of the solution?

(a) 0.010 M (b) 0.10 M (c) 99.99 M (d) 0.0010 M

(e) Cannot be determined from the information given.

_____ 11. We wish to dilute some 18.4 M H_2SO_4 solution to make 600 mL of 0.100 M H_2SO_4 solution. How much of the 18.4 M solution should we start with?

(a) 1.8 mL (b) 2.7 mL (c) 3.3 mL (d) 4.0 mL (e) 4.6 mL

_____ 12. What mass of a solution that is 10.0% by mass H_2SO_4 will be required to react with 24.0 grams of NaOH according to the following equation?

$$H_2SO_4 + 2NaOH \rightarrow Na_2SO_4 + 2H_2O$$

(a) 29.4 g (b) 294 g (c) 586 g (d) 58.6 g (e) 1980 g

_____ 13. What volume of 3.00 molar sulfuric acid, H_2SO_4, is required to react with 250 grams of calcium carbonate, $CaCO_3$?

$$CaCO_3 + H_2SO_4 \rightarrow CaSO_4 + CO_2 + H_2O$$

(a) 208 mL (b) 462 mL (c) 767 mL (d) 833 mL (e) 946 mL

_____ 14. Refer to the balanced equation in Short Answer Question 8. If 20.0 mL of 0.010 M H_3PO_4 solution is just neutralized by 60.0 mL of a $Ca(OH)_2$ solution, what is the molarity of the $Ca(OH)_2$ solution?

(a) 0.010 M (b) 0.200 M (c) 0.030 M (d) 0.040 M (e) 0.0050 M

Answers to Preliminary Test

Do not just look up answers. Think about the reasons for the answers.

True-False

1. True
2. False
3. True. This is one way to state the Law of Conservation of Matter.
4. True. This also follows from the Law of Conservation of Matter.
5. True
6. False
7. False. Count the number of oxygen atoms on the two sides of this equation. Count the number of hydrogen atoms. You should be able to balance this equation.
8. False. We must not change the formula of any substance in the equation; this would mean changing the substances described. O_2 is ordinary oxygen, which we all breathe. O_3 is ozone, a

33

very reactive, poisonous gas; it is not produced in the reaction of $KClO_3$. The balanced equation is

$$2KClO_3 \rightarrow 2KCl + 3O_2$$

9. True
10. False. 36 g of water would be required.
11. False. Remember that the reactants can react only in the mole ratio indicated by the balanced equation, and therefore in a weight ratio that we can calculate. Study the limiting reactant type of problems in Section 3-3.
12. False. Remember that 20.0% is taken to mean "by mass". We add enough compound to the solvent so that 10.0%, or 1/5, of the mass of any amount of solution would be due to the compound.
13. True. This method of expressing solution concentration deals with the number of moles. But we cannot directly measure how many moles of solute we are adding. Knowing the formula weight of the solute, we can convert number of moles needed to number of grams.
14. False. 1.00 M means that the solution contains 1.00 moles of solute (NaCl) per liter of *solution* (not per liter of solvent). Study the description of Figure 3-2.

Short Answer

1. Conservation of Matter
2. 1. The balanced equation is $Ca(OH)_2 + 2HNO_3 \rightarrow Ca(NO_3)_2 + 2H_2O$.
3. 5. The balanced equation is $C_3H_8 + 5O_2 \rightarrow 3CO_2 + 4H_2O$.
4. 3. The balanced equation is $2Al + 3CuSO_4 \rightarrow Al_2(SO_4)_3 + 3Cu$. It may be convenient to treat SO_4 (actually the sulfate ion, SO_4^{2-}) as a single group in balancing this equation.
5. 6. The balanced equation is $CS_2 + 3Cl_2 \rightarrow CCl_4 + S_2Cl_2$
6. 27. The balanced equation is $C_7H_{16} + 11O_2 \rightarrow 7CO_2 + 8H_2O$
7. 11. The balanced equation is $4Ag + 2H_2S + O_2 \rightarrow 2Ag_2S + 2H_2O$
8. $\dfrac{1 \text{ mole } Ca_3(PO_4)_2}{3 \text{ mole } Ca(OH)_2}$
9. $\dfrac{310.3 \text{ g } Ca_3(PO_4)_2}{3 \times 74.1 \text{ g } Ca(OH)_2}$
10. limiting reactant. This concept is discussed in Section 3-3.
11. CS_2. Each mole of CS_2 requires 3 moles of O_2, so the 2.0 moles of CS_2 would require 6.0 moles of O_2. Extra O_2 is supplied, so CS_2 is the limiting reactant. Study Section 3-3.
12. O_2; 1.5
13. 2.0
14. 4.0
15. 27.4. $Ca(OH)_2$ is the limiting reactant (can you show this?).

$$? \text{ g } Ca_3(PO_4)_2 = 19.6 \text{ g } Ca(OH)_2 \times \frac{1 \text{ mol } Ca(OH)_2}{74.1 \text{ g } Ca(OH)_2} \times \frac{1 \text{ mol } Ca_3(PO_4)_2}{3 \text{ mol } Ca(OH)_2}$$

$$\times \frac{310.3 \text{ g } Ca_3(PO_4)_2}{1 \text{ mol } Ca_3(PO_4)_2}$$

$$= 27.4 \text{ g } Ca_3(PO_4)_2$$

16. theoretical yield

17. actual yield

18. $\dfrac{\text{actual yield of product}}{\text{theoretical yield of product}}$ x 100%. Study Section 3-4, and be sure you know how to carry out related calculations.

19. 100%. This would mean that the reaction went to completion as written, stopping only when one of the reactants was completely used up, that there were no side-reactions to consume reactants, and that we were able to recover *all* of the product. You can imagine that such laboratory procedures are few.

20. 206 g.

$$\text{theoretical yield PI}_3 = 250 \text{ g I}_2 \text{ x } \frac{1 \text{ mol I}_2}{254 \text{ g I}_2} \text{ x } \frac{4 \text{ mol PI}_3 \text{ formed}}{6 \text{ mol I}_2 \text{ that react}} \text{ x } \frac{412 \text{ g PI}_3 \text{ formed}}{1 \text{ mol PI}_3}$$
$$= 270 \text{ g PI}_3 \text{ formed}$$

$$\% \text{ yield} = \frac{\text{actual yield}}{\text{theoretical yield}} \text{ x } 100\%$$

$$\text{actual yield} = \frac{\% \text{ yield x theoretical yield}}{100 \%} = \frac{76.3 \% \text{ x } 270 \text{ g}}{100\%} = 206 \text{ g}$$

21. solvent.

22. 2000 mL. Calculate this as shown in Section 3-7, Example 3-20. It is not necessary to know the identity of the solvent to work this problem.

23. 0.0496 *M*. You do not need to know what the solute is. See Section 3-6.

24. density. See Example 3-19, which is a similar problem stated in a different way. Be sure you can work such problems. Work Exercises 68 and 69 at the end of the text chapter.

Multiple Choice

1. (d). Equation (a) shows oxygen as atoms instead of as diatomic molecules, and is not balanced. Equation (b) is not balanced. Equation (c) is balanced, but shows carbon monoxide (CO) rather than carbon dioxide as a product. The reaction in equation (c) could occur if there were a shortage of oxygen in the combustion process. Reaction (c) requires less O_2 per mole of C_2H_4 than does the "complete" combustion reaction to CO_2 given in (d). See Example 3-6 and the first marginal note "Sometimes it is not possible..." on page 90 for a brief discussion of burning in the presence of excess O_2.

2. (a)

3. (b). Find this out by the method shown in Section 3-3. Be sure to convert the mole interpretation of the balanced equation to grams.

4. (d). Sulfur is the limiting reactant, so it would be consumed first. Consuming 8.0 grams of sulfur by this reaction requires 12.0 grams of O_2, so there would be 2.0 grams O_2 left over. Study Section 3-3.

5. (b). Using the unit factor approach, you can show that 1200 moles of CO must go into the second reaction to produce 800 moles of Fe. For the first reaction to produce those 1200 moles of CO, 600 moles of O_2 would be required.

6. (a)

7. (c)

8. (e)

9. (b)
10. (b)
11. (c)
12. (b)
13. (d)
14. (e)

4 Some Types of Chemical Reactions

Chapter Summary

Any study of chemistry should include discussion of the reactions that elements and compounds undergo. This part of the study is termed "descriptive chemistry." There are more than a hundred elements and several million known compounds, each able to undergo many reactions. A study of this huge mass of information might seem hopeless. Fortunately, we can systematize this information in several ways to make it manageable. This chapter will help you start your study of descriptive chemistry by showing you several ways to organize the information and introducing you to some general terminology of descriptive chemistry.

We organize properties and reactions with the aid of the classification scheme known as the **periodic table**. We can use this table (1) to organize our understanding of many properties and (2) to predict properties or reactions of elements with which we may not be familiar or for which the property or reaction may not have been studied. The term "periodic" arises because elements with similar properties recur somewhat regularly as we go through the elements by increasing atomic number. As you study this chapter, you should become increasingly aware of the regularities that many properties of the elements exhibit and with the similarities within groups of elements. Later (Chapters 5 and 6) we shall relate the periodic table to the electronic structures of atoms.

The development of this simple yet powerful correlation, described in Section 4-1, is one of the major intellectual achievements in the history of science. This section also explains the terminology we use to discuss the periodic table. Each horizontal row is referred to as a **period** and each vertical column is called a **group**. Here you are also introduced to the common names of some of the groups of elements—the **alkali metals**, **alkaline earth metals**, **halogens**, and **noble gases**.

The elements are often broadly classified as **metals**, **nonmetals**, or **metalloids**. The general physical and chemical properties that distinguish among these classes of elements are also discussed in Section 4-1. Metalloids have properties that are intermediate between those of the other two classes.

Section 4-2 introduces you to many aspects of reactions and properties in **aqueous** solutions. Many substances dissolve in water to give solutions that contain ions. Due to the mobility of these ions, such solutions are able to conduct electricity; these substances are called **electrolytes**. They range in this behavior from **strong** to **weak**. Aqueous solutions of other substances, **nonelectrolytes**, do not conduct electricity to any appreciable extent.

The second part of Section 4-2 is an introduction to the nature of the important substances known as **acids** and **bases**. We shall study acids, bases and their reactions in more detail in Chapters 10, 11, 18, and 19. According to the view presented here, in aqueous solution an **acid** acts as a source of **hydrogen ions**, H^+, and a **base** is a source of **hydroxide ions**, OH^-. The fundamental reaction that acids undergo is **ionization**,

$$H_nX(aq) \rightarrow nH^+(aq) + X^{n-}(aq)$$

where n is usually 1, but sometimes 2 (e.g., sulfuric acid, H_2SO_4) or even 3 (e.g., phosphoric acid, H_3PO_4). Some acids undergo this reaction essentially completely in dilute aqueous solutions; these are referred to as **strong acids**. The seven common strong acids and their anions (H_nX and X^{n-} in the above general representation) are listed in Table 4-5. You must learn the names and formulas of these acids and their anions. Some other acids undergo this reaction only slightly in dilute aqueous solution; these are referred to as **weak acids**. These are much more numerous, but a brief list of common weak acids and their anions appears in Table 4-6. In solutions of strong acids, very little of the acid exists in solution, nearly all of it having been converted to the anion. In weak acid solutions, the acid exists almost entirely as nonionized acid molecules, with relatively little of the anion formed; this is an example of a reaction that is strongly **reversible**. This chapter introduces a color convention that we use for representing acids and bases throughout the book

Similarly, **bases**, which are compounds that produce OH^- in aqueous solutions, can be strong or weak. Many common bases are metal hydroxides. Most of these are not very soluble in water, so only a few common bases need to be considered. You must learn the list of **strong bases** in Table 4-7. The fundamental reaction for the dissociation of strong bases can be written as

$$M(OH)_n(s) \rightarrow M^{n+}(aq) + nOH^-(aq),$$

where n is an integer, 1 for the alkali metals (Group IA) and 2 for the alkaline earth metals (Group IIA). Ammonia, NH_3, is the only common **weak base**.

Do not confuse the terminology of "weak" and "strong" acids and bases with that of "dilute" and "concentrated" solutions. The terms "dilute" and "concentrated" refer to *how much* of a substance we put into the solution. We have already learned about two methods for describing the concentrations of solutions (Section 3-6). The terms "strong" and "weak" refer to the *fraction* of acid or base that ionizes according to the reaction to give H^+ or OH^-. In Chapter 18 we will learn about more quantitative ways to describe the strengths of acids and bases.

The guidelines that summarize the solubilities of various common types of compounds in water (Section 4-2 part 5, Table 4-8) are very important. You should learn these guidelines.

In Section 4-3, we learn several useful ways of writing chemical equations, depending on what feature of the reaction we wish to emphasize. The three ways are: (a) the **formula unit equation**, in which all compounds are represented by complete formulas; (b) the **total ionic equation**, in which the formulas are written to show the form in which each substance exists in solution; and (c) the **net ionic equation**, which includes only those species that actually react. In this last version, the **spectator ions** have been eliminated from the equation. This formalism for writing equations is useful for many types of reactions.

Many reactions involve electron transfer. A formal way of keeping track of electron transfer is by the tabulation of **oxidation number** or **oxidation state** (Section 4-4). The rules for assigning oxidation numbers to elements in compounds will be quite useful in later study. You will use oxidation numbers to recognize electron transfer reactions, to help balance chemical equations for such reactions (Chapter 11), to study the electron-transfer reactions that take place in batteries and in electrolysis cells (Chapter 21), and to systematize many kinds of reactions (Chapters 22 through 25).

The common oxidation states given in Table 4-10 should be learned. You should practice until you can assign oxidation numbers to all elements in any compound or ion whose formula you are given. Remember that the two terms *oxidation number* and *oxidation state* mean the same thing and are used interchangeably. Also be sure that you understand that the oxidation number is a quantity that is determined "per atom." It is represented in the text by the notation $+n$ or $-n$ in a circle directly above the element symbol. By contrast, the ionic charge is the actual net electrical charge (in units of electronic charge) on the total ion, even if the ion is polyatomic. Ionic charge is represented in the text by the notation $n+$ or $n-$ above and to the right of the formula for the ion.

Oxidation-reduction reactions (Section 4-5) are those in which some substances undergo changes in oxidation number. For short, these are often called **redox** reactions. It is essential that you learn the terminology—**oxidation, reduction, oxidizing agent, reducing agent**—that is presented early in the section. An oxidizing agent (1) gains (or appears to gain) electrons, (2) undergoes a decrease in oxidation number (that is, is reduced), and (3) oxidizes other substances. A reducing agent (1) loses (or appears to lose) electrons, (2) undergoes an increase in oxidation number (that is, is oxidized), and (3) reduces other substances. Oxidation cannot occur without a corresponding reduction, and vice versa. You have already encountered many oxidation-reduction reactions earlier in this chapter. Throughout the text, we use the code colors blue for oxidation and red for reduction as a memory aid (remember *red* for *red*uction). A reaction in which one element is both oxidized and reduced is termed a **disproportionation reaction.** Acid-base neutralization reactions are *not* oxidation-reduction reactions, but many of the other types of reactions we have seen are frequently also redox reactions. The study of electron-transfer reactions (electrochemistry, electrolysis, voltaic cells, batteries, etc.), which is the subject of Chapter 21, also involves oxidation-reduction reactions. In addition to being able to recognize such reactions, you should learn to identify which substance is oxidized and which is reduced. Be sure that you understand the terminology. We sometimes call the *compound* or the *polyatomic ion* that contains the element undergoing a *decrease* in oxidation number the oxidizing agent; the compound that contains the element undergoing an oxidation number *increase* is called the reducing agent.

Combination reactions, those in which two or more substances combine to form a single compound, are discussed in Section 4-6. This class of reactions can be further subdivided according to the nature of the reactants. Remember (Chapter 3) that the reactants are the substances that are consumed as the reaction proceeds (to the left of the reaction arrow) and the products are those substances being formed (to the right of the arrow). (1) In some such reactions, the two reactants are both elements. Typical reactions are the formation of **hydrides** by reaction of metals or nonmetals with hydrogen, formation of **oxides** by reaction of metals or nonmetals with oxygen, and other reactions of metals with nonmetals. (2) Some other combination reactions have an element and a compound as reactants. (3) Two compounds can combine to form a single new compound.

Decomposition reactions (Section 4-7) are just the reverse of combination reactions; they contain a single reactant and more than one product. They can also be subclassified into three groups, this time in terms of the nature of the products. (1) In some decomposition reactions, both products are elements. (2) In other decomposition reactions, the products are an element and one or more compounds, although this type of reaction is not common. (3) A compound can decompose to form two or more compounds.

Displacement reactions (Section 4-8) are those in which one element displaces another from a compound. All displacement reactions are redox reactions. Examples of such reactions are the displacement of one metal from a compound by another, of hydrogen from a compound by a metal, and of one nonmetal by another. We call the metal (or nonmetal) most able to displace hydrogen or other metals (or other nonmetals) from a compound the *most active*. A ranking of the ability of metals or hydrogen to displace one another from compounds is known as the **activity series**, Table 4-12. The activities of the halogens decrease going down the group. You should be able to use these rankings to predict whether one particular element will displace another. In Chapter 21, we will see a quantitative approach to these rankings.

In metathesis reactions (Section 4-9), two compounds react to form two new compounds. The terminology of "changing partners" will help you to recognize metathesis reactions.

Acid-base reactions represent a very important kind of metathesis reaction. These are also called **neutralization reactions**, because the typical properties of acids and bases are neutralized.

In such a reaction, *an acid reacts with a base to produce a salt and (usually) water*. The essence of a neutralization reaction is the combination of H^+ and OH^- ions to form water molecules. This is represented by the following net ionic equation for any acid-base neutralization reaction between a strong acid and a strong base to form a soluble salt and water:

$$H^+(aq) + OH^-(aq) \rightarrow H_2O(l)$$

The spectator ions (and hence the salt that is formed) are different depending on the identities of the acid and the base involved in the reaction. Similar net ionic equations for neutralizations involving weak acids, weak or insoluble bases, or insoluble salts are also discussed in this section. Some metathesis reactions result in the formation of **precipitate** (an insoluble solid); these are called **precipitation reactions** (Section 4-9.2). The solubility guidelines help us recognize such reactions. A third type of metathesis reaction involves the formation of a gas (Section 4-9.3).

One of your principal goals of this chapter should be to classify reactions according to the types presented here. The summary and examples in Table 4-14 in Section 4-10 will be very helpful. Remember that some reactions can be classified in more than one way; for instance, some acid-base neutralization reactions are also precipitation reactions, and displacement reactions, many combination reactions, and many decomposition reactions are also oxidation-reduction reactions.

We must name compounds unambiguously so that we can communicate about chemistry. In Chapter 2, we saw an introduction to naming compounds. Now, in the final sections of Chapter 4, we learn about a more systematic way to name compounds. Many of the rules for naming inorganic compounds require knowing the oxidation number of one or more of the elements in the compound (Section 4-4). You should learn the names, charges, and formulas for the common ions appearing in Table 4-15. (You already know many of these from Chapter 2). Section 4-11 discusses the naming of **binary** compounds—those consisting of two elements. Most of the naming of compounds that you learn will be according to the systematic nomenclature of the IUPAC convention. But many nonsystematic traditional names are still widely used, and it will be necessary to learn to recognize some of these.

Ternary acids contain H, O, and one other element, usually a nonmetal. In Section 4-12, you will learn how to name ternary acids and their salts. You must learn the names of the "-ic" acids in Table 4-16. Then you can use these as a starting point to name many other ternary acids and their salts.

This chapter brings together many of the concepts and skills of earlier chapters and applies them to some important types of reactions. In addition, it introduces you to ways of viewing, describing, and systematizing such reactions and to terminology commonly used in discussing the reactions. As you study the chapter, you must become familiar with the terminology by being aware of the precise meanings of each term. Try also to relate this material to what you already know from earlier chapters.

Study Goals

Use these goals to help you organize your study. As in other chapters, some textbook Exercises related to each study goal are indicated.

1. Describe and understand what is meant by periodicity. Know and use the terminology of the periodic table. (Section 4-1; Exercises 1 through 4 and 9)

2. Know the contrasting physical and chemical properties of metals and nonmetals. Know where the metals, nonmetals, and metalloids are located in the periodic table. (Section 4-1; Exercise 12)

3. Know which elements are examples of each of the following: (a) alkali metals, (b) alkaline earth metals, (c) halogens, (d) noble gases. Relate each of these classes to position in the periodic table. (Section 4-1; Exercises 9 through 12)

4. Be able to use position in the periodic table to predict properties of an element and the formulas of compounds it forms. (Section 4-1; Exercises 5 through 8)

5. Describe solutions of strong electrolytes, weak electrolytes, and nonelectrolytes, giving examples of each. (Section 4-2.1; Exercises 23 through 25, 27 through 29, 31, 33, 34, and 36)

6. Be familiar with the characteristics of acids, bases, salts, and their solution. (Sections 4-2.1 through 4-2.4; Exercises 15, 18, 21 through 23, and 26)

7. Distinguish between strong acids and weak acids. Distinguish among strong bases, weak bases, and insoluble bases. (Sections 4-2.2 through 4-2.4; Exercises 17 through 23, 25, 28, and 34)

8. Remember the names and formulas of the seven common strong acids and their anions listed in Table 4-5. Remember the names and formulas of the common weak acids and their anions listed in Table 4-6. Remember the names and formulas of the eight strong bases in Table 4-7. (Sections 4-2.2 and 4-2.4; Exercises 17 through 24, 32, and 34)

9. Know the solubility guidelines (Table 4-8) and apply them to identify relatively soluble and relatively insoluble compounds. (Section 4-2.5; Exercises 30 and 82 through 92)

10. Distinguish among molecular, total ionic, and net ionic equations. (Section 4-3; Exercises 35 through 38 and 41 through 48)

11. Be able to write balanced molecular equations, total ionic equations, and net ionic equations as appropriate for reaction types discussed in this chapter. (Sections 4-3 and 4-5 th rough 4-9; Exercises 57 through 92)

12. Know the rules for assigning oxidation numbers and be able to assign oxidation numbers to elements, whether free or in molecules, ions, or compounds. (Section 4-4; Exercises 37 through 40)

13. Know what is meant by each of the following terms: (a) oxidation, (b) reduction, (c) oxidizing agent, (d) reducing agent. Identify oxidizing agents and reducing agents in equations for redox reactions. (Section 4-5; Exercises 41 through 48)

14. Classify chemical reactions as one or more of these types: (a) oxidation-reduction (redox), (b) combination, (c) decomposition, (d) displacement, (e) metathesis, (f) acid-base neutralization, (f) precipitation, (g) gas-formation. (Sections 4-5 through 4-9; Exercises 49 through 102)

15. Be able to use the activity series (Table 4-12) to predict whether or not one element would displace another from a compound. Write equations for displacement reactions. (Section 4-8; Exercises 57 through 68)

16. Given their formulas, name binary and other common inorganic compounds by the IUPAC system. Given their IUPAC names, write the formulas of common inorganic compounds. Name common inorganic compounds by other commonly used traditional methods. (Sections 4-11 and 4-12; Exercises 104 through 120)

17. Answer conceptual questions based on this chapter. (Conceptual Exercises 121 through 130)

18. Apply concepts and skills from earlier chapters to the ideas of this chapter. (Building Your Knowledge exercises, 131 through 134)

Some Important Terms in This Chapter

Write the meanings *in your own words*. Check the Key Terms list and the chapter reading. Then rewrite your definitions, still in your own words, to improve them. Study other new terms and review terms from preceding chapters if necessary.

periodicity

periodic table

metal

nonmetal

metalloid

noble gases

halogens

alkali metals

alkaline earth metals

electrolyte

nonelectrolyte

strong electrolyte

weak electrolyte

acid

base

salt

formula unit equation

total ionic equation

net ionic equation

oxidation number

oxidation

reduction

oxidizing agent

reducing agent

oxidation-reduction reaction

combination reaction

decomposition reaction

displacement reaction

metathesis reaction

neutralization reaction

precipitation reaction

gas-formation reaction

activity series

binary compound

ternary compound

Preliminary Test

As in other chapters, this test will check your understanding of basic concepts and types of calculations. Be sure to practice *many* of the additional textbook exercises, including those indicated with the Study Goals.

True-False

Mark each statement as true (T) or false (F).

_____ 1. There are more metallic elements than nonmetallic ones.
_____ 2. Any element is either a metal or a nonmetal.
_____ 3. The properties of all elements in a group of the periodic table are identical.
_____ 4. Calcium is more metallic than iron.
_____ 5. All solutions in water are good conductors of electricity.
_____ 6. There are more weak acids than strong acids.
_____ 7. All chemical reactions go to completion.
_____ 8. All of the common strong bases are metal hydroxides.
_____ 9. All metal hydroxides are strong bases.
_____ 10. A reaction in solution that forms an insoluble solid product is called a precipitation reaction.
_____ 11. Acid-base neutralization reactions can also be precipitation reactions.
_____ 12. Lighter metals can always displace heavier ones.
_____ 13. Oxidation number and oxidation state are the same thing.
_____ 14. Each element has a characteristic oxidation number in all its compounds.
_____ 15. Halogens have an oxidation number of -1 in all their compounds.

Short Answer

Answer with a word, a phrase, a formula, or a number with units as necessary.

1. A statement of the periodic law is _____
   _____.
2. The two scientists who are credited with the development of the ideas of periodicity of elemental properties were _____ and _____.
3. The elements in Group IA are referred to as the _____.
4. Calcium and barium are members of Group ____, known as the _____ group.
5. The members of the halogen group are _____.
6. An acid is a substance that produces _____ (name, formula) in aqueous solution.
7. An acid that does not very efficiently produce H^+ in aqueous solutions is called a (an) _____.
8. The anion of nitric acid is _____ (name, formula).
9. The anion of acetic acid is _____ (name, formula).
10. The acid of which cyanide ion is the anion is _____ (name, formula).
11. All of the strong bases listed in the text chapter are hydroxides of the elements of _____ and _____.

12. Organic acids contain the atom grouping _____.
13. A base, when dissolved in water, produces _____ (name, formula).
14. When a strong acid is dissolved in water, its ionization takes place _____.
15. When a weak acid is dissolved in water, its ionization takes place _____.
16. Classify each of the following substances as soluble or insoluble in water.

(a) KNO_3 _____
(b) $SrCl_3$ _____
(c) AgI _____
(d) $Al(OH)_3$ _____
(e) $(NH_4)_2S$ _____
(f) FeS _____
(g) $CaCO_3$ _____
(h) HBr _____
(i) $PbCl_2$ _____
(j) $NaSCN$ _____

(k) $Ca(OH)_2$ _____
(j) $AgNO_3$ _____
(m) $Mg(CH_3COO)_2$ _____
(n) $Zn(ClO_4)_2$ _____
(o) H_3PO_4 _____
(p) $Ba_3(PO_4)_2$ _____
(q) MgS _____
(r) $(NH_4)_2CO_3$ _____
(s) KOH _____
(t) Rb_2SO_4 _____

17. A reaction in which the elements or ions may be described as "changing partners" is called a _____ reaction.
18. A reaction in which one element reacts with a compound to give a new compound and an element is called a _____ reaction.
19. A reaction in which one element breaks apart to produce one or more elements or compounds is called a _____ reaction.
20. A reaction in which two or more elements of compounds come together to form a single new compound is called a _____ reaction.
21. An insoluble solid that is formed in a reaction is called a _____.
22. A reaction in which an insoluble solid is formed is called a _____ reaction.
23. When silver nitrate solution and sodium sulfate solution are mixed, silver sulfate precipitates, leaving a solution containing sodium ions and nitrate ions. The balanced molecular equation for this reaction is _____. This reaction may be classified as a _____ reaction. This reaction _____ (is, is not) an oxidation–reduction reaction.
24. Solid lead(IV) chloride, when heated to 100°C, reacts to give solid lead(II) chloride and chlorine gas. The balanced molecular equation for this reaction is _____. This reaction may be classified as a _____ reaction. This reaction _____ (is, is not) an oxidation-reduction reaction.
25. When the ammonia fumes above a solution are allowed to mix with the hydrogen chloride fumes above a hydrochloric acid solution, a "smoke" (actually a colloid of finely divided solid particles -- see Chapter 14) of ammonium chloride forms. The balanced formula unit equation for this reaction is _____. This reaction may be classified as a _____ reaction. This reaction _____ (is, is not) an oxidation-reduction reaction.
26. In the following equations, A, B, X, and Y represent hypothetical elements, while H and O are hydrogen and oxygen, respectively. Classify each of the following as one or more of the following: a combination reaction, a displacement reaction, a decomposition reaction, a metathesis reaction, an acid-base neutralization reaction.

(a) $XO + H_2O \rightarrow H_2SO_2$ _____
(b) $A + X_2 \rightarrow AX_2$ _____

46

(c) $A + BX \rightarrow AX + B$ _____

(d) $2AX + Y_2 \rightarrow 2AY + B$ _____

(e) $HX + BOH \rightarrow H_2O + BX$ _____

(f) $AX \rightarrow A + X$ _____

(g) $A + B \rightarrow AB$ _____

(h) $AX + BX \rightarrow ABX_2$ _____

(i) $2A_2O \rightarrow 2A_2 + O_2$ _____

(j) $2HX + B(OH)_2 \rightarrow 2H_2O + BX_2$ _____

(k) $AX + BY \rightarrow AY + BX$ _____

(l) $AXO_3 \rightarrow AO + XO_2$ _____

(m) $2A + 2HX \rightarrow 2AX + H_2$ _____

(n) $2A + X_2 \rightarrow 2AX$ _____

(o) $2AX + X_2 \rightarrow 2AX_2$ _____

(p) $A_2 + BX \rightarrow 2AX + B_2$ _____

(q) $AO + BO \rightarrow ABO_2$ _____

(r) $2AXO_3 \rightarrow 2AX + 3O_2$ _____

27. When an acid and a base act to neutralize one another, the products are _____ and _____.

28. The name and formula of the salt that results from the neutralization of calcium hydroxide and nitric acid are _____.

29. The formula of magnesium acetate is _____.

30. The net ionic reaction for all neutralization reactions involving strong acids and strong bases to form soluble salts in water is _____.

31. For the neutralization reaction of sulfuric acid and potassium hydroxide, the balanced formula unit equation is _____, the total ionic equation is _____, and the net ionic equation is _____. The spectator ions are _____.

32. The sum of the oxidation numbers of all of the atoms in a neutral compound is equal to ____.

33. The sum of the oxidation numbers of all of the atoms in a polyatomic ion is equal to _____.

34. In most of its compounds, the oxidation number of hydrogen is ____, whereas oxygen usually exhibits the oxidation number ____.

35. The oxidation numbers of the elements in the compound K_2S are K ____ and S ____.

36. In the compound NaCl, the oxidation numbers of the elements are Na ____ and Cl ____.

37. In the compound $NaClO_3$, the oxidation numbers of the elements are Na ____, Cl ____, and O ____.

38. In the sulfate ion, SO_4^{2-}, the oxidation numbers of the elements are S ____ and O ____.

39. A process in which electrons are lost is termed _____.

40. A substance that loses electrons in a reaction is termed a (an) _____.

41. A substance that undergoes an increase in oxidation number is a reaction is termed a (an) _____.

42. The substance that is reduced in a reaction is termed a (an) _____.

43. In any reaction in which reduction occurs, _____ must also take place.

44. Another name for an oxidation-reduction reaction is a _____.

- -

The reaction in which nitrogen and hydrogen react to give ammonia is an oxidation-reduction reaction in which all substances are gases. Questions 45 through 48 refer to this reaction.

45. The unbalanced formula unit equation that describes this reaction is _____.
46. In this reaction, nitrogen undergoes a change in oxidation number from _____ to _____; hydrogen undergoes a change in oxidation number from _____ to _____.
47. The balanced formula unit equation that describes this reaction is _____.
48. This reaction can also be classified as a _____ reaction.

- -

49. An oxidizing agent is a species that _____ electrons.
50. When an element in a substance undergoes an algebraic decrease in oxidation number, we say that the substance is being _____. In such a reaction it acts as the _____ agent.
51. The usual product formed when sulfuric acid acts as an oxidizing agent is SO_2; in this process, the oxidation number of sulfur changes from ____ to ____.
52. When permanganate ion acts as an oxidizing agent in acidic solution, the usual product formed is manganese(II) or manganous ion, Mn^{2+}; in this process, the oxidation number of manganese changes from ____ to ____.
53. When permanganate ion acts as an oxidizing agent in basic solution, the usual product formed is manganese dioxide, MnO_2; in this process, the oxidation number of manganese changes from ____ to ____.
54. When $Sn(OH)_3^-$ is converted to $Sn(OH)_6^{2-}$, the tin is being _____, from oxidation number ____ to ____.
55. A binary compound is one that consists of _____ elements.
56. A ternary compound is one that consists of _____ elements.
57. A ternary acid is one that contains H, O, and _____.

- -

Write the formulas for the compounds or ions named in Questions 58 through 83.

58. cupric chloride _____
59. copper(II) chloride _____
60. sodium acetate _____
61. sulfate ion _____
62. ammonium ion _____
63. permanganate ion _____
64. sodium hydrogen sulfate _____
65. sodium sulfate _____
66. sodium sulfite _____
67. sodium nitrate _____
68. sodium nitrite _____
69. lithium cyanide _____

70. iron(III) sulfite_____

71. sodium dichromate_____

72. sodium chromate _____

73. sulfurous acid _____

74. sulfuric acid _____

75. hypobromous acid _____

76. bromous acid _____

77. bromic acid _____

78. perbromic acid _____

79. hydrobromic acid _____

80. phosphoric acid _____

81. carbonic acid _____

82. magnesium carbonate _____

83. potassium hydrogen carbonate _____

- -

Name the compounds or ions whose formulas appear in Questions 83 through 109.

84. $CaSO_4$ _____

85. $FeSO_4$_____

86. $Fe_2(SO_4)_3$_____

87. $NaMnO_4$_____

88. $Fe(CN)_3$_____

89. $Ca(NO_2)_2$ _____

90. CdS _____

91. CrF_3 _____

92. AsO_4^{3-} _____

93. ClO_4^- _____

94. Zn^{2+} _____

95. $CoSO_3$ _____

96. $Ba(SCN)_2$ _____

97. CH_3COO^- _____

98. $NaHCO_3$ _____

99. $Fe_3(PO_4)_2$ _____

100. $(NH_4)_2CO_3$ _____

101. NiO _____

102. $AgCl$ _____

103. $Hg_2(CN)_2$ _____

104. HF (aqueous solution)_____

105. HI (aqueous solution) _____

106. $Hg(NO_2)_2$ _____

107. H_3PO_3 _____

108. HIO_3 _____

109. HNO_3 _____

Multiple Choice

_____ 1. Which of the following pairs of elements would be expected to be most similar chemically?
(a) Sn and Sr   (b) S and Se   (c) Ca and Na   (d) Ne and Na   (e) Ga and Ge

_____ 2. The formulas of the oxides of some second period elements are Li_2O, BeO, B_2O_3, and CO_2. Which of the following is likely to be the correct formula for aluminum sulfide?
(a) Al_2S (b) AlS   (c) Al_2S_3 (d) AlS_2
(e) Cannot predict without knowing the percent by mass of the compound.

_____ 3. Which of the following is *not* a strong acid?
(a) HF   (b) HCl   (c) HBr   (d) HI   (e) $HClO_4$

_____ 4. Which of the following is *not* a weak acid?
(a) HNO_3 (b) $(COOH)_2$ (c) H_2SO_3 (d) HSO_3^- (e) CH_3COOH

_____ 5. The salt that would be formed by the complete neutralization of $Mg(OH)_2$ with HCl is
(a) MgCl.   (b) $MgCl_2$.   (c) Mg_2Cl.   (d) HMg.   (e) ClOH.

_____ 6. The salt that would result from the complete neutralization reaction between sulfuric acid and potassium hydroxide is
(a) K_2SO_4.   (b) KSO_4.   (c) $K_2(SO_4)_2$.   (d) KS.   (e) K_2S.

_____ 7. The reaction

$$2HCl + Ca(OH)_2 \rightarrow 2H_2O + CaCl_2$$

can be described as
(a) a neutralization reaction.   (b) a displacement reaction.
(c) a disproportionation reaction.   (d) a precipitation reaction.
(e) an oxidation–reduction reaction.

_____ 8. The salt that would result from the neutralization reaction between phosphoric acid and calcium hydroxide is
(a) $CaPO_4$.   (b) $Ca_3(PO_4)_2$.   (c) $Ca_2(PO_4)_3$.   (d) $Ca(PO_4)_2$ (e) Ca_3PO_4.

_____ 9. The following describe properties of substances. Which one is not a typical acid property?
(a) It has a sour taste.
(b) It reacts with metal oxides to form salts and water.
(c) It reacts with acids to form salts and water.
(d) Its aqueous solutions conduct an electric current.
(e) It reacts with active metals to liberate H_2.

_____ 10. When heated in the presence of oxygen, O_2, the gas carbon monoxide, CO, is converted to carbon dioxide gas, CO_2. This reaction is best described as
(a) a displacement reaction.   (b) a neutralization reaction.
(c) a precipitation reaction.   (d) a redox reaction.

_____ 11. The oxidation numbers (oxidation states) of the elements in $KClO_3$ are
(a) K = +3, Cl = +3, O = –2.   (b) K = –1, Cl = –5, O = +2.

(c) K = +1, Cl = +5, O = –2. (d) K = +1, Cl = +3, O = –1.
(e) K = +1, Cl = +2, O = –3.

_____ 12. In the reaction

$$2Na(s) + 2H_2O(l) \rightarrow 2Na^+(aq) + H_2(g) + 2OH^-(aq)$$

sodium metal
(a) is acting as an acid.
(b) is the oxidizing agent and is being reduced.
(c) is the oxidizing agent and is being oxidized.
(d) is the reducing agent and is being oxidized.
(e) is the reducing agent and is being reduced.

_____ 13. In the reaction

$$Cr_2O_7^{2-}(aq) + 3H_2S(aq) + 8H^+(aq) \rightarrow 2Cr^{3+}(aq) + 2S(s) + 7H_2O(l)$$

which element is being oxidized?
(a) Cr   (b) H   (c) O   (d) S
(e) None, because this is not an oxidation-reduction reaction.

_____ 14. In the reaction of Question 13, the reducing agent is
(a) $Cr_2O_7^{2-}$.   (b) H_2S.   (c) H^+.   (d) H_2O.
(e) There is no reducing agent, because this is not a redox reaction.

_____ 15. In the reaction

$$Mg(s) + Cl_2(g) \rightarrow MgCl_2(s)$$

chlorine gas acts as
(a) an anion. (b) a cation. (c) an oxidizing agent. (d) an acid. (e) a reducing agent.

_____ 16. The reaction of Question 15 could be classified as
(a) a precipitation reaction.   (b) a neutralization reaction.
(c) a displacement reaction.   (d) a redox reaction.

_____ 17. In the reaction

$$Ca_3(PO_4)_2(s) + 3H_2SO_4(aq) \rightarrow 2H_3PO_4(aq) + 3CaSO_4(s)$$

which element is being oxidized?
(a) Ca   (b) P   (c) S   (d) O
(e) None, because this is not an oxidation–reduction reaction.

_____ 18. The mineral fluorite is an ionic substance consisting only of calcium and fluorine. What are
the oxidation numbers of Ca and F in fluorite?
(a) Ca = –1, F = +1   (b) Ca = +1, F = +2   (c) Ca = +2, F = –2
(d) Ca = +2, F = –1   (e) Ca = +2, F = +9

_____ 19. Consider the following equation that is involved in the rusting of iron:

$$4Fe + O_2 \rightarrow 2Fe_2O_3$$

Which one of the following statements about this equation is *not* correct?
(a) This is an example of an oxidation–reduction reaction.
(b) Metallic iron is a reducing agent.
(c) O_2 is an oxidizing agent.
(d) Metallic iron is reduced.

51

(e) The total gain in oxidation number of one element equals the total loss of oxidation number of another element.

_____ 20. Which of the following compounds contains an element with an oxidation number of +5?
(a) $HClO$ (b) HNO_2 (c) $HClO_3$ (d) $HBrO_4$ (e) H_2SO_4

_____ 21. Which of the following compounds contains an element with an oxidation number of +3?
(a) $HClO$ (b) HNO_2 (c) $HClO_3$ (d) $HBrO_4$ (e) H_2SO_4

_____ 22. In the net ionic equation

$$5SO_3^{2-} + 6H^+ + 2MnO_4^- \rightarrow 5SO_4^{2-} + 2Mn^{2+} + 3H_2O$$

which element is being reduced?
(a) S   (b) O   (c) H   (d) Mn
(e) None, because this is not a redox reaction.

_____ 23. Which of the following compounds contains an element with an oxidation number of +6?
(a) $HClO$ (b) HNO_3 (c) $HClO_3$ (d) $HBrO_4$ (e) H_2SO_4

_____ 24. Each of the following is a known acid containing chlorine. In which one does chlorine have oxidation number +7?
(a) HCl (b) $HClO$ (c) $HClO_2$ (d) $HClO_3$ (e) $HClO_4$

_____ 25. Manganese (Mn) exhibits a larger number of oxidation numbers in its compounds than any other of the Period 4 transition metals. The most important oxidation numbers of manganese are illustrated by the following oxides; MnO, MnO_2, and Mn_2O_7. What oxidation numbers of manganese are illustrated here?
(a) –2, –4, –7   (b) +2 and +7   (c) +2, +3, +9   (d) +2, +4, +7   (e) +2, +4, +14

_____ 26. In the reaction

$$2SO_2 + O_2 \rightarrow 2SO_3$$

what is the oxidizing agent?
(a) SO_2 (b) O_2 (c) SO_3
(d) None, because this is not a redox reaction.

_____ 27. In the reaction

$$Mg + H_2O \rightarrow MgO + H_2$$

what is the reducing agent?
(a) Mg (b) H_2O (c) MgO (d) H_2
(e) None, because this is not a redox reaction.

_____ 28. One of the reactions that takes place in a copper smelter to remove unwanted compounds from the ore is

$$CaCO_3 + SiO_2 \rightarrow CaSiO_3 + CO_2$$

In this reaction, what is the oxidizing agent?
(a) $CaCO_3$ (b) SiO_2 (c) $CaSiO_3$ (d) CO_2
(e) None, because this is not a redox reaction.

_____ 29. Copper(I) oxide can be prepared by boiling copper(I) chloride with a metal hydroxide. In this reaction, what is the oxidizing agent? The net ionic equation for this reaction is

$$2CuCl + 2OH^- \rightarrow Cu_2O + 2Cl^- + H_2O$$

(a) $CuCl$ (b) OH^- (c) Cu_2O (d) Cl^- (e) None, because this is not a redox reaction.

_____ 30. The following reaction (not balanced) is carried out in acidic solution. What is the oxidizing agent?

$$SO_3^{2-} + MnO_4^- \rightarrow SO_4^{2-} + Mn^{2+}$$

(a) SO_3^{2-} (b) MnO_4^- (c) SO_4^{2-} (d) Mn^{2+} (e) None (not a redox reaction)

_____ 31. In the reaction of Question 30, which species is described as the reducing agent?

(a) SO_3^{2-} (b) MnO_4^- (c) SO_4^{2-} (d) Mn^{2+}

(e) None, because this is not a redox reaction.

_____ 32. The reaction

$$Cr_2O_7^{2-} + H_2S \rightarrow Cr^{3+} + S$$

is carried out in acidic solution. What element is being oxidized?

(a) Cr   (b) H   (c) O   (d) S   (e) None, because this is not a redox reaction.

_____ 33. In the reaction of Question 32, which substance is described as the oxidizing agent?

(a) $Cr_2O_7^{2-}$ (b) H_2S (c) Cr^{3+} (d) S   (e) None, because this is not a redox reaction.

_____ 34. Which of the following is the formula for cobalt(II) nitrite?

(a) Co_2N (b) Co_2NO_2 (c) $Co(NO_2)_2$ (d) $CoNO_2$ (e) $Co(NO_3)_2$

_____ 35. Which of the following is the formula for iron(II) sulfite?

(a) $FeSO_4$ (b) FeS   (c) Fe_2SO_3 (d) $FeSO_3$ (e) $Fe(SO_3)_2$

_____ 36. What is the name of the ion SO_4^{2-}?

(a) sulfide   (b) sulfite   (c) sulfate   (d) sulfur tetroxide

(e) None of the preceding answers is correct.

_____ 37. Which of the following is the formula for iron(II) oxide?

(a) FeO   (b) FeO_3 (c) Fe_2O (d) Fe_2O_3 (e) Fe_3O_2

_____ 38. What is the name of the ion SO_3^{2-}?

(a) sulfide   (b) sulfite   (c) sulfate   (d) sulfur trioxide

(e) None of the preceding answers is correct.

_____ 39. Which of the following is the formula for nickel(II) sulfide?

(a) NiS   (b) Ni_2S (c) NiS_2 (d) $NiSO_4$ (e) Ni_2SO_4

_____ 40. What is the name of the ion NO_3^-?

(a) nitride   (b) nitrite   (c) nitrate   (d) nitrogen trioxide

(e) None of the preceding answers is correct.

_____ 41. Which of the following is the formula for manganese(II) hydroxide?

(a) MnO   (b) MnOH   (c) $Mn(OH)_2$ (d) Mn_2OH (e) $Mn_3(OH)_2$

_____ 42. Which of the following is the formula for iron(III) sulfate?

(a) $Fe_2(SO_4)_3$ (b) Fe_3SO_4 (c) $Fe(SO_3)_2$ (d) Fe_3SO_3 (e) $Fe_3(SO_4)_2$

_____ 43. Which of the following is the formula for chromium(III) nitrate?

(a) $CrNO_3$ (b) Cr_3NO_3 (c) $Cr(NO_3)_3$ (d) $Cr_2(NO_2)_3$ (e) $Cr(NO_2)_2$

NOTE: It takes a great deal of practice to learn to deal comfortably with the kind of material in this chapter. This is especially true for recognizing reaction types and identifying oxidizing and reducing agents. You should work many of the text exercises suggested with the Study Goals.

Answers to Preliminary Test

Do not just look up answers. Think about the reasons for the answers.

True-False

1. True. Look at Table 4-2.
2. False. The metalloids have properties that are intermediate between those of metals and nonmetals.
3. False. Properties of the elements in a group are similar, but not identical.
4. True. Metallic properties decrease from left to right across a row of the periodic table.
5. False. Only solutions that contain significant numbers of ions conduct electricity. Read about electrolytes and nonelectrolytes, Section 4-2.1.
6. True. There are seven common strong acids (Table 4-5) and many more weak acids, a *few* of which are listed in Table 4-6. You should know the seven common strong acids.
7. False. Many reactions can occur in both directions, so they do not proceed completely to form products. Such reactions are called reversible reactions (Section 4-2.4).
8. True. These are listed in Table 4-7. You should know this list of bases.
9. False. Many metal hydroxides are quite insoluble in water, so they cannot produce significant amounts of OH^- in solution.
10. True.
11. True. If the salt that is formed is insoluble, it forms a precipitate. An example is the neutralization reaction between sulfuric acid, H_2SO_4, and barium hydroxide, $Ba(OH)_2$. This reaction forms $BaSO_4$, which is an insoluble salt (refer to the solubility guidelines, pp. 124-125 and Table 4-8). Can you think of other examples?
12. False. See the activity series, Table 4-13 and study Section 4-8.
13. True.
14. False. Most elements can exhibit various oxidation numbers in their compounds, and some elements can have many.
15. False. This is true only for their binary compounds with metals, H, or NH_4^+. See Table 4-10.

Short Answer

1. Physical and chemical properties of the elements vary in a periodic fashion with the atomic numbers of the elements.
2. Mendeleev; Meyer
3. alkali metals. Be sure you remember the common group names (Section 4-1).
4. IIA; alkaline earth metals
5. F, Cl, Br, I, and At. Of these, only the first four are common. Notice the terminology by which a *group* is sometimes referred to as a *family* of elements.
6. hydrogen ion, H^+
7. weak acid
8. nitrate, NO_3^-. Remember that the anion of an acid is what is left when the acid gives up H^+ to water. See Section 4-2.2.
9. acetate, CH_3COO^-. Formulas for ions written without the proper charge are incorrect.
10. hydrocyanic acid, HCN. You can find out the acid(s) corresponding to an ion by adding one (or more) H^+. Remember to adjust the charge.

11. Group IA (alkali metals); the heavier members of Group IIA (alkaline earth metals).
12. —COOH.
13. hydroxide, OH⁻.
14. completely
15. very slightly
16. You should *learn* and use the solubility guidelines in Section 4-2.5, Table 4-8.

(a)	soluble	(f)	insoluble	(k)	soluble	(p)	insoluble
(b)	soluble	(g)	insoluble	(l)	soluble	(r)	soluble
(c)	insoluble	(h)	soluble	(m)	soluble	(q)	soluble
(d)	insoluble	(i)	insoluble	(n)	soluble	(s)	soluble
(e)	soluble	(j)	soluble	(o)	soluble	(t)	soluble

17. metathesis
18. displacement
19. decomposition
20. combination
21. precipitate
22. precipitation
23. $2AgNO_3(aq) + Na_2SO_4(aq) \rightarrow Ag_2SO_4(s) + 2NaNO_3(aq)$; precipitation; is not. You can always check to find out whether a reaction is a redox reaction by determining the oxidation numbers of all elements on both sides of the equation. In this equation, there are no oxidation number changes.
24. $PbCl_4(s) \rightarrow PbCl_2(s) + Cl_2(g)$; decomposition; is. Lead and part of the chlorine change in oxidation number.
25. $NH_3(g) + HCl(g) \rightarrow NH_4Cl(s)$; combination; is not.
26. (a) combination
    (b) combination
    (c) displacement
    (d) displacement
    (e) metathesis (a neutralization reaction)
    (f) decompostion
    (g) combination
    (h) combination
    (i) decomposition
    (j) metathesis (a neutralization reaction)
    (k) metathesis
    (l) decomposition
    (m) displacement
    (n) combination
    (o) combination
    (p) displacement
    (q) combination
    (r) decompostion

    It would be helpful for you to find one or more examples of each of these reactions in the textbook. You might need to look somewhere other than Chapter 4 for some of them.
27. water; a salt. Some salts are insoluble and precipitate from solution; others are soluble, remaining in solution in ionized form. NaCl, a compound that we commonly call "salt," is only

one member of an entire class of compounds that are collectively called salts in correct chemical usage.

28. calcium nitrate, $Ca(NO_3)_2$. Review the names of the anions of the common strong acids (Table 4-5) and the common weak acids (Table 4-6). You must be familiar with the names of some common ions that you studied in Table 2-3, Section 2-4.

29. $Mg(CH_3COO)_2$

30. $H^+(aq) + OH^-(aq) \rightarrow H_2O(l)$

31. molecular: $H_2SO_4(aq) + 2KOH(aq) \rightarrow 2H_2O(l) + K_2SO_4(aq)$

    total ionic: $[2H^+(aq) + SO_4^{2-}(aq)] + [2K^+(aq) + OH^-(aq)] \rightarrow$
    $$2H_2O(l) + [2K^+(aq) + SO_4^{2-}(aq)]$$

    net ionic: $H^+(aq) + OH^-(aq) \rightarrow H_2O(l)$
    spectator ions: K^+, SO_4^{2-}

32. 0. See the oxidation number rules in Section 4-6.

33. the charge on the ion. See Section 4-6.

34. +1; –2. The rare exceptions are pointed out in Table 4-11.

35. +1; –2. Remember that in a binary ionic compound, the oxidation numbers are equal to the charges on the ions. You should recall the rules in Section 4-6 and the common oxidation numbers in Table 4-11. The Group IA elements always have oxidation numbers of +1 in compounds; in binary compounds with less electronegative elements, the Group VIA elements have oxidation number –2.

36. +1; –1

37. +1; +5; –2. First figure out Na (+1 in compounds) and O (–2). Then the sum of all oxidation numbers in a compound must be 0, so Cl must be +5.

38. +6; –2. As in the preceding question, recall that O has oxidation number -2. Then the sum of oxidation numbers in a polyatomic ion is equal to the charge on the ion, so S must have oxidation number +6.

39. oxidation

40. reducing agent

41. reducing agent. Remember that a substance that undergoes an increase in oxidation number is said to be oxidized, and that the substance that is oxidized is the reducing agent.

42. oxidizing agent

43. oxidation

44. redox reaction

45. $N_2(g) + H_2(g) \rightarrow NH_3(g)$. It is incorrect to represent nitrogen gas as N or hydrogen as as H, because they do not exist in those forms. Each is diatomic.

46. 0, –3; 0, +1.

47. $N_2(g) + 3H_2(g) \rightarrow 2NH_3(g)$

48. combination

49. gains

50. reduced; oxidizing

51. +6; +4

52. +7; +2

53. +7; +4

54. oxidized; +2; +4

55. two

56. three

58. another element, usually a nonmetal

---

58. $CuCl_2$	67. $NaNO_3$	76. $HBrO_2$
59. $CuCl_2$	68. $NaNO_2$	77. $HBrO_3$
60. $NaCH_3COO$	69. $LiCN$	78. $HBrO_4$
61. SO_4^{2-}	70. $Fe_2(SO_4)_3$	79. $HBr(aq)$
62. NH_4^+	71. $Na_2Cr_2O_7$	70. H_3PO_4
63. MnO_4^-	72. Na_2CrO_4	71. H_2CO_3
64. $NaHSO_4$	73. H_2SO_3	72. $MgCO_3$
65. Na_2SO_4	74. H_2SO_4	73. $KHCO_3$
66. Na_2SO_3	75. $HBrO$	

NOTE: Where two answers are given for Questions 84 through 109, the first is the IUPAC name, which is usually the preferred one.

84. calcium sulfate	97. acetate ion
85. iron(II) sulfate or ferrous sulfate	98. iron(II) phosphate or ferrous phosphate
86. iron(III) sulfate or ferric sulfate	99. sodium hydrogen carbonate or sodium bicarbonate
87. sodium permanganate	100. ammonium carbonate
88. iron(III) cyanide or ferric cyanide	101. nickel(II) oxide or nickelous oxide
89. calcium nitrite	102. silver chloride
90. cadmium sulfide	103. mercury(I) cyanide or mercurous cyanide
91. chromium(III) fluoride or chromic fluoride	104. hydrofluoric acid
92. arsenate ion	105. hydroiodic acid
93. perchlorate ion	106. mercury(II) nitrite or mercuric nitrite
94. zinc ion	107. phosphorous acid
95. cobalt(II) sulfite or cobaltous sulfite	108. iodic acid
96. barium thiocyanate	109. nitric acid

Multiple Choice

1. (b). Remember that the most pronounced similarities are for elements in the same group (column) of the periodic table.
2. (c). Al is in the same column as B, and S is in the same column as O.
3. (a). Remember that HF is the only one of the hydrohalic acids (binary acids that involve halogens) that is not a strong acid.
4. (a). Notice that (b), (c), and (e) are listed as common weak acids. Sulfite ion, SO_3^{2-}, is listed as one of the anions derived from the weak acid H_2SO_3; this means that HSO_3^- must also be able to act as a weak acid.
5. (b)

6. (a)
7. (a). An acid, HCl, reacts with a base, Ca(OH)$_2$, to form water and a salt, CaCl$_2$.
8. (b). Calcium ion is Ca^{2+} and phosphate ion is PO$_4^{3-}$. These combine in a 3:2 ratio to form the salt Ca$_3$(PO$_4$)$_2$.
9. (c)
10. (d). The oxidation number of C changes from +2 to +4, while the oxidation number of O changes from 0 (in the element O$_2$) to –2.
11. (c)
12. (d)
13. (d). Determine the oxidation number of each element on both sides of the equation. Sulfur goes from –2 in H$_2$S to 0 in S (the free element).
14. (b). Remember that the compound containing the element that is oxidized is called the reducing agent.
15. (c)
16. (d). A solid product is formed, but the reaction does not take place in solution, so this is not a precipitation reaction.
17. (e). No element undergoes a change in oxidation number. You can always check oxidation numbers to find out whether a reaction is a redox reaction.
18. (d)
19. (d)
20. (c)
21. (b)
22. (d). Look for changes in oxidation numbers.
23. (e). Sulfur in this compound has an oxidation number of +6.
24. (e)
25. (d)
26. (b)
27. (a)
28. (e). Write down the oxidation numbers of all elements on both sides of the equation. No element changes oxidation number, so this is not a redox reaction.
29. (e)
30. (c). Remember that the compound or polyatomic ion that contains the element that is reduced is the oxidizing agent. Permanganate ion, MnO$_4^-$, is a common laboratory oxidizing agent.
31. (a)
32. (d)
33. (a). Dichromate ion, Cr$_2$O$_7^{2-}$, is another common laboratory oxidizing agent.
34. (c)
35. (d)
36. (c)
37. (a)
38. (b)
39. (a)
40. (c)
41. (c)
42. (a)
43. (c)

5 The Structure of Atoms

Chapter Summary

In this chapter, we learn about the structure of atoms by studying questions such as the following. Of what particles are atoms composed? How are these particles arranged in atoms? Are all atoms of a given element identical? If not, how do they differ and how are they alike? What evidence do we have for these descriptions? For a given element, how do we describe how many electrons the atom has, where these electrons are, and how strongly they are held? (This last question is an especially important one, because electrons are the particles in atoms that are involved in chemical reactions.)

First, let us see how the ideas presented in this chapter were developed. It is not our intent to learn the history of chemistry for its own sake. But an appreciation of how the ideas about atoms and their architecture have been developed can help us to develop our own understanding in a gradual and logical way. If the text or your instructor were to begin by telling you all information currently known or believed about atoms and subatomic particles, you would soon be overwhelmed, and would learn little. A presentation along historical lines can help us appreciate that chemistry is a living, developing science, with new evidence being found and old evidence being continually reassessed or reinterpreted. This presentation helps us to see that we describe the atom as we do, not because many of these properties of atoms or their particles can be directly observed, but because these descriptions are the best available explanations of our observations.

The first important modern idea about atoms was Dalton's theory, which was discussed in Section 2-1. Some features of the Dalton theory are still useful to us in chemistry; other features have required modification in light of later observations. As you study this chapter, watch for this pattern of development of ideas.

Section 5-1 introduces us to the three fundamental particles that we will work with in our understanding of chemistry—the basic particle of positive charge (**proton**), the fundamental particle of negative charge (**electron**), and the neutral particle (**neutron**). Much of our study of chemistry will be concerned with understanding the arrangements of these particles in atoms and how these arrangements are related to chemical and physical properties. We first encounter the properties of the two charged subatomic particles (Sections 5-2 and 5-3) through a discussion of the experiments that led to the discovery and study of the electron and the proton. Section 5-4 discusses the classic experiment that led to the idea of the nuclear atom, the Rutherford gold foil experiment. In this description of the atom, the protons and neutrons (Section 5-6) are in the **nucleus,** an extremely small volume in the center of the atom. Thus, the nucleus of the atom contains all of the positive charge and nearly all of the mass of the atom. The electrons account for all of the negative charge but for only a tiny fraction of the mass of the atom; they are distributed throughout the essentially empty space comprising the rest of the atom. The **atomic number** (Section 5-6) tells the number of protons in the nucleus of an atom of the element. It is important to remember that the atomic number characterizes each element uniquely; no two elements have the same atomic number, and two atoms with different numbers of protons in their nuclei must be different elements.

In Section 5-7, we learn how to describe different kinds of atoms according to the numbers of protons, neutrons, and electrons they contain. All atoms except the common form of hydrogen contain both protons and neutrons in their nuclei. The neutrons contribute to the total mass of the atom but do not alter its electrical properties, so they do not affect chemical properties. The existence of different forms of a given element that have different numbers of neutrons in the nucleus thus gives rise to atoms that are chemically identical but have different masses. These are called **isotopes**. The total number of particles in the nucleus of a particular atom is characterized by the **mass number**, which tells the total number of protons and neutrons in the nucleus. The **mass spectrometer**, which aided in determining this aspect of atomic structure, is helpful in understanding these concepts (Section 5-8). Most elements occur naturally as mixtures of isotopes, which behave the same chemically. Any chemical determination of the "atomic weight" can therefore measure only the **average** atomic weight of all isotopes present. This idea, discussed in Section 5-9, is the basis for our present atomic weight scale. On this scale, the mass of one atom of the most abundant isotope of carbon is assigned a mass of exactly 12 **atomic mass units (amu)**.

The most important aspect of atomic structure as it relates to chemistry is the arrangement of the electrons around the nucleus. The remainder of Chapter 5 is devoted to this topic. Most of our knowledge about electrons in atoms comes from a study of the ways in which matter interacts with light. Section 5-10 deals with the nature of light, which can sometimes be considered as a **wave** phenomenon (periodically varying electrical and magnetic fields—hence the name **electromagnetic radiation**). We learn that different colors of light correspond to different wavelengths or different frequencies of alternation of these fields. The other, equally important interpretation of light, applicable to other kinds of experiments, is in terms of a **particle** model. In this view, light consists of streams of particles called **photons**. For a given color of light, each photon consists of a particular amount of energy, called a **quantum**. Einstein's explanation of the **photoelectric effect** (Section 5-11) was an important early application of this view of light.

When excited atoms give off light (atomic emission spectra, Section 5-12), they give off only particular colors of light, i.e., only certain energies. This important observation shows that the atoms can undergo only certain specific energy changes. The **Bohr theory** of the atom introduces two ideas that are of central importance in our understanding of atoms: (1) an electron in an atom can be only in certain allowed places and (2) when it is in one of these allowed places, it possesses a definite amount of energy. We shall modify and extend these ideas later. The simple Bohr theory of the atom assumed that the electrons travel in well-defined orbits; each orbit is characterized by a positive integer, *n*, known as a **quantum number**. The Bohr theory actually only works well for simple atoms like hydrogen. The Enrichment feature on pages 201-203 emphasizes how the spectral wavelengths depend on *differences* in energies within the atom.

Very small particles such as the electron can also be described in terms of both wave and particle nature, Section 5-13. The modern **quantum-mechanical picture of the atom**, which we shall use throughout our study of chemistry, is introduced in Section 5-14. This theory extends the ideas of quantization of electronic energy and describes them in a highly mathematical approach, called **quantum theory**. In this theory, each electron in the atom is treated mathematically as though it were a wave. The electron is described by a wave equation called the Schrödinger equation, later made more general by Dirac. Even though we do not need to be concerned with the mathematical details, we must describe the solution to these wave equations—the wave function. We will use the solutions to these equations frequently in our study of chemistry.

In your study of the important Sections 5-14 through 5-17, keep in mind that this more refined and elegant theory, even though it looks quite complex, is really just concerned with two features of an electron in the atom: *where the electron is* and *what its energy is (how stable it is)*. Each electron in the atom corresponds to one of the solutions of the wave equation. Each solution is described by a particular set of four numbers, n, l, m_l and m_s known as the quantum numbers. You must become familiar with these numbers, their interpretations, and the rules for assigning them (Section 5-14).

Each valid set of n, l, and m_l defines one **atomic orbital**, which may be interpreted as a region in which there is a high probability of finding the electron. In Section 5-16 we see that each atomic orbitals are identified by a valid set of the quantum numbers n (distance from nucleus), l (orbital shape), and m_l (orbital orientation). In a given atom, orbitals that have the same values of n and l but that differ in their m_l values have the same energy; such orbitals are described as **equivalent** or **degenerate** orbitals.

You must learn to use rules based on the values of the four quantum numbers to tell the arrangement of the electrons in any specified atom, the **electron configuration** (Section 5-17). This is the *most important topic you have yet encountered in chemistry*. It serves as a basis for further understanding of chemical properties and bonding. The application of the rules and procedures that you learn here will later enable you to understand and predict properties of elements, chemical bonding, formulas, shapes of molecules, reactivities, and properties of compounds. Practice writing electron configurations of atoms, in several different notations.

We can figure out the electron configuration of an element by imagining that we add the electrons one by one, putting each into the available orbital of lowest energy. This imaginary procedure is called the **Aufbau procedure**, from the German word meaning "to build up." It is important to know the usual order that the orbitals fill: $1s$, $2s$, $2p$, $3s$, $3p$, $4s$, $3d$, $4p$, $5s$, $4d$, $5p$, $6s$, $4f$, $5d$, $6p$, $7s$, $5f$, $6d$, $7p$, ... A memory aid to help you reconstruct this order is given in Figure 5-29. But remember that this order is only a guide; the electron configurations of many atoms are not predicted by this order. Whatever the order of the orbitals, two additional rules that govern the order of filling and the number of electrons in the orbitals are the **Pauli Exclusion Principle** (Section 5-15) and **Hund's Rule** (Section 5-17).

Finally, we see in Section 5-18 that the periodic table, which was first introduced in Section 4-1 on the basis of chemical and physical similarities, corresponds to grouping of elements with similar electron configurations. In later chapters, we will see that elements with similar outer electron configurations are very similar in their chemical and physical properties. The periodic table is therefore an invaluable aid in classifying elements, their electron configurations, and their properties.

Study Goals

As in other chapters, some text Exercises related to the Study Goals are indicated.

1. Describe the experiments or ideas of the following scientists and their contributions to our knowledge of the structure of atoms: (a) Thomson, (b) Millikan, (c) Rutherford, (d) Moseley, (e) Rydberg, (f) Bohr, (g) Planck, (h) deBroglie, (i) Davisson and Germer, (j) Schrödinger, and (k) Dirac.
2. Describe the three fundamental particles that make up atoms. Describe the experimental evidence that led to the discovery of these three fundamental particles. (Sections 5-1 through 5-3 and 5-6; Exercises 1 through 3)
3. Describe the behavior of charged and uncharged particles and rays in electrical and magnetic fields. (Sections 5-2, 5-3, and 5-8; Exercises 2 through 4, 10, and 11)
4. Describe the nuclear atom. (Section 5-4; Exercises 5 through 9)
5. Distinguish between isotopes and atoms. (Section 5-7; Exercise 14)

6. Distinguish among atomic number, mass number, atomic weight, and isotopic weight. Carry out calculations based on these quantities. (Sections 5-5 through 5-8; Exercises 12 through 36)

7. Describe and illustrate the basic ideas of mass spectrometry. (Sections 5-7 and 5-8; Exercises 10, 11, 33, and 34)

8. From its nuclide symbol, tell the number and type of fundamental particles present in any atom. (Section 5-7; Exercises 15 through 18 and 20 through 23)

9. Understand the wave and particle descriptions of light and why both descriptions are useful. (Sections 5-10 and 5-11; Exercises 49 and 52)

10. Be able to carry out calculations to relate the descriptions of light to its wavelength, frequency, and energy. (Section 5-10; Exercises 37 through 48, 51, and 52)

11. Understand the relationship between the observation of atomic line spectra and the idea of distinct energy levels in atoms. Describe how the wavelength of light emitted or absorbed by an atom depends on the separations between its electronic energy levels. (Section 5-12; Exercises 53 through 62)

12. Understand the basis for the description of particles, such as the electron, either as a wave or as a particle. (Section 5-13; Exercises 63 through 66)

13. State and understand the basic ideas of quantum mechanics. (Section 5-14; Exercise 67)

14. Learn the four quantum numbers, what they represent, and what values they can have. (Sections 5-14 and 5-15; Exercises 68 and 69)

15. Understand the relationship between atomic orbitals and the corresponding values of the quantum numbers. Be familiar with the notation for identifying orbitals. (Sections 5-15 and 5-16; Exercises 70 through 79)

16. Describe atomic orbitals. Sketch the shapes of s, p, and d orbitals. Describe how the same kinds of orbitals in different shells differ. (Section 5-16; Exercises 69, 70, and 80)

17. Learn how the energies of orbitals depend on the quantum numbers. (Section 5-16)

18. Learn and apply the rules for the filling of electrons in atomic orbitals. Write out the electron configurations of specified atoms. Write out the electron configurations of especially the first 36 elements, both with arrow notation and with condensed notation. (Section 5-17; Exercises 81 through 91, 93 through 98, and 104 through 109)

19. Understand how the magnetic properties of an element depend on the number of unpaired electrons in an atom of the element. (Section 5-17; Exercises 92 and 99 through 102)

20. Relate the electron configuration of an atom to its position in the periodic table. Use Table 5-5 to write electron configurations of elements. (Section 5-18; Exercises 103, 110, and 111)

21. Answer conceptual questions based on this chapter. (Conceptual Exercises 112 through 117)

22. Apply concepts and skills from earlier chapters to the ideas of this chapter. (Building Your Knowledge exercises, 118 through 123)

Some Important Terms in This Chapter

Write the meanings *in your own words*. Check the Key Terms list and the chapter reading. Then rewrite your definitions, still in your own words, to improve them. Study other new terms, and review terms from preceding chapters if necessary.

electron

proton

neutron

atomic number

mass number

atomic mass unit

nucleus

isotope

nuclide symbol

electromagnetic radiation

frequency

wavelength

photon

quantum

quantum numbers

principal quantum number

subsidiary quantum number

magnetic quantum number

spin quantum number

atomic orbital

s orbital

p orbital

d orbital

f orbital

electron configuration

Aufbau Principle

Pauli Exclusion Principle

Hund's Rule

pairing

group

period

Preliminary Test

As in other chapters, this test will check your understanding of basic concepts and types of calculations. Be sure to practice *many* of the additional textbook Exercises, including those indicated with the Study Goals.

True-False

Mark each statement as true (T) or false (F).

_____ 1. Most of the mass of an atom is due to the presence of electrons.

_____ 2. The figure "56" in the symbol $^{56}_{26}\text{Fe}$ represents the total number of electrons and protons in the nucleus of an iron atom.

_____ 3. If we know that the atomic number of an element is 81, then the element must be thallium (Tl).

_____ 4. If we know that an element is thallium, then its atomic number must be 81.

_____ 5. If we know that an atom has mass number 101, then the element must be ruthenium (Ru).

_____ 6. In an atom, the protons and neutrons are in the nucleus, and the electrons are clustered together at some distance from the nucleus.

_____ 7. The characterization of cathode rays showed them to consist of streams of protons.

_____ 8. All cathode ray particles have the same charge/mass (e/m) ratio.
_____ 9. All canal ray particles have the same charge/mass (e/m) ratio.
_____ 10. In the Rutherford experiment, the scattering of streams of alpha particles by the atoms in a metal foil was studied.
_____ 11. The feature that distinguishes the atoms of various elements is their mass number.
_____ 12. All naturally occurring elements exist as mixture of isotopes.
_____ 13. Atomic numbers are always integers.
_____ 14. Mass numbers are always integers.
_____ 15. Masses of individual atoms are always integers.
_____ 16. Atomic weights are always integers.
_____ 17. The experimentally observed atomic weight of carbon is listed as 12.011, but this is due to a slight error in the definition of the atomic weight scale, because it should be exactly 12.
_____ 18. There are many colors of light that are not visible to the human eye.
_____ 19. The longer the wavelength of light, the higher is its frequency.
_____ 20. The higher the frequency of light, the higher is the energy of one photon of the light.
_____ 21. "A particle is a particle and a wave is a wave, and never the twain shall meet." (Apologies to Rudyard Kipling)
_____ 22. The number of lines in the emission spectrum of an element is equal to the number of electronic energy levels in the atom.
_____ 23. Each combination of n, l, and m_l corresponds to an orbital.
_____ 24. When two electrons occupy the same orbital, they must have opposite spins.
_____ 25. In a given atom, no more than two electrons can occupy a single orbital.
_____ 26. The second main energy level in an atom can have any number of electrons up to 10.
_____ 27. Electrons can exist only in atoms.
_____ 28. All neutral atoms of any given element are identical in all respects.
_____ 29. The higher its value of n, the higher is the energy of an atomic orbital.

Short Answer

Answer with a word, a phrase, a formula, or a number with units as necessary.

1. Fill in the following chart to indicate the effect of positively and negatively charges electrical plates on protons, electrons, and neutrons. Use the notation A for attracted, R for repelled, and U for unaffected.

	proton	electron	neutron
positive plate	(a) ____	(b) ____	(c) ____
negative plate	(d) ____	(e) ____	(f) ____

2. The charge on the nucleus of a carbon atom is _____.
3. The fundamental particle with positive charge is the _____.
4. The fundamental particle with no charge is the _____.
5. In an atom, the _____ and the _____ are in the nucleus.
6. The two fundamental particles that have nearly the same mass are the _____ and the _____.

7. The two fundamental particles that have equal but opposite charges are the _____ and the _____.

8. Think about the statement, "The nucleus is a combination of protons and neutrons." If we carefully consider the table of atomic numbers and observed atomic weight, we can see that the element _____ must have one isotope that is an exception to this statement.

9. The three kinds of radiation given off by radioactive substances are called _____, _____, and _____.

10. The symbol for an alpha particle is _____; this particle is shown to be identical to the ion that is designated by _____.

11. If a beam containing two kinds of particles having the same mass but different charge is sent through a magnetic field, the particles with the _____ charge would be deflected most.

12. If a beam containing two kinds of particles having the same charge but different mass is sent through a magnetic field, the particles with the _____ mass would be deflected most.

13. Atoms of the same element having different masses are called _____.

14. The symbol $^{23}_{11}Na$ is called as _____ symbol.

15. In the symbol $^{23}_{11}Na$, the number 11 is called the _____ and represents the number of _____; it also represents the number of _____ for the neutral atom.

16. In the symbol $^{23}_{11}Na$, the number 23 is called the _____ and represents the _____.

17. The element indium, In, consists of two isotopes. These two isotopes are designated by the symbols $^{113}_{49}In$ and $^{115}_{49}In$. The observed atomic weight of naturally occurring indium is 114.8. As a very good approximation, take the mass of a proton and that of a neutron to be 1 amu. The natural abundance of the two isotopes is _____ percent $^{113}_{49}In$ and _____ percent $^{115}_{49}In$.

18. The numbers of subatomic particles in a neutral atom of palladium (Pd) with mass number 106 are _____ protons, _____ neutrons, and _____ electrons.

19. The quantum number that designates the main energy level an electron occupies is _____, called the _____ quantum number.

20. The quantum number that designates the different possible subshells within a main shell is _____, called the _____ quantum number.

21. The value of _____, called the _____ designates the shape of the orbital.

22. The value of _____, called the _____ quantum number, designates the spatial orientation of an orbital.

23. The quantum number that designates the spin of the electron is _____.

24. An orbital that is spherically symmetrical has $l =$ _____.

25. An orbital that is spherically symmetrical is designated as a(an) _____ orbital.

26. An orbital with $l = 1$ is designated as a(an) _____ orbital.

27. A p orbital has _____ regions of high electron density.

28. In a single atom, the maximum number of electrons possible in a fourth main energy level is _____.

29. If two electrons in the same atom have the same values of n, l, and m_l, they must have _____.

30. For any value of the principal quantum number (except $n = 1$) there are three p orbitals. These three p orbitals correspond to regions of high electron density in three different directions from the atomic center. The three p orbitals on the same atom make angles of _____ with each other.

31. In the normal order of occupancy of electronic energy levels, the sublevel occupied just after $3p$ is ____.

32. In the normal order of occupancy of electronic energy levels, the sublevel occupied just after $4s$ is ____.

33. The ground state electron configuration of bromine (Br) is _____.

34. The ground state electron configuration of arsenic (As) is _____.

35. The number of unpaired electrons in a neutral atom of sulfur is ____.

36. The number of unpaired electrons in a neutral atom of iron is ____.

37. We describe a particular orbital as a $3d$ orbital. The possible values of the magnetic quantum number for this orbital are _____.

38. Substances that are strongly attracted into magnetic fields are termed _____ substances. The atomic feature that is responsible for this behavior is the presence of _____ in atoms.

39. In the periodic table, a vertical column of elements is called a _____.

40. In the periodic table, a horizontal row of elements is called a _____.

41. For the elements in Group IIIA, the configuration of the outermost electrons is _____.

42. For the elements in Group VIA, the number of unpaired electrons in each atom is _____.

43. In the periodic table, the nth period begins with the element whose outer electron configuration is _____ and ends with the element whose outer electron configuration is _____.

Multiple Choice

Select the one best answer for each question.

_____ 1. Two isotopes of the same element must differ in their
(a) atomic number.   (b) mass number.   (c) total charge.
(d) number of protons.   (e) number of electrons.

_____ 2. Given the unidentified isotope $^{61}_{27}X$, which of the following is true?
(a) The atom contains 61 neutrons.
(b) The neutral atom contains 27 electrons.
(c) The element must be promethium, Pm.
(d) The atom contains 34 protons
(e) The atom contains a number of protons plus electrons equal to 61.

_____ 3. Which of the following statements about magnesium atoms is *not correct*?
(a) All magnesium atoms have an atomic number of 12.
(b) All magnesium atoms have the same number of protons.
(c) All neutral magnesium atoms have 12 protons and 12 electrons.
(d) All magnesium atoms have the same number of neutrons.
(e) All atoms with an atomic number of 12 are magnesium atoms.

_____ 4. If a neutral atom has an atomic number of 28 and a mass number of 60, it contains
(a) 28 neutrons.   (b) 32 electrons.   (c) 60 electrons.
(d) 32 protons.   (e) a total number of nuclear particles equal to 60.

_____ 5. Cathode rays are _____. These particles have the _____ charge/mass ratio of the three fundamental particles found in the atom.
(a) neutrons, largest   (b) protons, smallest   (c) electrons, largest   (d) protons, largest
(e) None of the preceding answers is correct.

_____ 6. The discovery and characterization of cathode rays was important in the development of atomic theory because
  (a) it indicated that all matter contains protons.
  (b) it indicated that all matter contains alpha particles.
  (c) it indicated that all matter contained electrons.
  (d) it provided a basis for weight calculations in chemical reactions.
  (e) it led to the suggestion of the existence of the neutron.

_____ 7. Which of the following provided evidence that most of the mass of the atom, as well as all of the positive charge, is concentrated in a very small core, the nucleus?
  (a) the analysis of X-ray wavelengths
  (b) the scattering of alpha particles by a metal foil
  (c) the deflection of ions in a mass spectrometer
  (d) the results of the Millikan oil-drop experiment
  (e) the existence of elements with noninteger values for atomic weights.

_____ 8. What is the number of electrons in 1 mole of carbon atoms?
  (a) 6   (b) 12   (c) 6 x 6.022 x 10^{23}
  (d) 12 x 6.022 x 10^{23} (e) depends on the isotope of carbon

_____ 9. Suppose that a hypothetical element consists of a mixture of two isotopes. One isotope, having a mass of 44, is present in 70.0% atomic abundance, while the other isotope, having a mass of 46, accounts for the other 30.0%. We expect the experimentally determined atomic weight for this hypothetical element to be
  (a) 44.0   (b) 46.0   (c) 45.0   (d) 44.6   (e) 45.4

_____ 10. In an atom
  (a) the electrons contain most of the mass of the atom.
  (b) the nucleus occupies most of the volume of the atom.
  (c) all of the electrons and protons are in the nucleus.
  (d) the neutrons are in the nucleus.
  (e) the neutrons carry all of the negative charge.

_____ 11. When a beam of charged particles is sent through an electrical or magnetic field, the path of the particles becomes curved as the particles are deflected by the field. Which of the following statements correctly relates the direction and the amount of this curvature to the properties of the particles?
  (a) The direction of curvature depends on the charge; the amount of curvature depends on the e/m ratio.
  (b) The direction of the curvature depends on the e/m ratio; the amount of curvature depends on the charge.
  (c) The direction of curvature is independent of the charge; the amount of curvature depends on the e/m ratio.
  (d) The direction of curvature depends on the e/m ratio; the amount of curvature is does not depend on the e/m ratio (the sign of the charge does not affect these properties).
  (e) None of the above statements is correct.

_____ 12. The number of neutrons in one atom of the iron isotope $^{56}_{26}$Fe is _____.
  (a) 56   (b) 26   (c) 56 − 26   (d) 56 + 26
  (e) indeterminate unless we know the observed atomic weight of naturally occurring iron

_____ 13. The following are all proposals of Dalton's atomic theory. Which part of Dalton's atomic theory was shown to be wrong by the discovery of natural radioactivity?
   (a) Atoms of a given element all weigh the same.
   (b) Two or more kinds of atoms may combine in different ways to form more than one kind of chemical compound.
   (c) Atoms combine in simple numerical ratios to form compounds.
   (d) Atoms are permanent, unchanging, indivisible bodies.
   (e) Compounds consist of collections of molecules made up of atoms bonded together.

_____ 14. Which of the following must represent an element that is *different* from the others?
   (a) something with 12 protons, 24 neutrons, and 12 electrons
   (b) something with 12 protons, 24 neutrons, and 10 electrons
   (c) something with 12 protons, 26 neutrons, and 12 electrons
   (d) something with 10 protons, 24 neutrons, and 12 electrons
   (e) All of the preceding are isotopes of the same element.

_____ 15. The positive rays that are sometimes found in cathode ray tubes are characterized as particles having a smaller charge-to-mass ratio for heavier elements. The positive ray particles that are formed when the gas in the tube is hydrogen have a larger observed charge-to-mass ratio than those for any other gas. Which of these assumptions explains these observations?
   (a) The particles are all alpha particles.
   (b) The particles are all identical, regardless of the gas in the tube.
   (c) The particles are all ions formed by the removal of one or more electrons from neutral atoms of gases in the tube.
   (d) Each gas consists of a mixture of at least two kinds of atoms.
   (e) The positive ray particles are neutrons.

_____ 16. If a beam of each of the following were sent through a mass spectrometer, which would follow the most curved path (that is, be deflected most)?
   (a) neutral Ti atoms, mass number 47   (b) neutral Ti atoms, mass number 48
   (c) Ti^+ ions, mass number 47   (d) Ti^{2+} ions, mass number 47
   (e) Ti^{2+} ions, mass number 48

_____ 17. Which of the following provided evidence that the electrons in atoms are arranged in distinct energy levels?
   (a) the observation of line spectra from gas discharge tubes
   (b) the scattering of alpha particles by a metal foil
   (c) the deflection of ions in a mass spectrometer
   (d) the results of the Millikan oil-drop experiment
   (e) the existence of elements with noninteger values for atomic weights

_____ 18. Suppose that an atom had available to it *only* the six energy levels indicated below. What would be the total number of lines in the emission spectrum of this element?

$$
\begin{array}{ll}
\text{————} & n = 6 \\
\text{————} & n = 5 \\
\text{————} & n = 4 \\
\text{————} & n = 3 \\
\text{————} & n = 2 \\
\text{————} & n = 1
\end{array}
$$

   (a) 5   (b) 6   (c) 15   (d) 21   (e) 36

_____ 19. Bohr's theory of the hydrogen atom assumed that
  (a) electromagnetic radiation is given off when the electron moves in an orbit around the nucleus.
  (b) the electron can exist in any one of a set of discrete states (energy levels) without emitting radiation.
  (c) because a hydrogen atom has only one electron, the emission spectrum of hydrogen should consist of only one line.
  (d) the electron in a hydrogen atom can jump from one energy level to another without gain or loss of energy.
  (e) energy, in the form of radiation, must be continually supplied to keep the electron moving in its orbit.

_____ 20. An electron in a hydrogen atom could undergo any of the transitions listed below by changing its energy. Which transition would give off (emit) light of the longest wavelength. (You will not need to calculate wavelengths, but think about the Rydberg equation or the Planck relationship.)
  (a) $n = 3$ to $n = 2$ (b) $n = 3$ to $n = 1$ (c) $n = 2$ to $n = 1$
  (d) $n = 2$ to $n = 3$ (e) $n = 1$ to $n = 3$

_____ 21. An orbital that has $n = 4$ and $l = 2$ is identified as a _____ orbital.
  (a) $2f$ (b) $3s$ (c) $4s$ (d) $4p$ (e) $4d$

_____ 22. Which of the following is *not* a valid combination of quantum numbers?
  (a) $n = 3$, $l = 4$, $m_l = 0$, $m_s = 1/2$ (b) $n = 5$, $l = 4$, $m_l = 0$, $m_s = -1/2$
  (c) $n = 3$, $l = 2$, $m_l = -2$, $m_s = -1/2$ (d) $n = 2$, $l = 1$, $m_l = 1$, $m_s = 1/2$
  (e) $n = 4$, $l = 0$, $m_l = 0$, $m_s = 1/2$

_____ 23. Can an electron in an atom be described by the set of quantum numbers described below?
$$n = 5, l = 3, m_l = -2, m_s = -1/2$$
  (a) No, because l must be equal to $(n - 1)$.
  (b) No, because m_l cannot be negative.
  (c) No, because n cannot be as large as 5.
  (d) No, because m_s must be positive.
  (e) Yes.

_____ 24. How many orbitals are there in a set of equivalent p orbitals?
  (a) 1   (b) 2   (c) 3   (d) 5
  (e) The answer depends on the value of n (2p, 3p, etc.)

_____ 25. Which of the following orbitals has a shape that we describe as "spherically symmetrical"?
  (a) an s orbital   (b) a p orbital   (c) a d orbital
  (d) an f orbital   (e) all of the preceding

_____ 26. In a given atom, how many electrons can occupy the 3p set of orbitals?
  (a) 0   (b) 2   (c) 3   (d) 6   (e) 10

_____ 27. In a given atom, how many electrons can occupy the 4d set of orbitals?
  (a) 0   (b) 2   (c) 3   (d) 6   (e) 10

_____ 28. In a given atom, how many electrons can occupy the 2d set of orbitals?
  (a) 0   (b) 2   (c) 3   (d) 6   (e) 10

_____ 29. Which of the following atoms would be most strongly paramagnetic?
  (a) S   (b) Co   (c) Cr   (d) Cl   (e) Ne

_____ 30. Which of the following elements has three half-filled 2p orbitals in the ground state electron configuration of the neutral atom?
  (a) C   (b) N   (c) O   (d) F   (e) Ne

_____ 31. A neutral atom has a ground state electron configuration that we can designate $1s^2\,2s^2$. Which of the following statements about this atom is (or are) correct?
   (a) The element has an atomic number of 4 and is beryllium.
   (b) The atom contains four protons.
   (c) The atom has no unpaired electrons.
   (d) The atom has a total of two orbitals occupied.
   (e) All of the preceding statements are correct.

Answers to Preliminary Test

True-False

1. False. The electrons are the lightest of the three subatomic particles and contribute little to the total mass.
2. False. There are two reasons. First, the electrons are not in the nucleus, and second, 56 is the mass number, which represents the sum of the number of protons and the number of neutrons in the nucleus.
3. True. The atomic number uniquely defines the element.
4. True
5. False. It could be an isotope of Ru or of another element. Mass number does not uniquely define the element. There are many cases of more than one element having isotopes with the same mass number.
6. False. The electrons are distributed throughout a relatively large empty space surrounding the nucleus.
7. False. Cathode ray particles are electrons.
8. True
9. False. The canal ray particles are the positive ions that remain when an atom loses one or more electrons. These have different masses, depending on the identity of the original atom.
10. True
11. False. The atomic number (i.e., number of protons in the nucleus) determines the identity of the element. This is contrary to the earlier Daltonian idea of the atom.
12. False. See Table 5-3 and the accompanying discussion in Section 5-7.
13. True
14. True
15. False. See Sections 5-6 through 5-8.
16. False. The observed atomic weight of an element is the weighted average of the masses of the constituent isotopes. See the sample calculations in Section 5-9.
17. False. The atomic weight is defined so that the mass of the most common isotope of carbon is exactly 12 amu. Naturally occurring carbon also contains small amounts of another isotope, so the average atomic weight is not exactly 12. See Section 5-9.
18. True. You have probably already heard of ultraviolet and infrared light. The light visible to the human eye is only a very small part of the entire electromagnetic spectrum.
19. False. See Section 5-10. Wavelength and frequency are inversely proportional to one another. This point is illustrated by Example 5-3.
20. True. See Section 5-10. Energy is directly proportional to frequency. Be sure you understand the significance of Example 5-5.

21. False. Both light and particles such as the electron can exhibit either wave or particle behavior. See Sections 5-10 through 5-13.

22. False. Each line in the spectrum is the result of a transition between two electronic energy levels in the atom.

23. True. But remember that orbitals with the same values of n and l but different m_l values have the same energy in an isolated atom.

24. True

25. True

26. False. The second main energy level can have a maximum of eight electrons.

27. False. Recall that cathode rays are streams of electrons (Section 5-2) and that atoms can lose electrons to form ions (Section 5-3). Diffraction and scattering experiments on beams of electrons are mentioned in Section 5-13.

28. False. It is true that all atoms of a given element have the same number of protons; therefore to be neutral they must also have the same number of electrons. However, it is possible for the atoms to have different numbers of neutrons; thus they represent different isotopes of the same element.

29. False. You should know the normal order of electron energy levels, Figures 5-28 and 5-29. Remember, though, that the order of energies is sometimes different, depending on the occupancy of the orbitals. See Section 5-17, especially the discussion concerning the configurations of Cr and Cu.

Short Answer

1. (a) R; (b) A; (c) U; (d) A; (e) R; (f) U. Like charges repel, unlike charges attract, and particles having no charge are not affected by an electrical (or a magnetic) field.

2. 6+. Remember that the atomic number gives the number of protons in the nucleus. The only other particles in the nucleus are the neutrons, which are uncharged.

3. proton

4. neutron

5. protons; neutrons

6. proton; neutron

7. proton; electron

8. hydrogen. The experimentally observed atomic weight is only slightly greater than 1, so hydrogen must consist at least partially (in fact mostly) of an isotope with mass number 1. This isotope would have only one proton and no neutrons in the nucleus. In fact, this isotope of hydrogen is the only exception to the statement.

9. alpha rays (or particles); beta rays (or particles); gamma rays. Gamma rays, which consist of very high energy light, are usually not called gamma particles.

10. α; He^{2+}

11. larger. The deflection depends on the charge/mass ratio.

12. smaller

13. isotopes

14. nuclide

15. atomic number; protons; electrons

16. mass number; sum of number of protons and number of protons (or total number of nuclear particles)

73

17. 10; 90. Let x = fraction of atoms that have mass number 113; then $(1 - x)$ equals the fraction that have a mass number of 115. So

$$x(113) + (1 - x)(115) = 114.8$$
$$113x + 115 - 115x = 114.8$$
$$2x = 0.2 \quad \text{or} \quad x = 0.10 \text{ and } (1 - x) = 0.90$$

18. 46; 60; 46
19. n; principal (note the spelling—not principle!)
20. l; subsidiary (or azimuthal)
21. l, subsidiary (or azimuthal)
22. m_l, magnetic
23. m_s
24. 0
25. s
26. p
27. two
28. 32. For $n = 4$, the possible values of l are 0 (one s orbital), 1 (three p orbitals), 2 (five d orbitals), and 3 (seven f orbitals) for a total of 16 orbitals. Each orbital can hold 2 electrons for a total of 32 electrons possible in this energy level.
29. opposite values of m_s. Study the Pauli Exclusion Principle (Section 5-15).
30. 90°
31. $4s$
32. $3d$
33. $1s^2\, 2s^2\, 2p^6\, 3s^2\, 3p^6\, 4s^2\, 3d^{10}\, 4p^5$. Remember that it would also be correct to write the electron configuration with all orbitals having the same value of principal quantum number appearing together. Thus, we could have written $1s^2\, 2s^2\, 2p^6\, 3s^2\, 3p^6\, 3d^{10}\, 4s^2\, 4p^5$. This shows $3s$, $3p$, and $3d$ together, even though in the normal order $4s$ fills before $3d$.
34. $1s^2\, 2s^2\, 2p^6\, 3s^2\, 3p^6\, 4s^2\, 3d^{10}\, 4p^3$. See the additional comments in the answer to Question 33.
35. 2. The electron configuration is $1s^2\, 2s^2\, 2p^6\, 3s^2\, 3p^4$. In the $3p$ orbitals, the electrons are arranged as ↑↓ ↑ ↑ . All other electrons are paired.
36. 4. The electron configuration is $1s^2\, 2s^2\, 2p^6\, 3s^2\, 3p^6\, 4s^2\, 3d^6$. In the arrow notation, $3d^6$ represents ↑↓ ↑ ↑ ↑ ↑ . All other electrons are paired.
37. $-2, -1, 0, +1, +2$. This is a d orbital, so $l = 2$. Thus m_l can range from $-l$ to $+l$ by integers.
38. paramagnetic; unpaired electrons
39. group
40. period
41. $ns^2\, np^1$
42. 2. Each of these elements has the electron configuration $ns^2\, np^4$. The two s electrons are paired in a single orbital. The four p electrons are arranged as ↑↓ ↑ ↑ .
43. ns^1; $ns^2\, np^6$

Multiple Choice

1. (b)
2. (b)
3. (d). The observed atomic weight of magnesium is greater than 24, so there must be an isotope of magnesium that has a mass number greater than 24. This isotope would have more neutrons than the principal isotope, $^{24}_{12}\text{Mg}$.

4. (e)
5. (c)
6. (c)
7. (b)
8. (c). The atomic number of carbon is 6, which tells us that *each* atom of C contains 6 electrons, regardless of which isotope we are considering). But 1 mole of carbon contains 6.022×10^{23} carbon atoms, so the total number of electrons is

$$1 \text{ mol C atoms} \times \frac{6.022 \times 10^{23} \text{ C atoms}}{\text{mol C atoms}} \times \frac{6 \text{ electrons}}{\text{C atom}}$$

9. (d). Calculate as in Examples 5-1 and 5-2 of the text. Study Example 5-3, which uses the same ideas to solve for a different quantity.
10. (d)
11. (a)
12. (c)
13. (d)
14. (d). The number of protons (atomic number) determines the element. Answer (d) has a different atomic, so it must be a different element. Answers (a) and (c) are isotopes of the same element. Answer (b) is an ion formed from (a) by loss of two electrons, but it's still the same element.
15. (c)
16. (d). The larger the charge-to-mass ratio (e/m), the greater the deflection. Neutral atoms would not be deflected, because they have no charge. Answer (d) has twice as great a charge as (c), but the same mass, so (d) would be deflected more. Answer (e) also has a 2+ positive charge, but it has a larger mass than answer (d), so it has a *smaller* e/m ratio.
17. (a)
18. (c). Each transition from a higher to a lower energy level gives rise to one line in the emission spectrum. An electron could go from level $n = 6$ to any of five lower-energy levels (five possible transitions), from $n = 6$ to any of four lower-energy levels (four possible transitions), and so on. There are 15 such downward transitions.
19. (b)
20. (a). Answers (d) and (e) correspond to absorption of light, not emission. Of the other answers, (a) corresponds to the smallest energy difference (remember that the energy levels get closer together at higher n.). Wavelength is inversely proportional to energy of the light.
21. (e)
22. (a). l can be no larger than $(n - 1)$.
23. (e). Review the rules for the allowed values of the quantum numbers, Sections 5-15 and 5-16.
24. (c)
25. (a). See Section 5-16 regarding orbital shapes.
26. (d)
27. (e)
28. (a). There is no such thing as a 2d orbital This would mean that $n = 2$ and $l = 2$, which would violate the rules for the allowed values of quantum numbers.
29. (c). Cr has 6 unpaired electrons, the largest number of any of the atoms listed.
30. (b)
31. (e)

6 Chemical Periodicity

Chapter Summary

In the beginning of Chapter 4, we were introduced to the periodic table as a way of representing similarities in reactions and compounds of groups of elements. As we saw in Section 5-18, the marvelous utility of the periodic table in chemistry arises from the relationship of properties of the elements to their electron configurations. As you study this chapter, you will see many correlations between the regularities that many properties that elements exhibit and the regularities of their electron configurations in the periodic table. When you finish, you should be able to interpret the periodic table to remind you of electron configurations and to predict or correlate properties of substances.

Before you begin your study of this chapter, review the basic terminology of the periodic table that was presented in Section 4-1. Each horizontal row is referred to as a **period**, and each vertical column is called a **group**. You were also introduced to the common names of some of the groups of elements—the **alkali metals, alkaline earth metals, halogens,** and **noble gases**—and the classification of elements as **metals, nonmetals,** or **metalloids**. Another way of classifying elements for discussion is in terms of the "last" electron added to the electron configuration in the *Aufbau* process we learned in Chapter 5. (Review Section 5-18, especially Table 5-5. You should be able to tell quickly by looking at the position of an element in the periodic table into which kind of orbital the last electron went.) Thus, elements in which this last electron is in an *s* or a *p* orbital (excepting the noble gases) are termed **representative elements**, those in which the last electron is in a *d* orbital are **transition elements** or **transition metals**, and those with the last electron in an *f* orbital are **inner transition elements** (Section 6-1).

Many chemical and physical properties are related to the positions of the elements in the periodic table. Some of these are discussed in Sections 6-2 through 6-6. In turn, these properties will help us to understand bonding between elements and the properties of the resulting compounds. Keep two goals in mind as you study these sections. The first is to be able to explain how each of these observed trends results from changes in electron configurations across a period and down a group. The second is the reverse—using its position in the periodic table to predict the properties of an element.

One property that is easily correlated with position in the periodic table is the size of the atom, usually described in terms of the **atomic radii** of the elements (Section 6-2). The correlations that emerge are that the atomic sizes of the (representative) elements within a period decrease from left to right across a period, while within a group the sizes increase from top to bottom. Be sure that you understand how these trends in atomic radii are related to regular changes in electron configuration.

Chemical reactions result from the interaction of electrons of the elements involved. It is important to understand how tightly the electrons of different atoms are held, and how easily each kind of atom gains or loses electrons. When an atom loses one or more electrons, it forms a **positive ion (cation)**. When an atom loses one or more electrons, it forms a **negative ion (anion)**. The ease with which these two processes can occur can be measured as the **ionization energy** and the **electron affinity**, respectively, of the element. Sections 6-3 and 6-4 correlate trends in these two properties with position in the table, and hence with electron configurations of the elements. These correlations reveal that the ease with which the electrons are lost or gained depends both on the size of the atoms and on how effectively the nucleus can attract the outermost electrons. The ions that are thus formed are characteristically smaller (for cations) or larger (for anions) than the corresponding neutral atoms. These trends in ionic radii can also be related to the types of orbitals that the outermost electrons occupy (Section 6-5).

Section 6-6 introduces a very important concept that we shall use often for predicting and understanding chemical bonding. This is the **electronegativity scale**, a relative scale that expresses the tendency of an atom to attract electrons. You must remember the correlations of electronegativity with position in the periodic table. Electronegativities generally increase from left to right across a period and decrease from top to bottom within a group. We do not apply this terminology to the noble gases; because of their extremely stable configurations, these gases have practically no tendency to gain or lose electrons.

Oxygen and hydrogen combine with more elements than any others. For this reason, a study of the reactions and compounds of these two elements provides an illustration of the concepts and classifications embodied in the periodic table. Section 6-7 describes the properties of the element hydrogen and some methods of its preparation. In its reactions, hydrogen combines with metals or with nonmetals to form binary (two element) compounds called **hydrides.** These range in their properties from the ionic hydrides (with the more active metals) to the molecular hydrides (with nonmetals), through some compounds of intermediate character. The periodic table is the key to keeping track of the formulas and properties of these compounds.

Section 6-8 takes a similar approach to the properties, preparation, reactions, and compounds of oxygen. Most of this free element exists in the familiar diatomic form, O_2, sometimes called **dioxygen,** but usually just referred to as **oxygen.** The other **allotropic** form of this element, O_3, called ozone, is a very reactive gas. Oxygen combines with more elements than any other element, forming binary compounds (**oxides**) by direct reaction with all elements except the noble gases and a few of the less reactive metals, such as silver and gold. Oxygen does form compounds indirectly with several of these. You will study more about the reactions of this important element when we discuss the descriptive chemistry of other elements. The binary compounds of oxygen range in character from ionic (with metals) to molecular (with nonmetals). In the binary compounds with metals, oxygen can exist as the **oxide ion** (O^{2-}), as the **peroxide ion** (O_2^{2-}), or as the **superoxide ion** (O_2^{-}). Though the first of these is the most common, the tendency to form the others is correlated to trends in the periodic table (Section 6-8). The three types of oxide reactions discussed here are (1) the reaction of metal oxides (basic anhydrides) with water to form bases, (2) the reaction of nonmetal oxides (acid anhydrides) with water to form **ternary acids** (containing three elements—H, O, and a nonmetal), and (3) the reaction of these two types of oxides with one another to form salts. Note that all three of these occur without change of oxidation number.

The last two parts of Section 6-8 discuss **combustion reactions**; these are redox reactions in which oxygen combines rapidly with oxidizable materials in highly exothermic reactions with a visible flame (we call this "burning"). Some aspects of the atmospheric pollution that results from the combustion of fossil fuels are discussed.

Study Goals

Be sure to practice many of the indicated end-of-chapter Exercises, in addition to answering the Preliminary Test in this guide.

1. Relate the electron configuration of an atom to the position of the element in the periodic table (also Goal 20 of Chapter 5). (Section 6-1; Exercises 2 through 5, 7, 8, and 11 through 13)

2. Give examples of each of the following: (a) alkali metals, (b) alkaline earth metals, (c) halogens, (d) noble gases, (e) representative elements, (f) *d*-transition elements, (g) inner transition elements or *f*-transition elements, (h) lanthanides, and (i) actinides. Relate each of these classes to electron configuration and to position in the periodic table. (Review Section 4-1; Section 6-1; Exercises 1, 6, 9, and 10)

3. Summarize horizontal and vertical trends in the periodic table for each of the following properties. Relate each of these properties and their trends to electron configurations:
   (a) atomic radius (Section 6-2; Exercises 14 through 19)
   (b) ionization energy (Section 6-3; Exercises 20 through 31)
   (c) electron affinity (Section 6-4; Exercises 32 through 36)
   (d) ionic radius (Section 6-5; Exercises 37 through 42)
   (e) electronegativity (Section 6-6; Exercises 43 through 48)

4. Know some preparations and properties of hydrogen (H_2) and oxygen (O_2). (Sections 6-7 and 6-8; Exercises 56, 63 through 65, and 67)

5. Know some of the reactions of (a) hydrogen and the hydrides and (b) oxygen and the oxides. (Sections 6-7 and 6-8; Exercises 57, 58, 61, 68, and 69)

6. Recognize and give examples of the following classes of compounds: (a) metal hydrides, (b) nonmetal hydrides, (c) metal oxides (basic anhydrides), peroxides, and superoxides, (d) nonmetal oxides (acidic anhydrides). Give the preparations and reactions of these classes of compounds. (Sections 6-7 and 6-8; Exercises 59, 60, 62, 66, and 70 through 75

7. Know the basic features of combustion reactions. (Sections 6-8, last two subsections); Exercises 76 through 82)

8. Use the periodic table as a basis to recognize and answer a variety of types of questions about elements and their properties. (Exercises 49 through 55)

9. Answer conceptual questions based on this chapter. (Conceptual Exercises 83 through 87)

10. Apply concepts and skills from earlier chapters to the ideas of this chapter. (Building Your Knowledge exercises, 88 through 92)

Some Important Terms in This Chapter

Write the meanings *in your own words*. Check the Key Terms list and the chapter reading. Then rewrite your definitions, still in your own words, to improve them. Study other new terms, and review terms from preceding chapters if necessary. (Some of these terms have already been introduced in Chapters 4 or 5).

representative elements

d-transition elements

inner transition elements

metal

nonmetal

metalloid

noble gases

halogens

alkali metals

alkaline earth metals

atomic radius

effective nuclear charge

screening effect

ionization energy

electron affinity

anion

cation

ionic radius

electronegativity

hydride

acid anhydride

basic anhydride

oxide

peroxide

superoxide

ternary acid

Preliminary Test

As in other chapters, this test will check your understanding of basic concepts and types of calculations. Be sure to practice *many* of the additional textbook Exercises, including those indicated with the Study Goals.

Short Answer

Answer with a word, a phrase, a formula, or a number with units as necessary.

1. The transition metals are elements that are characterized by electrons being added to the _____.
2. The inner transition elements are characterized by electrons being added to the _____.
3. An element having outer electron configuration ns^2np^3 would be found in Group ____.
4. The number of valence electrons in an atom of any Group VIA element is ____.
5. The lightest three elements of Group VA, arranged in order of increasing atomic size, are _____.
6. Because of the increasing nuclear charge going left to right across the representative elements of a period, the first ionization energies _____.
7. Because of the increasing nuclear charge going left to right across the representative elements of a period, the atomic radii _____.
8. The most stable ion formed by a member of the halogen group would have a charge of _____.
9. The number of electrons in the iodide ion, I⁻, is _____.
10. The number of electrons is the calcium ion, Ca^{2+}, is _____.
11. Barium (Ba) would be _____ (more, less) likely to form a stable anion than a stable cation.
12. A cation is _____ (larger than, smaller than, the same size as) the neutral atom from which it is formed.
13. Atoms or ions that have the same number of electrons are called _____.
14. The neutral atom that is isoelectronic with Sr^{2+} is _____.
15. When the oxides of nonmetals dissolve in water, they form _____ solutions.
16. Basic anhydrides are the oxides of_____.
17. Strong soluble bases, such as the Group IA metal hydroxides, can often be prepared by the reaction of_____ with water.
18. Many ternary acids can be prepared by dissolving the appropriate _____ _____ in water.
19. A hydride is a compound composed of_____ and _____.
20. A molecular hydride contains hydrogen and _____, whereas an ionic hydride contains hydrogen and _____.

21. The general formula of the hydrides of the halogens is _____ (denoting a halogen as X).

22. The general formula of the hydrides of the Group VIA nonmetals is _____ (denoting a Group VIA nonmetal as Y).

23. The general formula for the normal oxides of the alkali metals is _____ (denoting an alkali metal as M).

24. The two binary compounds that oxygen forms with hydrogen are _____ and _____. (Give name and formula of each.)

25. Of the two allotropic forms of oxygen, _____ is much more stable than _____. (Give name and formula of each.)

26. Metal peroxides contain the _____ ion, with formula _____; in this ion, the oxidation number of oxygen is _____.

27. Some nonmetallic oxides are referred to as _____ anhydrides, because they _____.

28. Many ionic metal oxides are called _____ anhydrides, because they _____.

29. Oxides that can dissolve in either acids or bases are termed _____.

30. The formula for sodium peroxide is _____.

31. The formula for hydrogen peroxide is _____.

32. The name of the ion O^{2-} is _____.

33. The name of the ion O_2^- is _____.

34. The name of the ion O_2^{2-} is _____.

35. The process in which oxygen combines rapidly with some materials in highly exothermic reactions, with a visible flame, is called _____.

36. The usual products of the complete combustion of hydrocarbons are _____ and _____. (Give name and formula of each.)

37. Combustion reactions always give off _____.

38. Combustion of fuels containing sulfur produces _____ (name, formula), which is a very harmful atmospheric pollutant.

39. The substance produced in Question 38 is slowly oxidized in air to _____, which then combines with moisture in air to form _____. (Give name and formula of each.)

Multiple Choice

_____ 1. For each of the noble gases except helium, the outermost occupied main energy level contains how many electrons?
   (a) 1   (b) 2   (c) 6   (d) 8   (e) 18

_____ 2. Which of the following statements concerning the element iodine is *not correct*?
   (a) Iodine gives up electrons readily.
   (b) Iodine is in the fifth period of the periodic table.
   (c) Iodine has chemical properties similar to those of chlorine.
   (d) Iodine is a halogen.
   (e) One molecule of iodine contains two atoms.

_____ 3. Zirconium (Zr) is classified as
  (a) a representative element.   (b) a transition element.
  (c) a lanthanide.   (d) an actinide.   (e) a noble gas.

_____ 4. Which of the following is a transition element?
  (a) P   (b) Pd   (c) Ar   (d) U   (e) Na

_____ 5. Which of the following is a representative element?
  (a) P   (b) Pd   (c) Eu   (d) U   (e) V

_____ 6. An element with an outer electron configuration ns^2 (excepting helium) is classified as
  (a) an alkali metal.   (b) an alkaline earth metal.   (c) a halogen.
  (d) a transition element.   (e) an inner transition element.

_____ 7. When the properties of the elements with atomic numbers 114 and 115 are well characterized, they will probably fit into
  (a) the actinide series of inner transition elements.
  (b) Groups IVA and VA.
  (c) a new section of the periodic table, not yet represented.
  (d) the noble gas group.
  (e) the halogen group.

_____ 8. Atoms of the elements in the same A group in the periodic table have
  (a) the same electronegativity.
  (b) the same total number of electrons.
  (c) the same size.
  (d) the same number of valence electrons.
  (e) the same number of protons.

_____ 9. Where in the periodic table are the least electronegative elements found? (Ignore the noble
  gases.)
  (a) upper left   (b) upper right   (c) lower left
  (d) lower right   (e) middle

_____ 10. Which of the following elements would have the highest first ionization energy?
  (a) Na   (b) Al   (c) Si   (d) S   (e) Cl

_____ 11. For which of the following elements would the size of the neutral atom (atomic radius) be
  smallest)?
  (a) K   (b) Na   (c) Mg   (d) Ca   (e) Sr

_____ 12. Cesium (Cs) would be expected to have a _____ first ionization energy and a _____
  electronegativity.
  (a) high, high   (b) high, low   (c) low, high   (d) low, low

_____ 13. The alkali metals have low first and high second ionization energies. The alkaline earth
  metals have relatively low values for both first and second ionization energies. These
  observations suggest that
  (a) both alkali metals and alkaline earth metals should form very stable ions with charge 1+.
  (b) both alkali metals and alkaline earth metals should form very stable ions with charge 2+.
  (c) alkali metals should form stable 1+ ions, while alkaline earth metals should form stable
     2+ ions.
  (d) alkali metals should form stable 1+ ions, while alkaline earth metals should form stable
     2– ions.
  (e) alkali metals should form stable 1– ions, while alkaline earth metals should form stable
     2– ions.

14. Which one of the following isoelectronic ions would be expected to have the largest radius?
    (a) S^{2-}　　　　(b) Cl^-　　　　(c) K^+　　　　(d) Ca^{2+}　　　(e) Sc^{3+}

15. Which of the following pairs would correctly be described as isoelectronic?
    (a) Na and Na^+　(b) Ca and Be　(c) Cu^+ and Zn^{2+}　(d) P^- and P^{2-}　(e) H and He

16. How many electrons are there in one Cr^{3+} ion?
    (a) 24　　　　　(b) 52　　　　　(c) 27　　　　　(d) 21
    (e) Cannot be answered without knowing the mass number.

17. Which of the Group VIA elements is not a solid at room temperature and atmospheric pressure?
    (a) oxygen　　(b) sulfur　(c) selenium　(d) tellurium　(e) polonium

18. The element oxygen, in the form O_2, accounts for which of the following percentages, by volume, of dry air?
    (a) about 20%　(b) about 40%　(c) about 50%　(d) about 60%　(e) about 80%

19. Which of the following Group VIA elements has the highest first ionization energy?
    (a) oxygen　　(b) sulfur　(c) selenium　(d) tellurium

20. Some soluble bases can be formed by the action of metals with water. For which of the following metals would this not be a likely way of preparing the corresponding hydroxide base?
    (a) K　　　　(b) Ca　　　　(c) Na　　　　(d) Ba　　　　(e) Fe

21. Each of the following ions forms a stable hydroxide. For which of the ions would the hydroxide *not* be amphoteric?
    (a) Sn^{2+}　　　(b) Sn^{4+}　　　(c) Cr^{3+}　　　(d) Be^{2+}　　　(e) Ba^{2+}

22. Which one of the following is a molecular hydride?
    (a) RbH　　　(b) SrH_2　　　(c) NaH　　　(d) PH_3　　　(e) CaH_2

23. Arrange the following oxides in order of increasing ionic character: CaO, SiO_2, SO_2.
    (a) $BaO < SiO_2 < SO_2$　　　　　　(b) $SO_2 < BaO < SiO_2$
    (c) $SiO_2 < SO_2 < BaO$　　　　　　(d) $BaO < SO_2 < SiO_2$
    (e) $SO_2 < SiO_2 < BaO$

24. Which of the following compounds is a superoxide?
    (a) Na_2O_2　　　(b) SrO　　　(c) KO_2　　　(d) FeO　　　(e) Cl_2O_7

25. For which of the following elements is the normal oxide classified as most acidic?
    (a) P　　　　　(b) Al　　　　(c) Ca　　　　(d) Na
    (e) No other answer is correct, because an acid must contain hydrogen.

26. Reaction between which of the following pairs is an example of a reaction between an acidic anhydride and a basic anhydride.
    (a) CaO and SiO_2　　　　(b) Mg and $CuBr_2$　　　　(c) CaO and H_2O
    (d) NH_3 and HNO_3　　　(e) KH and H_2O

Answers to Preliminary Test

Do not just look up answers. Think about the reasons for the answers.

Short Answer

1. *d* orbitals
2. *f* orbitals

3. VA. It might help you to remember that for the representative elements, the A group number is equal to the combined number of s and p electrons in the outer shell.

4. 6. Notice that the group number for representative elements also corresponds to the number of valence electrons, and hence to the number of dots in the Lewis dot representation of the neutral atom.

5. N, P, As. Remember that within a group, atomic radii increase going down the group because of the filling of energy levels with higher n.

6. increase. Keep in mind that higher ionization energy means increased difficulty of removing an electron.

7. decrease

8. 1-. Each halogen has a configuration of ns^2np^5 when neutral, so with a gain of one electron, it would attain the noble gas configuration of ns^2np^6.

9. 54. The atomic number of iodine is 53. It gains one electron to form the I⁻ ion.

10. 18. A calcium atom has 20 electrons. It loses two electrons to form Ca^{2+} ion.

11. less. Ba has rather low first and second ionization energies but positive electron affinity. You could also look at its electronegativity, which is low. This means that is has a good tendency to lose electrons (form cations) and very little tendency to gain electrons (form anions).

12. smaller than. Understand the explanation in Section 6-5.

13. isoelectronic. The prefix "iso" means "same." Do you remember *iso*topes? Watch for further occurrences of this prefix in your study of chemistry.

14. Kr. To form an ion with a charge of 2+, an Sr atom must lose two of its original 38 electrons, to give 36 electrons in Sr^{2+}. A neutral Kr atom has 36 electrons.

15. acidic. See Section 6-8.

16. metals. See Section 6-8.

17. the metallic element. Some such reactions occur at dangerously explosive rates.

18. nonmetal oxide. See Section 6-8.

19. hydrogen; another element

20. a nonmetal; an active metal. Remember that the hydrides of the less active metals are intermediate in character between the most ionic and the most covalent hydrides. See the classification of the hydrides of the representative elements in Figure 6-4.

21. HX

22. H_2Y

23. M_2O. But you should be aware that the alkali metals sometimes form peroxides or superoxides in their direct reactions with O_2. See Table 6-4.

24. water, H_2O; hydrogen peroxide, H_2O_2

25. dioxygen, O_2; ozone, O_3

26. peroxide; O_2^{2-}; −1. Keep in mind that we present an ionic charge as $n-$ or $n+$, while the oxidation number or oxidation state is designated as $-n$ or $+n$. The peroxides are one of the few classes of compounds in which oxygen exhibits an oxidation number other than −2.

27. acid; react with water to give acid solution. How many examples of this behavior can you list?

28. basic; react with water to form bases. These are primarily the oxides of the alkali metals and those of the heavier alkaline earth metals.

29. amphoteric

30. Na_2O_2

31. H_2O_2

32. oxide

33. superoxide

34. peroxide
35. combustion
36. water, H_2O; carbon dioxide, CO_2
37. heat. These are always exothermic. The usual reason for carrying out combustion reactions is to use the heat energy produced, rather than to make desirable products.
38. sulfur dioxide, SO_2
39. sulfur trioxide, SO_3; sulfuric acid, H_2SO_4

Multiple Choice

1. (d). All have outer configuration ns^2np^6.
2. (a). Iodine is near the right side of the periodic table. Thus, it has a fairly high electronegativity, and does not give up electrons easily, but gains them readily.
3. (b). Be clear about these classifications, Section 6-1.
4. (b). The term "transition element" without further qualification refers to the d-transition elements.
5. (a)
6. (b)
7. (b). To see this, you can look at the periodic table and count blocks to see where elements 114 and 115 would fall. Alternatively, you can figure out their configurations. Element 114 would have outer shell $7s^27p^2$, and would thus fall in Group IVA, with the other ns^2np^2 elements. Reason in the same way for element 115.
8. (d)
9. (c). Electronegativity increases left to right and decreases top to bottom.
10. (e). Higher ionization energy means greater difficulty of removing electrons. First ionization energy increases going left to right across a period. See Figure 6-2 and the accompanying discussion in terms of electron configurations.
11. (c). Atomic radii generally decrease left to right within a period and increase going down a group. But the changes in this property going down a group are much more pronounced than those going across a period. Thus, we know Mg would be smaller than Na because it is further to the right. We know that Mg would also be smaller than Ca, because it is further to the right. K is larger than Na (further down) and larger than Ca (further left). Sr is further down than any other of these, so it is the largest. Study Section 6-2.
12. (d). Remember the meanings of these terms. A low ionization energy implies ease of removal of electrons, while a low electronegativity means a poor tendency to attract electrons to it. Both of these descriptions fit Cs, with outer configuration $6s^1$.
13. (c)
14. (a). All have the same number of electrons (18). Sc has the highest nuclear charge (21+) to attract these electrons, and S has the lowest nuclear charge (16+). Thus, S^{2-} would be have its electrons most loosely held, and S^{2-} would be the largest of these ions.
15. (c)
16. (d). Neutral Cr has 24 electrons; Cr^{3+} has three fewer.
17. (a)
18. (a)
19. (a). This trend of decreasing first ionization energy going down the group prevails in all representative groups of the periodic table. Review Section 6-3.

20. (e). This method is applicable only to the Group IA metals and to the heavier Group IIA metals. Study Section 6-8.
21. (e)
22. (d). Hydrides of nonmetals are molecular (Section 6-7)
23. (e). Increasing ionic character of oxides is correlated with increasing metallic character of the other element.
24. (c). Remember how to identify oxides, peroxides, and superoxides by determining the oxidation state of oxygen in the compound.
25. (a)
26. (a)

7 Chemical Bonding

Chapter Summary

Chapter 7 begins our study of chemical bonding with a discussion of how atoms are held together in compounds. There are two extremes in the possible modes of bonding: (1) **ionic bonding**, in which one atom or group of atoms completely gives up or **transfers** electrons to another, resulting in charged atoms or groups of atoms (ions) that then attract one another, and (2) **covalent bonding**, in which electrons are **shared** between two atoms. We can correlate many properties of compounds with the predominant type of bonding that holds them together (page 271). We should recognize, however, that these are the extremes in a continuous gradation of bonding, with nearly all bonds being to some degree both ionic and covalent.

Chemical bonding usually involves only the outermost or incompletely filled shells of electrons. We often concentrate our attention on these **valence electrons**. A formalism for representing and keeping track of the numbers of valence electrons in atoms, the **Lewis dot representation**, is correlated with the position of the element in the periodic table (Section 7-1). Because this formalism is somewhat more cumbersome and less useful for *d*- or *f*-transition elements, it is usually used only for representative elements.

The important Section 7-2 contains a detailed description of **ionic bonding**. The main ideas are as follows: (1) The elements with low electronegativities (and low ionization energies) readily lose electrons, giving them up to the elements with high electronegativities and quite negative electron affinities. This forms ions that then attract one another because of their opposite charges. (2) The number of electrons an atom gains or loses to form a stable ion depends on its electron configuration. The gain or loss results in a more stable configuration, usually a noble gas configuration for representative elements. This lets us understand and predict the formulas of simple **binary** (two-element) compounds, as summarized in Table 7-2. Do not try to memorize this table; understand, in terms of the electron configurations of the elements involved, *why* the formulas of the compounds are as shown in this table. The electron configuration diagrams and Lewis dot representations that are used in the discussion of the section will help you learn to predict formulas.

The rest of this chapter presents an introduction to **covalent bonding**. This type of bonding occurs when the atoms have the same or similar electronegativities, so that they attain more stable configurations only by sharing electron pairs. When *one pair* of electrons is shared, we refer to the bond as a **single bond**, when *two pairs* are shared it is a **double bond**, and when *three pairs* are shared it is a **triple bond**.

The sharing of electron pairs to form covalent bonds results in a balance of attractive and repulsive forces. The drive to overlap orbitals to share electrons is the attractive force, as described in the discussion of Section 7-3, while the repulsion is due to the like charges of the two nuclei. For molecules more complicated than H_2, there is additional repulsion by the unshared electrons on the two atoms. The potential energy of the pair of atoms at various distances can be represented by curves such as those in Figure 7-4. The distance at which the curve becomes a minimum is the distance at which the bonded combination of atoms is most stable. The depth of the curve below the zero of potential energy represents the strength of the bond. This is the amount of energy that must be supplied to separate the bonded pair of atoms into two isolated atoms.

The **Lewis formula** is a convenient notation to help us count valence electrons and propose stable bonding arrangements. In this formula, shared and unshared electrons are each represented by a dot; alternatively, each shared electron pair can be represented by a single line. You should also be familiar with the following points of terminology. The terms **monatomic**, **diatomic**, **triatomic**, and **polyatomic** refer to *how many atoms* are in the molecules; for instance, H_2 and HF are diatomic molecules, O_3 (ozone) and SO_2 are triatomic, and so on. The term **homonuclear** means that the molecule is composed of all one kind of atom; O_2, H_2 and O_3 are each homonuclear. The term **heteronuclear** means that the molecule contains more than one kind of atom; HF and SO_2 are heteronuclear. The same terms can be applied to ions.

Sections 7-4 through 7-7 contain very important ideas. Here we discuss in detail the Lewis formulas for keeping track of the number and kind of bonds. Section 7-5 introduces the **octet rule**, a very useful aid to writing Lewis formulas. This rule can be expressed in the simple mathematical relationship $S = N - A$. It summarizes the observation that representative elements usually achieve a noble gas configuration in most of their compounds. This generalization is known as the octet rule, because all of the noble gases except helium have eight electrons in their outer shells. An obvious exception among the representative elements is hydrogen. It attains a noble gas configuration (that of He) by getting only two electrons in its valence shell. You should work through examples in the chapter to understand the application of this rule. Drawing Lewis formulas for molecules and ions is essential to understanding molecular structure and to predicting polarities of bonds and arrangement of atoms. Development of the skills presented in Section 7-5 is worth a great deal of practice.

As we develop our understanding of chemical bonding in molecules, we find that some molecules or ions cannot be properly described by a single Lewis formula. In such a case, we introduce the idea of **resonance** (Section 7-6), a concept that describes the true structure as the average of several Lewis structures. This is made possible by the **delocalization** of the bonding electrons over several atoms, rather than electron pairs being shared explicitly between two atoms. The Enrichment feature entitled "Formal Charges" presents a formalism that is helpful in selecting the correct Lewis formula.

We must not take the octet rule too strictly. There are many cases in which elements form stable compounds by attaining some number of electrons other than eight in their outer shell. Section 7-7 first lists elements and compounds for which such exceptions are to be expected, and then explains several such exceptions. One type of exception includes atoms that are so far from having eight electrons in their valence shell that they cannot easily attain that state by electron sharing and do not readily form ionic compounds. Elements at the tops of Groups IIA and IIIA behave in this way. In a second type of exception, the molecule or ion has an odd number of electrons. Elements that involve d or f electrons in their bonding constitute a third class of exceptions. The chemistry of transition elements, which involves mostly their d electrons, is discussed further in Chapters 23 and 25.

The atoms in a heteronuclear molecule may not attract the shared electrons equally (Section 7-8). This leads to an imbalance in the electron distribution in the bond. We refer to such a bond as a **polar** covalent bond, in contrast to **nonpolar** covalent bonds, in which the bonding electron pair is equally shared. Such an arrangement is referred to as a **dipole** (or **bond dipole**). Section 7-9 extends the idea of bond polarity to describe the **dipole moment** of the molecule. We shall see in Chapter 8 that the experimental measurement of this quantity gives valuable information about the three-dimensional arrangements of atoms in molecules. We should view pure (nonpolar) covalent bonding and ionic bonding as the two ideal extremes of chemical bonding, with all intermediate stages also possible (Section 7-10). Some compounds, K_2SO_4, for example, exhibit both ionic and strong covalent interactions.

Study Goals

Be sure you work many of the suggested Exercises. Those that involve writing Lewis representations of ions and molecules (Study Goals 7, 10, and 11 are especially important.)

1. Know how to write Lewis dot representations of atoms of representative elements. Understand how these are related to periodic table group numbers. (Section 7-1; Exercises 4 and 5)
2. Distinguish between the major aspects of ionic and covalent bonding. (Chapter introduction; Exercises 1 through 3)
3. Know some of the major differences in physical properties of ionic and covalent compounds. (Chapter introduction; Exercises 12 and 13)
4. Predict whether bonding between a given pair of elements or in a given compound would be primarily covalent or ionic. (Section 7-2; Exercises 6 through 10)
5. Relate the stable ions formed by elements to their electron configurations and to their positions in the periodic table. (Section 7-2; Exercises 11 and 20 through 23)
6. Predict the formulas of binary ionic compounds from the positions of their constituent representative elements in the periodic table. (Section 7-2; Exercises 14 through 19)
7. Write equations for the reactions of representative elements to form binary ionic compounds. Be able to represent reactants and products with (a) simple chemical symbols and formulas, (b) electron configuration diagrams, and (c) Lewis formulas. (Section 7-2)
8. Understand what forces are responsible for the formation of covalent bonds. (Pages 278 and 279, and Section 7-3; Exercises 24 through 27)
9. Distinguish among single, double, and triple bonds. (Section 7-3; Exercise 26)
10. Draw Lewis formulas for molecules and polyatomic ions that contain only representative elements. Recognize which ones do not follow the octet rule. (Sections 7-4, 7-5, and 7-7; Exercises 32 through 62)
11. Recognize molecules and ions for which resonance structures are possible; draw appropriate Lewis formulas. (Section 7-6; Exercises 47 through 52)
12. Be able to write formal charges for the atoms in covalently bonded molecules and ions. Use these as a guide to assess whether Lewis formulas may be correct. (Enrichment feature following Section 7-6; Exercises 63 through 66)
13. Distinguish between polar covalent bonds and nonpolar covalent bonds. (Sections 7-8 through 7-10; Exercises 67, 68, 70, 72, and 73)
14. Be able to tell whether the bonding between two elements is ionic, polar covalent, or nonpolar covalent. You should be able to do this either with numerical values of electronegativity or on the basis of the positions of the elements in the periodic table. Understand the continuous range of bonding possibilities. (Sections 7-8 through 7-10; Exercises 67 through 81)
15. Answer conceptual questions based on this chapter. (Conceptual Exercises 82 through 85)
16. Apply concepts and skills from earlier chapters to the ideas of this chapter. (Building Your Knowledge exercises, 86 through 93)

Some Important Terms in This Chapter

Write the meanings *in your own words*. Check the Key Terms list and the chapter reading. Then rewrite your definitions, still in your own words, to improve them. Study other new terms, and review terms from preceding chapters if necessary.

Lewis dot representation (for atoms)

chemical bonding

covalent bonding

ionic bonding

octet rule

single bond

double bond

triple bond

polar bond

nonpolar bond

dipole moment

Lewis formula (for molecules or ions)

dash formula

polyatomic ion

resonance

formal charge

Preliminary Test

These preliminary test questions are quite basic. They are intended to check your understanding of the main ideas and terminology of the chapter. You will need to work many additional Exercises at the end of Chapter 7.

Short Answer

Answer with a word, a phrase, a formula, or a number with units as necessary.

1. The Lewis dot representation of an atom of calcium is _____.
2. The Lewis dot representation of an atom of phosphorus is _____.
3. The Lewis dot representation for any halogen atom (represent it as X) is _____.
4. The formula of the ionic compound of magnesium and oxygen is _____.
5. The formula of the ionic compound of rubidium and sulfur is _____.
6. The formula of the ionic compound of calcium and nitrogen is _____.
7. In the formation of a covalent bond, the principal "attractive force" is _____ _____.
8. In the formation of a covalent bond, the principal "repulsive forces" are due to _____ _____ and to _____.
9. When two atoms combine to form a stable chemical bond, the _____ of the bonded pair of atoms is lower than that of the isolated atoms.
10. When two atoms are joined by a single bond, they share _____ electrons.
11. When two atoms are joined by a double bond, they share _____ electrons.
12. Which of the following molecules would be described as diatomic: He, H_2, Cl_2, HCl, CS_2, NH_3, H_2O, S_8, P_4? _____

13. Which of the molecules in Question 19 would be described as homonuclear? _____ Which as heteronuclear? _____

14. When the bonding pair of electrons in a covalent bond is equally shared between the two atoms, the bond is called _____; when one atom attracts the shared pair of electrons more strongly than the other, the bond is called _____.

15. A covalent bond is polar if the two atoms involved differ significantly in their _____ values.

16. In which are the covalent bonds more polar, CH_4 or CF_4? _____

17. Consider a series of bonds in which carbon is bonded to another second period element: C–O, C–F, C–N, and C–C. Place these bonds in order of decreasing polar (ionic) character. _____.

Each of the following compounds or ions obeys the octet rule. Apply the relationship $S = N - A$ to each of them and draw the Lewis formula; draw each Lewis formula using only dots and again using dashes to represent bonding electrons.

Compound or ion	N	A	S	Lewis formula
18. CH_4				
19. CCl_4				
20. NH_3				
21. NH_4^+				
22. CH_3Cl				
23. CS_2				
24. PF_3				

Compound or ion	N	A	S	Lewis formula

25. SO_4^{2-}

26. HCN (C is the central atom)

27. C_2H_2

28. N_2H_4

Multiple Choice

The valence electrons of an atom in its ground state are represented by the configuration $3s^2\ 3p^4$. Questions 1 through 3 refer to this atom.

_____ 1. To which group in the periodic table does the element belong?
      (a) IIA   (b) IIIA   (c) VIA   (d) VIIA   (e) noble gas

_____ 2. How many of the valence orbitals of the neutral atom are half-filled?
      (a) 1   (b) 2   (c) 3   (d) 4   (e) 0

_____ 3. What is this element?
      (a) O   (b) S   (c) P   (d) Cr   (e) Se

_____ 4. How many valence electrons are there in an isolated carbon atom?
      (a) 0   (b) 1   (c) 2   (d) 4   (e) 6

_____ 5. The electron configuration $1s^22s^22p^63s^23p^6$ could represent
      (a) an atom of Ar.   (b) a Cl^- ion.   (c) a K^+ ion.
      (d) any of the ions or atoms mentioned.   (e) none of the ions or atoms mentioned.

_____ 6. What is the ground state electron configuration of the S^{2-} ion?
      (a) $1s^22s^22p^63s^23p^4$ (b) $1s^22s^22p^63s^23d^6$ (c) $1s^22s^22p^63s^23p^2$
      (d) The same as the ground state configuration of a Ne atom.
      (e) The same as the ground state configuration of an Ar atom.

_____ 7. A barium (Ba) atom would form its most stable ion by the
      (a) gain of one electron.   (b) gain of two electrons.   (c) loss of one electron.
      (d) loss of two electrons.   (e) loss of 56 electrons.

_____ 8. How many electrons are in the $4s$ orbital of the Ca^{2+} ion?
      (a) 3   (b) 2   (c) 1   (d) 0   (e) None of the preceding answers is correct.

_____ 9. All of the following represent stable ions. Four of the five have a noble gas configuration. Which one does *not have a noble gas configuration*?
(a) Fe^{3+} (b) Br^- (c) S^{2-} (d) Ba^{2+} (e) K^+

_____ 10. The ion Cl^- can best be described as
(a) a metal ion.   (b) a monatomic anion.   (c) a monatomic cation.
(d) a neutral ion.   (e) a Lewis dot ion.

_____ 11. How many electrons are present in the valence shell of fluorine in the compound KF?
(a) 6   (b) 7   (c) 8   (d) 9   (e) 10

_____ 12. Predict the formula of an ionic compound containing potassium (K) and selenium (Se)?
(a) KSe   (b) K_2Se (c) KSe_2 (d) K_2Se_2 (e) K_3Se_2

_____ 13. Which of the following is the best Lewis dot representation of Na_2O?

(a) $2Na^+, [:\ddot{O}\cdot]^-$ (b) $2Na\cdot^+, [:\ddot{O}\cdot]^+$ (c) $2Na^-, [:\ddot{O}:]^+$ (d) $2Na^+, [:\ddot{O}:]^{2-}$

(e) None of the preceding answers is correct, because Na and O have similar electronegativities and probably do not form ions.

_____ 14. Which of these pairs of elements combine with the most covalent (least polar) bonding?
(a) Ca and N   (b) Fe and Cl   (c) Be and I   (d) C and O   (e) Ca and S

_____ 15. Which of the following might be expected to be a stable compound?
(a) ArH   (b) NaO   (c) BaSe   (d) CCl_2 (e) Ca_2O

_____ 16. A metal (call it X) in Group IA of the periodic table forms a compound with an element (call it Y) of Group VIA. The compound would most likely have the formula _____ and would be _____.
(a) X_2Y, ionic   (b) X_2Y, covalent   (c) XY_2, ionic
(d) XY_2, covalent   (e) XY_6, covalent

_____ 17. Which of the following is the best description of covalent bonding?
(a) Bonding that occurs between elements with very different electronegativities, and with transfer of electrons.
(b) Bonding that occurs between elements with very different electronegativities, and with sharing of electrons.
(c) Bonding that occurs between elements with similar electronegativities, and with transfer of electrons.
(d) Bonding that occurs between elements with similar electronegativities, and with sharing of electrons.
(e) Bonding that occurs between ions and elements, and with sharing of electrons.

_____ 18. Which of the following substances would be expected to be the best conductor of electricity in the liquid state (i.e., when melted or condensed if not already a liquid)?
(a) CH_4 (b) PCl_3 (c) SiO_2 (d) H_2O (e) BaF_2

_____ 19. Which of the following substances would be expected to have the highest melting point?
(a) NaF   (b) CH_4 (c) H_2O (d) HF   (e) BF_3

_____ 20. On the usual Pauling scale of electronegativities, the electronegativity of selenium (Se) is 2.4, while that of chlorine (Cl) is 3.0. Based on these values, we should expect the bonding between Se and Cl to be
(a) ionic.   (b) covalent and nonpolar.
(c) covalent and polar with the Se end of the bond a little negative.
(d) covalent and polar with the Se end of the bond a little positive.
(e) unstable in any circumstances.

_____ 21. Which of the following diagrams correctly indicates the polarity of the bond?

(a) $\overset{\longrightarrow}{C-H}$ (b) $\overset{\longrightarrow}{Cl-C}$ (c) $\overset{\longrightarrow}{F-H}$ (d) $\overset{\longrightarrow}{F-C}$

(e) None of the preceding answers is correct.

_____ 22. Nitrogen can form single covalent bonds with boron, carbon, nitrogen, oxygen, and fluorine. The bond would be expected to be polar with the bonding electrons displaced most strongly away from the nitrogen in the case when it is bonded to

(a) boron.   (b) carbon.   (c) nitrogen.   (d) oxygen.   (e) fluorine.

_____ 23. Nitrogen and hydrogen can each exist as two isotopes. This means that molecules of ammonia, NH_3, could exist having several different masses. How many different masses of ammonia molecules would be possible?

(a) 1   (b) 2   (c) 4   (d) 6   (e) more than 6

_____ 24. Which of the following contains both covalent and ionic bonding?

(a) KF   (b) CO_2 (c) CH_2 (d) NH_4F (e) C_2H_4

_____ 25. Which of the following contains *no* polar bonds?

(a) HF   (b) CCl_4 (c) O_2 (d) NH_3 (e) PCl_5

_____ 26. What is the total number of electrons shared in the NH_3 molecule?

(a) 2   (b) 3   (c) 4   (d) 6   (e) 8

_____ 27. What is the total number of valence electrons that should appear in a Lewis formula for C_3H_5ClO? (It is not necessary to draw the Lewis formula in order to figure this out!)

(a) 10   (b) 20   (c) 30   (d) 40   (e) 50

_____ 28. What is the number of shared electrons that should appear in the Lewis formula for C_3H_5ClO? (It is not necessary to draw the Lewis formula in order to figure this out, either!)

(a) 10   (b) 20   (c) 30   (d) 40   (e) 50

_____ 29. Which of the following covalent substances contains a triple bond?

(a) H_2S (b) NH_3 (c) C_2H_2

(d) H_2CO (Hint: C is the central atom)   (e) HOF (Hint: F and H are both bonded to O)

_____ 30. The ion NH_4^+ can best be described as

(a) an anion.   (b) a cation.   (c) a Lewis dot ion.

(d) a monatomic ion.   (e) a homonuclear ion.

_____ 31. Which of the following statements concerning resonance is *not* correct?

(a) Resonance involves more than one formal Lewis formula to represent the bonding.

(b) Resonance involves the sharing of electrons in delocalized orbitals.

(c) The true energy of a molecule in which resonance is important is lower than that of any of the hypothetical molecules described by the contributing Lewis formulas.

(d) Resonance is important in understanding the structures of some ions as well as the structures of neutral molecules.

(e) An example of the use of the concept of resonance in understanding bonding is the explanation why the two S–O bonds in SO_2 are observed to be nonequivalent.

_____ 32. Which of the following contains only one unshared pair of valence electrons?

(a) H_2O (b) NH_3 (c) CH_4 (d) NaCl   (e) BF_3

Answers to Preliminary Test

Short Answer

1. Ca: But remember that it is arbitrary on which side of the symbol we write the dots, so : Ca or Ca̤ or Ca̤ would also be correct representations. We usually do indicate each pair of electrons as a pair of dots.

2. :Ṗ·

3. :Ẍ·

4. MgO. Each magnesium atom gives up two electrons ($3s^2$) to form a Mg^{2+} ion. Each oxygen atom accepts two electrons (changing $2s^22p^4$ to $2s^22p^6$) to form an O^{2-} ion. Thus, there are equal numbers of Mg^{2+} and O^{2-} formed and the chemical formula is MgO.

5. Rb_2S. Reason as in Question 4. Rb is a Group IA element (alkali metal), and S is in Group VIA.

6. Ca_3N_2. Same reasoning as for Question 4.

7. the tendency of atoms to attain stable electron configurations by overlapping their orbitals so as to share electrons.

8. repulsion of the nuclei of the two atoms; repulsion of the unshared pairs of electrons of the two atoms.

9. potential energy

10. two (or one pair of)

11. four (or two pairs of)

12. H_2, Cl_2, HCl. Remember that the term *diatomic* means that the molecule consists of two atoms, whether or not they are the same kind of atom.

13. homonuclear: He, H_2, Cl_2, S_8, P_4; heteronuclear: HCl, CS_2, NH_3, H_2O. The term *homonuclear* means "consisting of only one kind of atom," whereas *heteronuclear* means "consisting of more than one kind of atom."

14. nonpolar; polar

15. electronegativity

16. CF_4. It will be helpful to remember that C and H have almost the same electronegativity values. You must also remember the trends in electronegativity with respect to position in the periodic table.

17. C–F > C–O > C–N > C–C. Bond polarity depends on the difference in electronegativity between the two bonded atoms. Remember the trends in electronegativities (Section 6-6). Each of the other elements mentioned here is more electronegative than carbon, so C–C is the least polar bond shown, and C–F is most polar.

	Compound or ion	N	A	S	Lewis formula	
18.	CH_4	16	8	8	H H : C : H H	H H–C–H H
19.	CCl_4	40	32	8	: Cl : : Cl : C : Cl : : Cl :	: Cl : : Cl–C–Cl : : Cl :
20.	NH_3	14	8	6	H : N : H H	H–N–H H
21.	$NH_4{}^+$ (Remember to account for the + charge by decreasing A by 1)	16	8	8	H + H : N : H H	H + H–N–H H
22.	CH_3Cl	22	14	8	H H : C : Cl : H	H H–C– Cl : H
23.	CS_2	24	16	8	S :: C :: S	S = C = S
24.	PF_3	32	26	6	: F : P : F : : F :	: F –P– F : : F :
25.	$SO_4{}^{2-}$ (Remember to account for the 2– charge by increasing A by 2)	40	32	8	: O : 2– : O : S : O : : O :	: O : 2– : O –S– O : : O :
26.	HCN	18	10	8	H : C ::: N :	H–C≡N :
27.	C_2H_2	20	10	10	H : C ::: C : H	H–C≡C–H
28.	N_2H_4	24	14	10	H : N : N : H H H	H–N–N–H H H

Multiple Choice

1. (c)
2. (b)
3. (b)
4. (d)
5. (d)
6. (e)
7. (d)
8. (d)
9. (a)
10. (b). Remember the meanings of the terms. An anion is a negative ion and a cation is a positive ion. The term "monatomic" means consisting of one atom. No ion could be neutral, by definition.
11. (c). In this compound, fluorine exists as the fluoride ion, F^-.
12. (b)
13. (d). One check you can perform is that the total charge in an ionic compound must be equal to zero, which eliminates answer (a). Answer (b) makes no sense because a sodium atom with one valence electron would not be an ion, nor would an oxygen with six valence electrons. Likewise, (c) can be eliminated because Na does not form a stable negative ion and O would not form a positive ion—and these Lewis formulas are wrong anyway. Answer (e) is wrong in its statement of properties.
14. (d). This is the pair of elements with the smallest electronegativity difference of any listed. Remember how trends in electronegativity are associated with positions in the periodic table.
15. (c). In none of the others do all of the atoms attain a stable configuration.
16. (a)
17. (d)
18. (e). BaF_2 is the only ionic substance listed. Pure water, which is covalent, is a *poor* conductor of electricity—the commonly observed ability of water to conduct electricity is due to small amounts of dissolved ions.
19. (a). NaF is the only ionic substance listed.
20. (d)
21. (e)
22. (e)
23. (e). There are eight possible combinations. Can you write all of them?
24. (d)
25. (c). Each of the other compounds listed contains covalent bonds between elements of unequal electronegativities.
26. (d). Use the formula $S = N - A$.
27. (c). This is A in the formula $S = N - A$.
28. (b). This is S in the formula $S = N - A$. $N = 50$ and $A = 30$ (Question 27).
29. (c)
30. (b)
31. (e). The two S–O bonds in SO_2 are equivalent.
32. (b)

8   Molecular Structure and Covalent Bonding Theories

Chapter Summary

This chapter deals with shapes of molecules and ions and with the orbitals that atoms use to share electrons. Several important ideas form a unifying theme throughout the chapter. Section 8-1 gives an overview of the main themes of this chapter. Pay special attention to the boxed Important Note on page 307, and consult your instructor, if necessary, about the order in which you should study the chapter material. The numbered steps and the flow chart on page 308 summarize the procedures you will use in this chapter to analyze the structure and bonding in any polyatomic molecule or ion. Writing the Lewis formula for a molecule or ion is an essential starting point for the material in this chapter, so review that procedure from Chapter 7.

Each atom in a polyatomic molecule or ion arranges the electron pairs in its valence shell in such a way as to minimize the repulsion of these electrons pairs. This is the central idea of the **valence shell electron pair repulsion (VSEPR) theory**, Section 8-2. Section 8-3 discusses one important consequence of the shape of a molecule—its polarity, or dipole moment. The experimental measurement of molecular polarity can give us valuable information about the three-dimensional arrangement of atoms in molecules. In some molecules, the polar bonds are arranged so that the bond dipoles cancel, to leave the entire molecule nonpolar (dipole moment equal to zero). In other molecules, the bond dipoles are arranged to they do not cancel, leaving the entire molecule polar (dipole moment not equal to zero). The conditions on page 312 and the flow chart in Figure 8-1 (page 313) summarize the approach for predicting whether a polyatomic molecule is polar or nonpolar. This approach will be used throughout this chapter.

To reach minimum energy (maximum stability), atoms in molecules often modify their orbitals to a different arrangement than they would have in isolated atoms. In the terminology of the **valence bond (VB) theory**, Section 8-4, this mixing or **hybridization** of atomic orbitals to form **hybrid orbitals** allows us to correlate the bonding with the observed molecular shapes.

We can consider the geometry of molecules and ions to fall into one of several cases, depending on the number of atoms to which a central atom is bonded and the number of unshared electron pairs that this central atom has (Table 8-1). For each of these cases, we can use the VSEPR theory to predict the three-dimensional arrangement of the electrons around the central atom. This, in turn, allows us to predict the arrangements of atoms around the central atom. The valence bond theory is used to discuss the number of electrons occupying each type of hybrid orbital and the overlap of these orbitals (Table 8-2). Sections 8-5 through 8-12 discuss the most common bonding arrangements in terms of these two theories.

As you study this material, look ahead to the summary of molecular and electronic geometries in Section 8-15 and in Table 8-4. You must be careful to distinguish clearly between **electronic geometry** ("Where are the *valence-shell electron pairs* around the central atom?") and the **molecular geometry** ("Where are the *other atoms* around the central atom?"). As you will see, there are only five common answers to the first question; the electronic geometry on the central atom is usually either linear with sp hybrid orbitals (Section 8-5), trigonal planar with sp^2 hybrid orbitals (Section 8-6), tetrahedral with sp^3 hybrid orbitals (Sections 8-7 through 8-9), trigonal bipyramidal with sp^3d hybrid orbitals (Section 8-11), or octahedral with sp^3d^2 hybrid orbitals (Section 8-12). After you have determined the electronic geometry for a molecule or ion, then you can determine its molecular geometry by considering which of the electronic locations attach the central atom to other atoms and which ones merely correspond to unshared electron pairs on that central atom. For example (Sections 8-7 through 8-9), molecules that have four regions of high electron density all have tetrahedral electronic geometry; their molecular geometries could be tetrahedral (e.g., CH_4), pyramidal (e.g., NH_3), or angular (e.g., H_2O). Table 8-3 summarizes how the presence of lone pairs on the central atom affects molecular geometry. Once you have found the overall geometries by using the two theories presented in this chapter, you can often understand variations by considering the stronger repulsion of unshared pairs for other electron pairs, and by taking bond polarities into account.

With the understanding about orbitals we have developed in this chapter, we can describe how multiple bonding occurs. Double bonds are discussed in detail in Section 8-13 and triple bonds in Section 8-14. Be sure you understand the difference between sigma and pi bonds, and which combinations of orbitals can form each kind of bond.

Study Goals

Much practice is required to master the bonding descriptions in this chapter. You should work *many* of the exercises at the end of the chapter in the text.

1. Outline the significance and main ideas of the valence shell electron pair repulsion (VSEPR) theory. (Section 8-2; Exercises 1 through 4, 9, and 10)
2. Outline the significance and main ideas of the valence bond (VB) theory. (Section 8-4; Exercises 11 through 17, 21, and 48)
3. Be able to identify a central atom and count its number of regions of high electron density. (Section 8-2)
4. Distinguish between electronic geometry and molecular geometry. (Sections 8-2 and 8-5 through 8-12; Exercises 5 through 8)
5. Use the VSEPR theory to describe or predict the electronic geometry *and* the molecular geometry of a molecule or polyatomic ion. (Parts B of Sections 8-5 through 8-12; Exercises 22 through 47)
6. Name and describe the hybrid orbitals that correspond to each possible number of regions of high electron density. (Section 8-4; Exercises 16 through 20)
7. Be able to use valence bond structures to describe the bonding in molecules and ions having one central atom. (Parts C of Sections 8-5 through 8-12 and Section 8-15; Exercises 42 through 57, 63, and 64)
8. Use valence bond structures to describe the bonding in molecules and ions that have more than one central atom. (Sections 8-6 through 8-14; Problem-Solving Tip on pages 322-323; Exercises 56 through 59)
9. Be able to use valence bond structures to describe how double and triple bonding occur. (Section 8-13 and 8-14; Exercises 60 through 64)
10. In the valence bond description, distinguish between sigma and pi bonds. (Sections 8-13 and 8-14; Exercises 14, 15, 60, 61, 63, and 64)

101

11. Knowing the electronic structure and shape of a molecule, determine whether it is polar or nonpolar. (Review Sections 7-8 and 7-9; Sections 8-3, 8-5 through 8-12; Exercises 37 through 26, 27, and 38 through 47)
12. Use your understanding of this chapter to recognize and solve a variety of types of questions about structures, bonding, and molecular polarities. (Mixed Exercises, 69 through 78)
13. Answer conceptual questions based on this chapter. (Conceptual Exercises 79 through 85)
14. Apply concepts and skills from earlier chapters to the ideas of this chapter. (Building Your Knowledge exercises, 85 through 89)

Some Important Terms in This Chapter

Write the meanings *in your own words*. Check the chapter and rewrite your definitions, to improve them. Study and review other new terms.

central atom

valence shell

valence shell electron pair repulsion theory

valence bond theory

regions of high electron density

electronic geometry

molecular geometry

linear

trigonal planar

tetrahedral

trigonal bipyramidal

octahedral

pyramidal

angular

hybridization

hybrid orbital

polar molecule

nonpolar molecule

sigma (σ) bond

pi (π) bond

Preliminary Test

These preliminary test questions are quite basic. They are intended to check your understanding of the main ideas and terminology of the chapter. You should practice many additional Exercises at the end of Chapter 8 in the textbook.

True-False

Mark each statement as true (T) or false (F).

_____ 1. The number of regions of high electron density on the central atom in H_2O is 4.
_____ 2. The number of regions of high electron density on the central atom in CCl_4 is 4.
_____ 3. The number of regions of high electron density on the central atom in NF_3 is 4.
_____ 4. Because all atoms satisfy the octet rule in their bonding, the number of regions of high electron density on the central atom is always 4 or some multiple of 4.
_____ 5. The number of regions of high electron density is equal to S in the formula $S = N - A$.
_____ 6. All molecules with four regions of high electron density about a central atom have tetrahedral electronic geometry.
_____ 7. All molecules with four regions of high electron density about a central atom have tetrahedral molecular geometry.
_____ 8. A molecule of the type AB_5 (where A is the central atom) could have trigonal bipyramidal molecular geometry.
_____ 9. A molecule of the type AB_5 (where A is the central atom) must have trigonal bipyramidal molecular geometry.
_____ 10. In the molecule BF_3, all F–B–F bond angles are equal.
_____ 11. In the molecule CF_4, all F–C–F bond angles are equal.
_____ 12. In the molecule NF_3, all F–N–F bond angles are equal.
_____ 13. All molecules with tetrahedral electronic geometry must have bond angles equal to 109.5°.
_____ 14. In the molecule NF_3, all F–N–F bond angles are equal to 109.5°.
_____ 15. In the molecule PF_5, all F–P–F bond angles are equal.
_____ 16. A molecule that contains polar bonds may be polar.
_____ 17. A molecule that contains polar bonds must be polar.
_____ 18. A three-atom molecule with linear molecular geometry must be nonpolar.
_____ 19. A three-atom molecule with linear molecular geometry must be polar.
_____ 20. A tetrahedral molecule must be polar.
_____ 21. A tetrahedral molecule must be nonpolar.

Short Answer

Answer with a word, a phrase, a formula, or a number with units as necessary.

1. The theory that accounts for the shapes of molecules by considering the repulsion of the outer-most electron pairs on a central atom is called the _____ theory.

2. The theory that describes the formation of covalent bonds as being due to the overlap of the atomic orbitals on two atoms is called the _____ theory.

3. The mixing or rearrangement of individual "pure" atomic orbitals to form new orbitals with different spatial orientations is called _____.

4. The angle between two sp hybrid orbitals on the same atom is _____.

5. The angle between two sp^2 hybrid orbitals on the same atom is _____.

6. The angle between two sp^3 hybrid orbitals on the same atom is _____.

7. The electronic geometry of the central atom in the compound HCN is _____; the hybrid orbitals on the central atom in this compound are _____ orbitals; the molecular shape is described as _____.

8. The electronic geometry of the central atom in the compound PCl_3 is _____; the hybrid orbitals on the central atom in this compound are _____ orbitals; the molecular shape is described as _____.

9. The electronic geometry of the central atom in the compound BCl_3 is _____; the hybrid orbitals on the central atom in this compound are _____ orbitals; the molecular shape is described as _____.

10. The electronic geometry of the central atom in the compound H_2O is _____; the hybrid orbitals on the central atom in this compound are _____ orbitals; the molecular shape is described as _____.

11. The electronic geometry of the central atom in the ion SO_4^{2-} is _____; the hybrid orbitals on the central atom in this ion are _____ orbitals; the ionic shape is described as _____.

12. The electronic geometry of the central atom in the ion AlF_6^{3-} is _____; the hybrid orbitals on the central atom in this ion are _____ orbitals; the ionic shape is described as _____.

Multiple Choice

_____ 1. Which of the following is a polar molecule?
(a) CO_2 (a linear molecule)   (b) BF_3 (a trigonal planar molecule)
(c) CH_4 (a tetrahedral molecule)   (d) NH_3 (a pyramidal molecule)   (e) H_2

_____ 2. Which one of the following statements regarding sp^3 hybrid orbitals is *not* correct?
(a) The result of sp^3 hybridization is a set of four orbitals with tetrahedral geometry.
(b) The result of sp^3 hybridization is a set of four orbitals, one of which is a lower energy, because one of them came from an s orbital.
(c) The angle between two sp^3 hybrid orbitals on an atom is about 109.5°.
(d) If an atom has four electrons in a set of sp^3 hybrid orbitals, the lowest energy configuration would have all four electrons unpaired.
(e) The result of sp^3 hybridization is four equivalent orbitals.

_____ 3. The main reason for describing the bonding around nitrogen in ammonia, NH_3, in terms of sp^3 hybrid orbitals on nitrogen is that
(a) the angle between two N–H bonds is closer to 109.5° than to any other ideal value.
(b) the angle between two N–H bonds is closer to 90° than to any other ideal value.
(c) the angle between two N–H bonds is closer to 120° than to any other ideal value.
(d) the hydrogen atoms attain the filled $1s^2$ configuration by bond formation.
(e) the ground state configuration of nitrogen has four unpaired electrons.

_____ 4. Which of the following contains at least one atom that uses sp^3 hybridization in its bonding?
     (a) H_2O (b) NH_3 (c) CH_4
     (d) All of the molecules listed.   (e) None of the molecules listed.

_____ 5. Which of the following contains at least one atom whose bonding may be described in terms of sp^2 hybridization?
     (a) H_2O (b) NH_3 (c) C_2H_2 (d) C_2H_4 (e) H_2

_____ 6. In the compound hexachloroethane, C_2Cl_6, the angle between two C–Cl bonds on the same carbon atom is closest to which of the following values?
     (a) 60°   (b) 90°   (c) 109.5°   (d) 120°   (e) 180°

_____ 7. The geometry of a molecule of silane, SiH_4, can be described as
     (a) trigonal planar.   (b) octahedral.   (c) tetrahedral.   (d) linear.   (e) angular.

_____ 8. The shape of a molecule of phosphorus trichloride, PCl_3, can be described as
     (a) tetrahedral.   (b) trigonal planar.   (c) trigonal bipyramidal
     (d) trigonal pyramidal.   (e) angular.

_____ 9. The hybrid orbitals on the central atom in PCl_3 are
     (a) sp.   (b) sp^2.   (c) sp^3.   (d) sp^3d.   (e) sp^3d^2.

_____ 10. The shape of a molecule of phosphorus pentachloride, PCl_5, can be described as
     (a) tetrahedral.   (b) trigonal planar.   (c) pyramidal.
     (d) trigonal bipyramidal.   (e) angular.

_____ 11. The hybrid orbitals on the central atom in PCl_5 are
     (a) sp.   (b) sp^2.   (c) sp^3.   (d) sp^3d.   (e) sp^3d^2.

_____ 12. In the compound SF_6, what is the arrangement of fluorine atoms around the sulfur atom, and what hybrid orbitals of the sulfur atom are used in bonding?
     (a) hexagonal, sp^3 (b) tetrahedral, sp^3 (c) octahedral, sp^3d
     (d) octahedral, sp^3d^2 (e) trigonal planar, sp^2

_____ 13. In the compound CH_3Cl, the bonding orbitals of the carbon atom are
     (a) p orbitals.   (b) sp hybrid orbitals.   (c) sp^2 hybrid orbitals.
     (d) sp^3 hybrid orbitals.   (e) sp^4 hybrid orbitals.

_____ 14. The compound tetrachloroethylene, C_2Cl_4, has a carbon-carbon double bond. The angle between two C–Cl bonds on the same C atom should be closest to which of the following?
     (a) 60°   (b) 90°   (c) 109.5°   (d) 120°   (e) 180°

_____ 15. If an atom is attached to two other atoms and has no valence-shell unshared electron pairs, then the molecule's shape would be described as
     (a) angular.   (b) linear.   (c) tetrahedral.   (d) trigonal planar.   (e) octahedral.

_____ 16. The water molecule, H_2O, is angular rather than linear, because
     (a) the two hydrogen atoms attract one another.
     (b) the bonding involves s orbitals of H.
     (c) the bonding involves sp^2 orbitals of O.
     (d) the bonding involves sp^3 orbitals of O.
     (e) the s orbital of one H atom overlaps with the s orbital of the other H atom.

_____ 17. Acetylene, C_2H_2, contains a carbon-carbon triple bond. Which one of the following statements about this molecule is _not_ true?
     (a) The four atoms are bonded together in a rectangular arrangement.
     (b) Each carbon-hydrogen bond is a single bond.
     (c) Each carbon atom is sp hybridized.
     (d) The bonding around each carbon atom is linear.
     (e) The atoms are all in the same straight line.

106

_____ 18. A triple bond includes
(a) one σ and two π bonds.   (b) two π and one σ bonds.   (c) three σ bonds.
(d) three π bonds.   (e) None of the preceding answers is correct.

_____ 19. The following five statements describe differences between a single bond and a double bond. Which one of these statements is *incorrect*?
(a) A double bond is the sharing of four electrons by two atoms, while a single bond is the sharing of two electrons by the two atoms.
(b) Atoms connected by a double bond are more difficult to rotate with respect to each other than atoms connected by a single bond.
(c) A double bond usually consists of a σ bond and a π bond, while a single bond is usually a π bond.
(d) In a single bond, the region of highest density of bonding electrons is directly between the two atomic centers, while in a double bond another region of high density of bonding electrons also exists on two sides of the line between the atomic centers.
(e) A double bond is stronger (harder to break) than a single bond between the same two atoms.

_____ 20. A π bond can be formed by the overlap of
(a) two *s* orbitals on different atoms.   (b) two *p* orbitals on the same atom.
(c) two *p* orbitals on different atoms, these *p* orbitals lying side-by-side in the same plane.
(d) two *p* orbitals on different atoms, these *p* orbitals lying end-to-end in the same plane.
(e) two *p* orbitals on different atoms, these *p* orbitals lying side-by-side in planes that are perpendicular to one another.

_____ 21. In order to understand the shape of H_2CO, we describe the hybridization of carbon (the central atom) as sp^2. In order to fully describe the double bond, we postulate that the hybridization of the oxygen atom is
(a) *sp*.   (b) sp^2.   (c) sp^3.   (d) sp^3d.   (e) sp^3d^2.

Answers to Preliminary Test

Do not just look up answers. Think about the reasons for the answers.

True-False

1. True. For all questions like this, you first need to draw the Lewis dot formula for the molecule or ion. Review this topic in Sections 7-4 through 7-7.
2. True
3. True
4. False. This is false for two reasons: (1) Not all atoms obey the octet rule in their covalent bonding. Be familiar with the main types of exceptions (Section 7-7). (2) Even if the octet rule is satisfied, the bonding may involve double or triple bonds, which each account for only *one* high electron density region.
5. False. This is false for two reasons: (1) We only count the regions of high electron density on one central atom at a time, and the formula $S = N - A$ refers to an entire molecule or ion. (2) Even if there is only one central atom, the term S in this formula tells the total number of shared electrons. These may be shared in single, double, or triple bonds, each of which counts as only *one* region of high electron density.

6. True
7. False. Such a molecule may have the central atom bonded to fewer than four other atoms, in which case the electronic geometry and the molecular geometry are different. For examples, study the discussions of NH_3 in Section 8-8 and H_2O in Section 8-9.
8. True. This can occur when all of the regions of high density are electrons that bond the central atom to others. An example is PF_5, discussed in Section 8-11.
9. False. This could occur only if all of the regions of high electron density were bonding sets to other atoms, but this is not necessary. Suppose there were *six* regions of high electron density (octahedral electronic geometry), with one of these an unshared electron pair. IF_5 or SF_5^- are examples of such species.
10. True
11. True
12. True
13. False. See the discussion in Sections 8-8 and 8-9.
14. False. The repulsion of the bonded pairs by the lone pairs causes the bond angles F–N–F to be less than the ideal tetrahedral value of 109.5°.
15. False. Even though there are no lone pairs to account for, the angles in a trigonal bipyramidal arrangement are not all equal. Six are 90° (axial-equatorial), three are 120° (equatorial-equatorial), and one is 180° (axial-axial).
16. True
17. False. If any dipoles due to bond polarity are arranged so that their effects cancel, the molecule would still be nonpolar. A simple example is CO_2.
18. False. HCN is an example of such a molecule that is polar.
19. False. CO_2 is an example of such a molecule that is nonpolar.
20. False. If all four atoms (B) attached to the central atom (A) in a tetrahedral species AB_4 are the same, then the molecule or ion would be nonpolar. Examples of this are CH_4, CF_4, and SO_4^{2-}.
21. False. If one (or more) of the atoms is different from the others, the molecule *could* still be polar. An example would be CH_3Cl.

Short Answer

1. valence shell electron pair repulsion. This is often referred to as the VSEPR theory. You should remember what those initials stand for, because it will help you to remember the main idea of the theory.
2. valence bond
3. hybridization
4. 180°
5. 120°
6. 109.5°
7. linear; *sp*; linear
8. tetrahedral; *sp*^{3}; pyramidal
9. trigonal planar; *sp*^{2}; trigonal planar
10. tetrahedral; *sp*^{3}; angular
11. tetrahedral; *sp*^{3}; tetrahedral
12. octahedral; *sp*$^3d^2$; octahedral

Multiple Choice

1. (d)
2. (b)
3. (a)
4. (d)
5. (d)
6. (c)
7. (c)
8. (d)
9. (c)
10. (d)
11. (d)
12. (d)
13. (d)
14. (d)
15. (b)
16. (d)
17. (a)
18. (a)
19. (c)
20. (c)
21. (b)

9 Molecular Orbitals in Chemical Bonding

Chapter Summary

In Chapter 7, we learned that valence bond (VB) theory describes covalent bonding as the result of the overlap of orbitals that remain on or between the two atoms involved. We also saw that we could use VSEPR theory to describe the molecular geometry as the result of repulsions among valence shell electron pairs on the atoms. In Chapter 9 we learn about an alternative view of covalent bonding, the **molecular orbital (MO) theory**. Recall that in VB theory, we proposed that the various orbitals on a *single* atom could mix and recombine (hybridize) to form hybrid orbitals. These have different energies and shapes than the "pure" or isolated atom orbitals. In the MO theory, the orbitals that are mixed are on *different* atoms, and the term "hybridization" is not used. The resulting **molecular orbitals** are no longer the "property" of just one atom, but extend over at least the pair of atoms involved, or in some cases over the entire molecule.

For many molecules, either approach gives a good description of the molecule. The VB approach has the advantage of being easily visualized. The MO theory has an advantage with respect to quantitative calculations of energies and geometry, description of electron distributions, and correlation with magnetic properties. In other cases, the MO theory can give a correct description of some molecular property (for example, the paramagnetism of O_2) when the VB approach fails.

When two atomic orbitals from different atoms combine, they produce two molecular orbitals. Of these molecular orbitals, one (the **bonding orbital**) is at lower energy than the original orbitals while the other (the **antibonding orbital**) is at higher energy than the pure atomic orbitals. The superscript star (*) means that the orbital is antibonding, while the absence of the superscript means that the orbital is bonding. Section 9-1 describes the various ways in which atomic orbitals can combine to give molecular orbitals. The relations between the shapes of the atomic orbitals and those of the bonding and antibonding orbitals are described. Be sure that you understand the notation used to describe molecular orbitals. The notation **sigma (σ)** indicates that the molecular orbital is cylindrically symmetrical about a line through the internuclear axis. The notation **pi (π)** means that the molecular orbital has mirror symmetry across a plane through the two nuclei. The *subscripts* indicate the type of atomic orbitals that combined to give this molecular orbital—*s* if both were *s* orbitals, *sp* if one was *s* and one was *p*, and *p* if both were *p*. The only way (at least using only *s* and *p* orbitals as we do in this chapter) that a π molecular orbital can be formed is by *side-on* overlap of *p* orbitals on two different atoms.

Section 9-2 extends the description of the molecular orbitals to include the relative energies of the various molecular orbitals. It is important to remember the **order** of the energy levels for the various homonuclear diatomic molecules of the first- and second-period elements (Figure 9-5). (Remember that for O_2, F_2, and Ne_2 and their ions, the order of energies of the σ_p and the π_p orbitals is reversed from the order of these orbitals for the other molecules and ions.) We fill the orbitals with the correct number of electrons for the molecule, following the Aufbau procedure, Hund's Rule, and the Pauli Principle. You will recognize that this is the same procedure for molecules that we used earlier with isolated atoms (Sections 5-15 and 5-16). The procedure for finding the molecular orbital description of a molecule or ion is summarized on pages 357-358. Once we have placed the correct number of electrons into the set of molecular orbitals, we can describe the stability of the resulting arrangement using the **bond order**, defined in Section 9-3. The higher the bond order, the more stable we predict the molecule or ion to be, and the shorter the **bond length** and the greater the **bond energy**.

Section 9-4 covers in detail the neutral homonuclear diatomics of the first and second periods. A careful study of this section will help you learn to apply the MO theory and to become familiar with the notation used. As you will see in this section, the MO theory can also be used to describe bonding in diatomic cations and anions by removing or adding the appropriate number of electrons. For many heteronuclear diatomic molecules, the general features of the molecular orbital diagrams already studied can be adjusted to take into account the electronegativity difference between the atoms. This is discussed in Section 9-5, with NO (a slightly polar molecule) and HF (a very polar molecule) as examples.

The molecular orbital theory is also very useful in discussions of molecules and ions in which electron delocalization is an important feature of the bonding. (These are species that the valence bond theory describes with resonance structures.) Applications of this approach to the carbonate ion (CO_3^{2-}) and the benzene molecule (C_6H_6), are discussed in Section 9-6.

Study Goals

As in other chapters, some typical textbook Exercises related to each Study Goal are indicated.

1. Distinguish between the main concepts of the valence bond (VB) theory and the molecular orbital (MO) theory. (Introduction to Chapter 9 and Section 9-1; Exercises 1 through 5 and 7)
2. Distinguish between molecular orbitals and atomic orbitals. (Section 9-1; Exercises 3 through 6)
3. Describe each of the various molecular orbitals discussed in the chapter. Understand how each of these orbitals is related to the "pure" atomic orbitals from which it is formed. (Sections 9-1 and 9-5; Exercises 5 through 9 and 12)
4. Distinguish between sigma (σ) and pi (π) molecular orbitals. Distinguish between bonding and antibonding molecular orbitals. (Section 9-1; Exercises 6 through 9 and 12)
5. Remember the order of energies of the molecular orbitals for simple homonuclear diatomic molecules and ions. Construct molecular orbital diagrams for these molecules and ions. (Section 9-2; Exercise 10)
6. Apply the concepts of molecular orbital theory to determine bond orders and to predict relative stabilities of simple homonuclear diatomic molecules and ions. (Sections 9-3 and 9-4; Exercises 14 through 28)
7. Remember the order of energies of the molecular orbitals for simple heteronuclear diatomic molecules and ions. Relate the appearance of molecular orbital diagrams to polarities of bonds. Construct molecular orbital diagrams for these molecules and ions. (Section 9-5; Exercises 29 through 36)
8. Sketch three-dimensional representations of molecular orbitals in some molecules and ions for which delocalization is important. (Section 9-6; Exercise 37 and 38)
9. Recognize and answer questions of various types about molecular orbitals. (Mixed Exercises, 38 through 44)
10. Relate what you know from earlier chapters to the new ideas in this chapter. (Building Your Knowledge exercises, 45 and 46)

Some Important Terms in This Chapter

Write the meanings *in your own words*. Check the Key Terms list and the chapter reading. Then rewrite your definitions, still in your own words, to improve them. Study other new terms, and review terms from preceding chapters if necessary.

molecular orbital

atomic orbital

bonding orbital

antibonding orbital

nonbonding orbital

sigma orbital

pi orbital

bond order

delocalization

Preliminary Test

As in other chapters, this test will check your understanding of basic concepts and their applications. Practice *many* of the additional textbook exercises, including those indicated with the Study Goals.

Short Answer

1. The maximum number of electrons that could occupy one molecular orbital is ____.
2. When two atomic orbitals combine, the number of molecular orbitals formed is ____.
3. When four atomic orbitals combine, the number of molecular orbitals formed is ____.
4. Head-on overlap of two p orbitals results in formation of the molecular orbitals _____ and _____.
5. Side-on overlap of two p orbitals results in formation of the molecular orbitals _____ and _____.
6. Molecular orbitals that are at lower energy than the atomic orbitals from which they are formed are called _____ orbitals.
7. Molecular orbitals that, when occupied, increase the stability of the bonding between the two atoms are called _____ orbitals.
8. The quantity that uses the relative numbers of electrons in bonding and antibonding orbitals to predict the stability of bonds is called the _____.
9. The definition of bond order is _____.
10. For a molecule to be stable, its bond order must be _____.
11. The amount of energy necessary to break a mole of bonds is called the _____.
12. Generally, the more stable a bond, the _____ is the bond order, the _____ is the bond length, and the _____. (Answer with words like greater, shorter, etc.)
13. The order of energies of the molecular orbitals of the homonuclear diatomic molecules and ions of the first- and second-period elements (atomic numbers up to and including that of N) is _____.
14. The order of energies of the molecular orbitals of the homonuclear diatomic molecules and ions of the second-period elements (atomic numbers greater than that of N) is _____.
15. A molecule or ion with an odd number of electrons must be _____.
16. In order for a molecule to be paramagnetic, it must have _____ electrons.
17. Of the homonuclear diatomic molecules whose electron distributions are shown in Table 9-1 in the text, the ones that are paramagnetic are _____.
18. The magnetic properties of F_2 would be described as _____.
19. Of the first- and second-row homonuclear diatomic molecules, the most stable one is predicted to be _____.
20. The electron configuration of Li_2 is _____.
21. The bond order of Li_2 is _____; thus, we predict that the molecule of Li_2 would be _____.
22. For heteronuclear diatomics, the atomic orbitals of the more electronegative element are _____ in energy than the corresponding orbitals of the less electronegative element.
23. According to the molecular orbital theory, we predict that the order of stability of NO, NO^-, and NO^+ would be (most stable to least stable) _____.

Multiple Choice

_____ 1. When two s orbitals combine to form molecular orbitals, the orbitals formed are
(a) one σ_s and one σ_s^*.   (b) two σ_s.   (c) two σ_s^*.
(d) one σ_s and one π_s.   (e) one π_s and one σ_s^*.

_____ 2. When an s orbital and a p orbital combine to form molecular orbitals, the orbitals formed are
(a) one σ_s and one σ_p.   (b) one π_s and one π_p.   (c) one σ_{sp} and one σ_{sp}^*.
(d) one p_{sp} and one p_{sp}^*.   (e) one σ_{sp} and one π_{sp}^*.

_____ 3. When two p orbitals overlap head-on to form molecular orbitals, the orbitals formed are
(a) one σ_s and one σ_p.   (b) one π_s and one π_p.   (c) one σ_p and one σ_p^*.
(d) one π_p and one π_p^*.   (e) one σ_{sp} and one π_{sp}^*.

_____ 4. When two p orbitals overlap side-on to form molecular orbitals, the orbitals formed are
(a) one σ_s and one σ_p.   (b) one π_s and one π_p.   (c) one σ_p and one σ_p^*.
(d) one π_p and one π_p^*.   (e) one σ_{sp} and one π_{sp}^*.

_____ 5. Which of the following describes a diatomic molecule with bond order equal to zero?
(a) More electrons are in bonding than in antibonding orbitals.
(b) More electrons are in antibonding than in bonding orbitals.
(c) No electrons are in bonding orbitals.
(d) No electrons are in antibonding orbitals.
(e) Equal numbers of electrons are in bonding and in antibonding orbitals.

_____ 6. Which of the following molecules or ions would have a bond order equal to $2\frac{1}{2}$?
(a) He_2 (b) O_2^- (c) NO   (d) O_2 (e) Li_2^+

_____ 7. Which of the following molecules would be predicted to be the most stable?
(a) H_2 (b) Li_2 (c) B_2 (d) Be_2 (e) C_2

- -

For questions 8 through 10, use the molecular orbital diagram in Figure 9-5(b) of the text. Put 16 electrons into this diagram.

_____ 8. What is the bond order of the molecule or ion described?
(a) 1   (b) $1\frac{1}{2}$ (c) 2   (d) $2\frac{1}{2}$ (e) 3

_____ 9. Which of the following statements is true about this molecule or ion?
(a) It is diamagnetic.
(b) It has a dipole moment.
(c) It is a homonuclear diatomic species.
(d) It is a heteronuclear diatomic species.
(e) None of the preceding statements is true.

_____ 10. The MO diagram best describes which of the following molecules or ions?
(a) NO^- (b) O_2 (c) CO^- (d) NO^+ (e) CN^-

Answers to Preliminary Test

Do not just look up answers. Think about the reasons for the answers.

Short Answer

1. two. Just as for pure atomic orbitals, a molecular orbital is a region in space that can hold a maximum of two electrons, and then only if they have opposite spin (that is, are paired).
2. two. The number of molecular orbitals formed is always the same as the number of atomic orbitals used to form them.
3. four. See the answer to Question 2.
4. σ_p and σ_p^*. When two orbitals combine, one of the resultant molecular orbitals is bonding and one is antibonding. These are σ orbitals because they have cylindrical symmetry about the internuclear line. See Figure 9-3.
5. π_p and π_p^*. See Figure 9-4.
6. bonding
7. bonding
8. bond order. See Section 9-3.
9. (number of electrons in bonding orbitals – number of electrons in antibonding orbitals)/2. See Section 9-3.
10. greater than 0
11. bond energy. See Section 9-3.
12. greater; shorter; greater. See Section 9-3.
13. See Figure 9-5(a) for the answer to this question. You should know this order of orbitals.
14. See Figure 9-5(b) for the answer to this question. You should know this order of orbitals.
15. paramagnetic. Remember that the presence of unpaired electrons leads to paramagnetism. If an atom or molecule has an odd number of electrons, they cannot all be paired.
16. unpaired
17. B_2 and O_2. Only these have unpaired electrons.
18. diamagnetic
19. N_2. This molecule has the highest bond order.
20. $\sigma_{1s}^2 \sigma_{1s}^{*2} \sigma_{2s}^2$.
21. 1; stable. The bond order is calculated as $(4 - 2)/2 = 1$.
22. lower. One way to remember this is to recall that the electrons are drawn preferentially to the more electronegative element.
23. NO^+, NO, NO^-. To figure this out, use the MO energy level diagram. You could use either diagram in Figure 9-5, and lower the O side, because O has the lower electronegativity. Then put in the appropriate number of electrons and determine the bond order. NO has 15 electrons and a bond order of $2\frac{1}{2}$; NO^+ has 14 electrons and a bond order of 3 (same as N_2); NO^- has 16 electrons and a bond order of 2 (same as O_2). Higher bond order means higher stability.

Multiple Choice

1. (a). See Figure 9-2 and the discussion in Section 9-1.
2. (c). See Figure 9-8 and the discussion in Section 9-5.
3. (c). See Figure 9-3 and the discussion in Section 9-1.

4. (d). See Figure 9-3 and the discussion in Section 9-1.
5. (e)
6. (c). Be sure you know how to fill the orbitals and then calculate bond order.
7. (e). The bond orders are $H_2 = 1$, $Li_2 = 1$, $B_2 = 1$, $Be_2 = 0$, $C_2 = 2$. C_2 has the highest bond order, so it is predicted to be the most stable of the molecules listed here. This would be true in the high-temperature gas phase. However, at room temperature, elemental carbon exists in other forms far more stable than C_2, such as graphite and diamond.
9. (c). $(10 - 6)/2$
10. (c). We can tell because the atomic orbitals of the two atoms are at the same energies. Remember that for heteronuclear diatomics, the atomic orbitals of the more electronegative element are lower in energy. See Figure 9-7 for such a diagram.
11. (b). Count the number of electrons in each of the species listed. Only O_2 has 16 electrons and is also homonuclear.

10 Reactions in Aqueous Solutions I: Acids, Bases, and Salts

Chapter Summary

This chapter continues and broadens the descriptive chemistry associated with acids, bases, and salts that was introduced in Chapter 4. Some important occurrences and practical uses of acids, bases, and salts are discussed in the introduction to Chapter 10. The general properties of aqueous solutions of protonic acids and soluble bases are summarized in Section 10-1. The list of properties presented there emphasizes several ways in which acids and bases are opposite in their behavior. You have already learned in Chapter 4 about the **Arrhenius** view of acids and bases (Section 10-2). In that view, acids are substances that produce H^+ in aqueous solution, whereas bases act to produce OH^- ions. We now know that H^+ exists in aqueous solution as the **hydrated hydrogen ion**, usually represented as the **hydronium ion, H_3O^+** (Section 10-3). The most significant overall feature of acids and bases, common to all of the studies in this chapter, is the essentially *opposite nature* of these two classes of compounds. The most important aspect of their opposite nature is their ability to neutralize one another. One of the drawbacks of the Arrhenius approach is that it does not emphasize sufficiently the "oppositeness" of acids and bases. Much of Chapter 10 is devoted to several other ways of looking at acids and bases. Each of these "theories" about acids, bases, and salts is a different way of systematically describing the behavior of these substances. The red/blue color coding for acids/bases that was introduced in Chapter 4 of the textbook is continued in this chapter.

One of the most useful descriptions of acid-base behavior is the **Brønsted-Lowry theory** (Section 10-4). The theory is used in this chapter to describe a broad range of reactions. An alternative view of acid-base behavior, the **Lewis theory**, is also presented. The chapter closes with some quantitative considerations of acid-base reactions in water solutions.

According to the Brønsted-Lowry view of acid-base behavior (Section 10-3), an acid is a substance that **donates H^+** (referred to as a proton, even though it is hydrated in aqueous solution), and a base is a substance that **accepts H^+**. Thus we see that this approach emphasizes the opposite nature of these two classes of substances. Pay close attention to the terminology used to describe behavior and properties so these terms will help you later in the chapter. Reactions are reversible, so we can write many acid-base reactions in the general form

$$\text{acid}_1 + \text{base}_2 \rightleftharpoons \text{acid}_2 + \text{base}_1.$$

When any Brønsted-Lowry acid gives up H^+, it becomes capable of accepting H^+, that is, of acting as a base. Thus, we identify **conjugate acid-base pairs** of substances—a pair of substances that differ by an H^+. Such apparently diverse reactions as the **autoionization** of water (Section 10-5), the dissolving of hydrogen chloride in water to form hydrochloric acid, and the dissolving of ammonia in water to give a basic solution may all be described in this way. Learning to recognize conjugate acid-base pairs will be helpful in your study of this chapter. As a memory aid, the text uses rectangles to indicate one acid-base (red-blue) pair and ovals for the other pair (see page 375). We also learn in Section 10-4 that the stronger an acid is, the weaker is its conjugate base. For any base to accept H^+, it must have available an unshared electron pair; this will be a basis for a different acid-base classification later in the chapter.

Amphoterism (Section 10-6) is the ability of a substance to act either as an acid or as a base. It will be helpful to remember the formulas of the **amphoteric hydroxides** and the complex ions to which they are related (Table 10-1).

One of our continuing goals in chemistry, beyond merely learning and classifying the reactions, is to understand trends in properties in terms of electronic and structural features of the molecules—in other words, in terms of the bonding. Section 10-7 emphasizes such an understanding. First we correlate relative strengths of **binary** protonic acids with their bonding properties. This section relates trends in strengths of acids (and of course the reverse trends in strengths of their conjugate bases) to such features as electronegativity differences and bond strengths. The important notion of the **leveling** effect is introduced in this section. We also can correlate the acid strengths of **ternary acids** with the bonding features of the acid molecules: structure, oxidation state of the central element, position in the periodic table, etc. Learn the generalizations presented in this section regarding trends in acid strength.

Section 10-8 deals in a systematic way with the neutralization reactions between acids and bases. You should review Section 4-5 before beginning your study of this section. You must also familiarize yourself with the writing of **total ionic equations** and **net ionic equations** (Sections 4-3 and 4-5) to represent reactions. A review of the general rules regarding solubilities of inorganic compounds in water (Section 4-2, part 5 and Table 4-8), will assist you in understanding this section. As you study the neutralization reactions in Section 10-6, you will see that the useful forms of the equations (that is, which aspects of the reaction that we wish to emphasize) depend on several factors: (1) the strength of the acid involved, (2) the strength and solubility of the base involved, (3) the solubility of the salt in water, and (4) the ionic or covalent nature of the soluble substances. Be sure you see how these factors affect the ways in which we write the chemical equation.

When the salts formed by neutralization of acid and base contain no remaining unreacted H^+ or OH^-, they are referred to as **normal salts**. But if less than a stoichiometric amount of base reacts with a **polyprotic acid** (more than one H^+ per molecule), the product is an **acidic salt** (i.e., it still has ionizable H^+). Similarly, a **basic salt** (still has ionizable OH^-) is formed when an acid partially neutralizes a **polyhydroxy base** (more than one OH^- per base formula unit). The acidic and basic salts are the subject of Section 10-9.

As we observed earlier, a Brønsted-Lowry base must have a lone pair available for bond formation with the H^+ that it accepts. The generalization of this behavior leads to the **Lewis theory** (Section 10-10), the broadest of the acid-base theories. In this theory, any substance that supplies an electron pair for bond formation, whether to H^+ or to any other species, is classified as a base; the substance that accepts a share in this lone pair is then classified as an acid. The Lewis theory includes all acid-base behavior previously discussed, but it extends the acid-base terminology to include reactions that do not involve H^+. This view of acid-base chemistry help to systematize additional types of reactions, such as those involving coordination compounds (Chapter 25).

Several methods of preparing common acids are discussed in Section 10-11.

Study Goals

Be sure to review the ideas and skills from indicated sections of earlier chapters. Several of the study goals in this chapter take much practice, so be sure you answer many of the Exercises in the text. Be systematic in your study.

1. Review the fundamental ideas of acids and bases. Memorize the lists of the common strong acids (Table 4-5) and the common strong soluble bases (Table 4-7). Classify common acids as strong or weak; classify common bases as strong soluble, insoluble, or weak. (Review Section 4-2.5; Section 10-1; Exercises 1 through 3 and 5 through 9)

2. Be familiar with the characteristics of acids, bases, and their solutions. (Section 10-1; Exercises 1, 5, and 31 through 35)

3. Define acid, base, and neutralization in terms of the Arrhenius theory. Give specific examples with formulas. (Section 10-2; Exercise 4)

4. Be familiar with the description of the hydrated hydrogen ion. (Section 10-3; Exercises 10 through 12)

5. Understand the Brønsted-Lowry acid-base theory. Be able to identify the Brønsted-Lowry acids and bases in reactions. (Section 10-4; Exercises 13 through 20)

6. Understand and apply the Brønsted-Lowry terminology of conjugate acid-base pairs and their relative strengths. (Section 10-4; Exercises 21 through 29)

7. Understand the important aspects of the ionization of water (Section 10-5)

8. Know the identity, properties, and characteristic reactions of the amphoteric hydroxides. Correlate amphoteric behavior of metal hydroxides with the relative strengths of (a) metal-oxygen and (b) oxygen-hydrogen bonds associated with coordinated water molecules. (Section 10-6; Exercises 36 and 38, 71, and 72)

9. Explain trends in acid strengths of the binary acids. Understand the correlations of strengths of ternary acids with oxidation number of the central atom, with position of the central atom in the periodic table, and with number of oxygen atoms. (Section 10-4; Exercises 39 through 53)

10. Review the solubility guidelines for ionic substances in water. (Review Section 4-2, part 5, and Table 4-8)

11. Be able write balanced molecular equations, total ionic equations, and net ionic equations for neutralization reactions of acids and bases to form salts. Be able to name all compounds involved in these equations. (Section 10-8; Exercises 54 through 64)

12. Tell what is meant by each of the following terms: (a) normal salt, (b) acidic salt, and (c) basic salt. Give examples of each and be able to tell by what kinds of reactions each type of salt could be formed. (Section 10-9; Exercises 65 through 74)

13. Understand the Lewis acid-base theory. Be able to recognize Lewis acid-base reactions and classify their acids and bases. (Section 10-10; Exercises 75 through 82)

14. Write equations illustrating the preparations of binary acids and of ternary acids. (Sections 10-9; Exercises 83 through 85)

15. Recognize and answer various types of questions about acids, bases, and salts. (Mixed exercises, 86 through 93)

16. Answer conceptual questions based on this chapter. (Conceptual Exercises 94 through 96)

17. Apply concepts and skills from earlier chapters to the ideas of this chapter. (Building Your Knowledge exercises, 97 through 101)

Some Important Terms in This Chapter

Write the meanings *in your own words*. Check the Key Terms list and the chapter reading. Then rewrite your definitions, still in your own words, to improve them. Study other new terms, and review terms from preceding chapters if necessary.

Arrhenius acid

Arrhenius base

salt

formula unit equation

total ionic equation

net ionic equation

Brønsted-Lowry acid

Brønsted-Lowry base

conjugate acid-base pair

ionization

autoionization

leveling effect

neutralization

normal salt

acidic salt

basic salt

binary acid

ternary acid

Lewis acid

Lewis base

amphoterism

Preliminary Test

As in other chapters, this test will check your understanding of basic concepts and types of calculations. Be sure to practice *many* of the additional textbook Exercises, including those indicated with the Study Goals.

Short Answer

Answer with a word, a phrase, a formula, or a number with units as necessary.

1. Write the definitions of acids and bases according to each of the following theories:

Theory	Definition
Arrhenius	(a) acid
	(b) base
Brønsted-Lowry	(c) acid
	(d) base
Lewis	(e) acid
	(f) base

2. The essential feature of bases, according to all theories considered here, is that they act to _____ acids.

3. In either the Arrhenius or the Brønsted-Lowry view of aqueous acid-base neutralization reactions, one of the products of the neutralization is molecules of _____. (Give name and formula.)

4. According to the Brønsted-Lowry description of acids and bases, when an acid acts to donate H^+, it is transformed into its _____.

5. The conjugate base of HCl is _____. (name, formula)

6. The conjugate base of acetic acid is _____. (name, formula)

7. The conjugate acid of ClO_4^- is _____. (name, formula)

8. The conjugate acid of HSO_4^- is _____. (name, formula)

9. The conjugate base of HSO_4^- is _____(name, formula)

10. The conjugate acid of water is _____(name, formula)

11. The conjugate base of water is _____(name, formula)

12. The process of water ionizing to produce an equal number of hydrated hydrogen ions (that is, hydronium ions, H_3O^+) and hydroxide ions OH^- is called the _____ of water.

13. A substance that can act as either an acid or a base is termed _____.

14. The stronger an acid, the _____ is its conjugate base.

15. HCl is a stronger acid than HCN, so we know that CN^- must be a _____ base than Cl^-.

16. Even though it is not included in the formal definition, we see that in order for a substance to act as a Brønsted-Lowry base, it must have _____.

17. We can attribute the weaker acidity of HF compared with the other hydrohalic acids to the _____ between H and F than between H and the other halogens.

18. "The strongest acid that can exist in aqueous solution is H_3O^+. Acids stronger than this react with water to produce H^+ and their conjugate bases." This is often referred to as the _____ of water.

19. All neutralization reactions in aqueous solution produce _____ and _____.

20. Nearly all soluble salts produce _____ when they dissolve in water.

21. The net ionic equation for the reaction of all strong acids with strong soluble bases to form soluble salts and water is _____.

22. An acid that contains only one acidic (ionizable) hydrogen atom per formula unit is called _____; one that contains two is called _____.

23. The most common soluble weak base is ammonia. The net ionic equation for the reaction of a monoprotic strong acid HX with aqueous ammonia is _____.

24. The usual driving force for neutralization reactions in aqueous solution is the formation of _____; a secondary factor favoring such neutralizations might be the formation of _____.

25. Salts that contain no unreacted H^+ or OH^- ions are referred to as _____ salts; those with excess available H^+ are termed _____ salts, whereas those with remaining OH^- are called _____ salts.

26. An acid that is composed of hydrogen, oxygen, and one other element is referred to as a _____ acid.

27. The four ternary acids of chlorine are (give name and formula for each):
    _____, _____,
    _____, and _____.

28. The salts of nitric acid contain the ion _____, and are called _____; the normal salts of sulfurous acid contain the ion _____, and are called _____.

29. For various ternary acids containing the same central element (i.e., the element other then H and O), strength of the acid increases as the _____ of the central atom _____.

30. For ternary acids containing different elements from the same periodic group in the same oxidation state, acid strength increases with _____ of the central element.

31. Due to the formation of complex hydroxyanions, many amphoteric metal hydroxides are soluble in _____.

32. Strong soluble bases, such as the Group IA metal hydroxides, can often be prepared by the reaction of _____ with water.

33. Many ternary acids can be prepared by dissolving the appropriate _____ _____ in water.

34. When CO_2 is dissolved in water, as in carbonated drinks, the acid formed is _____. (Give name, formula)

35. In the Lewis acid-base theory, a neutralization reaction always involves _____ bond formation.

Multiple Choice

___ 1. According to the Brønsted-Lowry concept of acids and bases, which of the following statements about a base is *not* correct?
   (a) A base accepts a hydrogen ion by sharing a previously unshared electron pair with H^+.
   (b) A base accepts a hydrogen ion by forming a coordinate covalent bond to H^+.
   (c) If a base is weak, its conjugate acid will be a strong acid.
   (d) A base can be formed only by ionization of a compound containing a hydroxide group.

___ 2. Which of the following could not be a Brønsted-Lowry acid?
   (a) H_2O (b) HN_3 (c) H_3O^+ (d) NH_4^+ (e) BF_3

___ 3. Which of the following could not act as a Lewis base?
   (a) PCl_3 (b) CN^- (c) I^- (d) CH_4 (e) H_2O

_____ 4. When a salt such as sodium acetate is dissolved in water, the acetate ion acts according to the reaction

$$CH_3COO^-(aq) + H_2O(l) \rightarrow CH_3COOH(aq) + OH^-(aq)$$

In this reaction, acetate ion is acting as
(a) an acid.   (b) a base.   (c) an amphoteric substance.
(d) an oxidizing agent.   (e) a reducing agent.

_____ 5. What substance is acting as a Lewis acid in the following reaction?

$$BF_3 + F^- \rightarrow BF_4^-$$

(a) BF_3 (b) F^- (c) BF_4^-
(d) All of substances listed.   (e) None of the substances listed

_____ 6. Cupric ion, Cu^{2+}, is pale blue in water solution. When ammonia is added to the solution, the deep-blue complex ion $[Cu(NH_3)_4]^{2+}$ is formed. In the formation of this complex ion, Cu^{2+} is acting as
(a) a Lewis acid.   (b) a Lewis base.   (c) an Arrhenius acid.
(d) a Brønsted-Lowry acid.   (e) a Brønsted-Lowry base.

_____ 7. In the reaction

$$CN^- + NH_4^+ \rightarrow HCN + NH_3$$

the cyanide ion acts as
(a) a base.   (b) an acid.   (c) an oxidizing agent
(d) a reducing agent.   (e) an amphoteric substance.

_____ 8. The following describe properties of substances. Which one is not a typical acid property?
(a) It has a sour taste.
(b) It reacts with metal oxides to form salts and water.
(c) It reacts with acids to form salts and water.
(d) Its aqueous solutions conduct an electrical current.
(e) It reacts with active metals to liberate H_2.

_____ 9. In water solution, HCN is only very slightly ionized in the reaction

$$HCN(aq) + H_2O(l) \rightarrow H_3O^+(aq) + CN^-(aq)$$

This observation shows that CN^- is a _____ than is water.
(a) stronger acid   (b) stronger base   (c) weaker acid   (d) weaker base

_____ 10. What is the conjugate base of HI?
(a) H^+ (b) HI   (c) I^- (d) HIO_3 (e) H_2I

_____ 11. HBr is a strong acid when dissolved in water. HF is a weak acid when dissolved in water. Which of the following is the strongest base?
(a) H^+ (b) Br^- (c) F^- (d) H_2O
(e) Insufficient information is given to answer the question.

_____ 12. Which of the following substances produces an acidic solution in water?
(a) SO_3 (b) NH_3 (c) NaO   (d) $Ca(OH)_2$ (e) LiH

_____ 13. Which of the following is an example of an acidic salt?
(a) KNO_3 (b) $KHSO_4$ (c) H_3PO_4 (d) $Mg(OH)_2$ (e) $Al(OH)Cl_2$

_____ 14. Which of the following is an example of a basic salt?
(a) KNO_3 (b) $KHSO_4$ (c) H_3PO_4 (d) $Mg(OH)_2$ (e) $Al(OH)Cl_2$

_____ 15. Which of the following is an example of a normal salt?
(a) KNO_3 (b) $KHSO_4$ (c) H_3PO_4 (d) $Mg(OH)_2$ (e) $Al(OH)Cl_2$

_____ 16. Each of the following ions forms a stable hydroxide. For which of the ions would the hydroxide *not* be amphoteric?

(a) Sn^{2+} (b) Sn^{4+} (c) Cr^{3+} (d) Be^{2+} (e) Ba^{2+}

_____ 17. In which of the following pairs is the order of acid strength *incorrect* ?

(a) $HF < HCl$ (b) $H_2SeO_4 < H_2SO_4$ (c) $HNO_2 < HBO_3$

(d) $H_2SO_4 < H_2SO_3$ (e) $HCl < HI$

Answers to Preliminary Test

Do not just look up answers. Think about the reasons for the answers.

Short Answer

1. (a) acts to produce H^+ (or H_3O^+)   (b) acts to produce OH^-
   (c) a proton (H^+) donor   (d) a proton (H^+) acceptor
   (e) accepts a share in an electron pair   (f) makes available a share in an electron pair
2. neutralize
3. water, H_2O
4. conjugate base
5. chloride, Cl^-
6. acetate, CH_3COO^-. Perhaps you should review the names, formulas, and anions (here called conjugate bases) of common acids, Tables 4-5 and 4-6.
7. perchloric acid, $HClO_4$
8. sulfuric acid, H_2SO_4
9. sulfate, $SO_4{}^{2-}$
10. hydronium, H_3O^+
11. hydroxide, OH^-
12. autoionization
13. amphoteric
14. weaker
15. stronger
16. an unshared electron pair
17. greater electronegativity difference. See Section 10-7.
18. leveling effect. See Section 10-7.
19. water, a salt
20. ions
21. $H^+ (aq) + OH^- (aq) \rightarrow H_2O\ (l)$. See Section 10-8.
22. monoprotic; diprotic. Be sure you understand the term polyprotic.
23. $H^+ (aq) + NH_3 (aq) \rightarrow NH_4{}^+ (aq)$
24. nonionized water molecules; an insoluble salt
25. normal; acidic; basic
26. ternary
27. perchloric acid, $HClO_4$; chloric acid, $HClO_3$; chlorous acid, $HClO_2$; hypochlorous acid, $HClO$
28. $NO_3{}^-$; nitrates; $SO_3{}^{2-}$; sulfites
29. oxidation state; increases. Notice that this also corresponds to an increasing number of oxygen atoms. For example, the order of increasing acid strengths of the *ternary* acids of chlorine is

$HClO < HClO_2 < HClO_3 < HClO_4$. Be sure that you remember that this applies only when comparing *ternary* acids of the same *element*.

30. increasing electronegativity
31. bases
32. the metallic element. Some of these reactions occur at dangerously explosive rates.
33. nonmetal oxide. See Section 10-11.
34. carbonic acid, H_2CO_3
35. coordinate covalent. Be sure that you know what this term means. It refers to the formation of a covalent bond by one atom supplying both electrons for the bond.

Multiple Choice

1. (d)
2. (e). A Brønsted-Lowry acid acts as a source of H^+; BF_3 does not contain H.
3. (d). A Lewis base provides an unshared electron pair for bond formation. Of the compounds listed, only CH_4 has no unshared valence electron pairs.
4. (b). It accepts H^+.
5. (a). It accepts the electron pair from F^- (the base) in the formation of a coordinate covalent bond.
6. (a). Ammonia (the Lewis base) provides electron pairs to Cu^{2+}.
7. (a). In this case, as a Brønsted-Lowry base, accepting H^+.
8. (c)
9. (b). The observation that most of the HCN is still present in undissociated form tells us that CN^- wins the competition for H^+ much more often than does H_2O.
10. (c)
11. (c). Reason as in Question 9. F^- must be a stronger base than H_2O, which must in turn be a stronger base than Br^-.
12. (a). Understand the relation of acidic oxides to positions in the periodic table.
13. (b)
14. (e)
15. (a)
16. (e)
17. (d). You should know the common strong acids. You should be able to use the generalizations about strengths of binary and ternary acids discussed in Section 10-7 to predict trends, even if you do not recognize the acids as strong or weak.

11 Reactions in Aqueous Solutions II: Calculations

Chapter Summary

Chapter 11 deals with some calculations about reactions in aqueous solutions. Two general classes of reactions are presented here—acid-base reactions and oxidation-reduction (redox) reactions. These calculations are based entirely on the stoichiometry calculations presented in Chapters 2 and 3, and you should review those calculations now—especially Sections 3-6 through 3-8.

The first four sections of Chapter 11 deal with some quantitative aspects of acid-base reactions in aqueous solutions. You will see that the terminology of the Brønsted-Lowry and Arrhenius theories (Chapter 10) is used here. The chemical reaction of all aqueous acid-base neutralizations involves

$$H^+(aq) + OH^-(aq) \rightarrow H_2O(l).$$

The central idea behind neutralization calculations is that equal numbers of moles of H^+ and OH^- react in such reactions, no matter what acid or base is involved. When this condition has been reached, the solution contains water, the anion of the acid, and the cation of the base. If the salt that is formed is insoluble, it precipitates from the reaction mixture as a solid; if it is soluble, it remains in solution as dissolved ions.

The essential criterion for neutralization is that equal numbers of moles of H^+ and OH^- (called **equivalent** amounts of acid and base) are required. But this criterion cannot be applied directly because we do not directly measure number of moles. Thus, a quantity to express solution concentration that relates *number of moles* (which is hard to measure) to *volume of solution* (which is easy to measure) is quite useful. Molarity is such a quantity.

Section 11-1 uses molarities to describe the amounts of acid and base solutions that are necessary to attain neutralization. Notice that some acids (e.g., HCl or HNO_3) supply 1 mole of H^+ per mole of acid, whereas others (e.g., H_2SO_4) supply 2 moles of H^+ per mole of acid. Comparable situations exist for bases with different numbers of moles of OH^- per mole of base. Be sure you are thoroughly familiar with the terminology of **titration, standard solutions, primary standards,** and **secondary standards** (Section 11-2). In the next two sections you will learn more about the calculations involved in acid-base neutralizations and in titrations. First, these calculations are carried out using the familiar mole method and molarity (Section 11-3).

An alternate approach is based on the concept of **equivalent weight** (Section 11-4). For any acid, this is defined as the weight of the acid, expressed in grams, that will furnish 1 mole of H^+. Similarly, the equivalent weight of a base is the amount of that base, expressed in grams, that will furnish 1 mole of OH^-. In these terms, one equivalent weight (usually called one **equivalent**) of *any acid* will neutralize one equivalent weight of *any base*. We then define the **normality** of a solution as the number of equivalents of solute per liter of solution. This provides a convenient basis for calculation of the volumes of acid and base solutions involved in a reaction. For convenience, we sometimes express normality in terms of **milliequivalents** per milliliter of solution. Study the examples in Section 11-4 carefully; note especially the relationships (1) between formula weight and equivalent weight for various acids and bases and (2) between molarity and normality for various acid or base solutions.

After you understand the ideas of these sections, get lots of practice by working appropriate problems at the end of the chapter—see Study Goals 4 and 6 for suggested exercises. Remember that the basis for these calculations is the following relationship at the point of neutralization (the **equivalence point**):

$$\text{no eq acid} = \text{no eq base}, \quad \text{or} \quad \text{no meq acid} = \text{no meq base}.$$

Another useful relation using solution concentrations is that

$$\text{no eq} = N \times L, \quad \text{or} \quad \text{no meq} = N \times mL.$$

These may be combined to get still another useful form:

$$\text{vol}_{acid} \times N_{acid} = \text{vol}_{base} \times N_{base},$$

where the two volumes must be expressed in the same units, i.e., both in L or both in mL.

The rest of this chapter deals with **oxidation-reduction (redox) reactions**. You have learned to use the formalism of oxidation numbers (Section 4-6) to recognize such reactions (Section 4-7). Review these two sections of Chapter 4.

Many applications of chemical reactions require that we write the equation in **balanced** form. Sometimes redox equations can be difficult to balance. These reactions involve electron transfer, and electrons are neither created nor destroyed in ordinary chemical reactions. These facts lead to the useful principle that the total increase in oxidation numbers must equal the total decrease in oxidation numbers. This, in turn, leads us to two very useful methods for balancing redox equations, discussed in Sections 11-5 through 11-7.

Of course, both methods for balancing redox equations give the same answer. It would be good to be familiar with both methods, though your instructor may prefer that you concentrate on becoming proficient in one or the other. In any case, be sure that you know how to carry out a completely reliable check on whether the resulting equation is balanced. Be sure you check both mass balance (for *each* element in the equation) and charge balance.

One method that many chemists use is the **half-reaction** method, frequently called the **ion-electron** method (Section 11-5). In this method we formally separate the oxidation (electron loss) and reduction (electron gain) portions of the process into half-reactions. We balance each half-reaction separately and then equalize the number of electrons lost and gained in the two half-reactions. We then add the balanced half-reactions to give the overall balanced equation. As you learn this method, always apply the foolproof check of whether the final equation is balanced. Check also to be sure that no electrons remain on either side of the equation. You can see that to apply this method you need to decide which species are changing in oxidation numbers, but the actual values of the oxidation numbers are not used in the balancing process.

Practice first with equations in which all reactants and products are given. This includes all equations given in Exercises 54 through 59 at the end of the text chapter. However, for many reactions in aqueous solutions, the equation as given may not specify all reactants and products. In such a case, it is necessary to add enough H and O to complete the mass balance. Section 11-6 discusses this step, which is accomplished by adding *two* of the three species H^+, OH^-, and H_2O to the appropriate sides of the equation. Use the following reasoning to remember which of these species you should add. (1) In an acidic solution, there is a lot of water (the solvent) and a large amount of H^+, but very little OH^- can be present. Thus, in an acidic solution there is essentially no OH^- to act as a reactant, and it cannot be written on the left-hand side of the equation; likewise OH^- should not be

128

written as a product in acidic solution, because it would immediately be consumed by the excess H^+ to form water. So we conclude that *in acidic solution we can add only H^+ or H_2O, but we never add OH^-*. (2) Similar reasoning (can you go through it?) regarding basic solutions (excess OH^-, very little H^+) leads to the conclusion that *in basic solution we can add only OH^- or H_2O, but we never add H^+*. Exercises 61 through 65 at the end of the text chapter will give you practice in this aspect of balancing redox equations.

Section 11-7 describes the **change-in-oxidation-number** method, which is based on the equality of total increase and decrease in oxidation numbers. Whenever you apply this method, be systematic, and follow the recommended procedures carefully until they are a habit. The same approach described earlier for adding H_2O and either H^+ or OH^- is applied in this method, but now to the overall equation.

The final section of this chapter deals with **redox titrations**. For these, we base our calculations on the quantitative relationships given by the balanced equations that describe oxidation-reduction reactions. As before, we can carry out these calculations with solution concentrations expressed as molarity; in Section 11-8 this approach is illustrated with several examples.

Study Goals

Be sure to review the ideas and skills from indicated sections of earlier chapters. Several of the study goals in this chapter take much practice, so be sure you answer many of the Exercises in the text. Be systematic in your study.

1. Review earlier calculations involving molarity. (Review Sections 3-6 through 3-8; Exercises 1 through 7, 20, and 21)
2. Using specified concentrations of solutions, be able to calculate (a) whether a specified mixture of acid and base solutions results in neutralization and (b) the amounts of salts formed and acid or base remaining when an acid and a base solution are mixed. (Section 11-1; Exercises 8 through 19 and 22)
3. Be familiar with the terminology and procedures of titrations and standardization of solutions of acids and bases. (Section 11-2; Exercises 23 through 27)
4. Using molarity, be able to calculate (a) the concentration of an acid (or base) solution from information about the amount of base (or acid) required to neutralize it (standardization), (b) the volume of an acid (or base) solution required to neutralize a specified amount of base (or acid), and (c) the amount of acid or base in an unknown sample, by titration with a solution of known concentration. (Section 11-3; Exercises 28 through 41)
5. Define equivalent weight and normality for an acid or base solution. Be able to carry out calculations relating these quantities (a) to one another and (b) to formula weight and molarity. (Section 11-4; Exercises 42 through 47)
6. Be able to carry out the calculations of Study Goal 4 using normality to express solution concentration. (Section 11-4; Exercises 48 through 51)
7. Balance molecular and net ionic equations for redox reactions by either the *change-in-oxidation-number* method or the *ion-electron* method. Be able to add H^+, OH^-, and H_2O, as appropriate, to achieve mass balance of H and O. (Sections 11-5 through 11-7; Exercises 52 through 65)
8. Using molarity, perform stoichiometric calculations involving redox reactions in which one or more of the reactants is in solution. (Section 11-8; Exercises 66 through 75)

9. Use the concepts of the chapter to recognize and solve a variety of types of questions. (Mixed Exercises, 76 through 86)
10. Answer conceptual questions based on this chapter. (Conceptual Exercises 87 through 88)
11. Apply concepts and skills from earlier chapters to the ideas of this chapter. (Building Your Knowledge exercises, 89 through 99)

Some Important Terms in This Chapter

Write the meanings *in your own words*. Check the Key Terms list and the chapter reading. Then rewrite your definitions, still in your own words, to improve them. Study other new terms, and review terms from preceding chapters if necessary.

titration

end point

equivalence point

indicator

standardization

standard solution

primary standard

molarity

normality

equivalent weight of an acid

equivalent weight of a base

oxidation

reduction

redox reaction

redox titration

Preliminary Test

As in other chapters, this test will check your understanding of basic concepts and types of calculations. Be sure to practice *many* of the additional textbook Exercises, including those indicated with the Study Goals.

Short Answer

Answer with a word, a phrase, a formula, or a number with units as necessary.

1. One millimole of HCl would neutralize exactly _____ millimole(s) of KOH.
2. One millimole of HCl would neutralize exactly _____ millimole(s) of $Ba(OH)_2$.
3. One meq of HCl would neutralize exactly _____ meq of KOH.
4. One meq of HCl would neutralize exactly _____ meq of $Ba(OH)_2$.
5. One meq of H_2SO_4 would neutralize exactly _____ meq of KOH.
6. In terms of concentration and volume of solution, the number of millimoles of solute can be calculated as _____.
7. In terms of concentration and volume of solution, the number of meq of solute can be calculated as _____.

8. The formula weight of nitric acid is 63.02. In the acid-base neutralization reaction

$$HNO_3 + NaOH \rightarrow H_2O + NaNO_3$$

the weight of one equivalent of nitric acid is _____ gram(s).

9. The formula weight of nitric acid is 63.02. In the acid-base neutralization reaction

$$2HNO_3 + Ca(OH)_2 \rightarrow 2H_2O + Ca(NO_3)_2$$

the weight of one equivalent of nitric acid is _____ gram(s).

10. The weight of 1 meq of nitric acid is _____.

11. The formula weight of oxalic acid is 90.04. The weight of one weight of oxalic acid is _____ gram(s).

12. The formula weight of calcium hydroxide is 74.10. The weight of one equivalent weight of calcium hydroxide is _____ gram(s).

13. A solution of hydrochloric acid is described as 0.15 molar (0.15 M). In terms of normality, the concentration of this solution would be described as _____.

14. A solution of phosphoric acid is described as 0.15 molar (0.15 M). In terms of normality, the concentration of this solution would be described as _____.

15. The volume of 0.200 N HNO_3 required to exactly neutralize 50.0 mL of a 0.150 N solution of $Ca(OH)_2$ is _____.

16. A solution whose concentration is accurately known is called a _____ solution.

17. The process by which we determine the volume of a solution required to react with a specific amount of a substance is called _____.

18. A substance that is of sufficient purity and ease of handling that the number of moles (or equivalents) present may be determined by direct weighing is referred to as a _____.

19. When we need additional hydrogen and oxygen to balance the equation for a redox reaction that is carried out in acidic solution, we may add _____ or _____ to either side of the equation as needed, but we must not add _____.

20. The "change-in-oxidation-number" method for balanced redox equations uses the idea that the _____ by all species being _____ must equal the _____ by all species being _____.

21. The "ion-electron" method for balancing redox equations is based on the idea that the _____ by all species being _____ must equal the _____ by all species being _____.

22. In applying the ion-electron method we formally separate the oxidation and the reduction processes, and write these as two separate _____, which we later recombine after we balance each one separately.

Multiple Choice

_____ 1. Which of the following statements is true?
  (a) The equivalent weight of an acid is always less than its formula weight.
  (b) The equivalent weight of an acid is always equal to its formula weight.
  (c) The equivalent weight of an acid is always greater than its formula weight.
  (d) The equivalent weight of an acid is always greater than or equal to its formula weight.
  (e) The equivalent weight of an acid is always less than or equal to its formula weight.

_____ 2. Which of the following statements is true?
    (a) The normality of an acid solution is always less than its molarity.
    (b) The normality of an acid solution is always equal to its molarity.
    (c) The normality of an acid solution is always greater than its molarity.
    (d) The normality of an acid solution is always greater than or equal to its molarity.
    (e) The normality of an acid solution is always less than or equal to its molarity.

_____ 3. The formula weight of NaOH is 40.0 g/mol. What is the equivalent weight of NaOH?
    (a) 40.0 g/eq   (b) 20.0 g/eq   (c) 80.0 g/eq
    (d) The answer depends on the reaction undergone by NaOH.

_____ 4. The formula weight of the base $Sr(OH)_2$ is 121.6 g/mol. What is the equivalent weight of $Sr(OH)_2$?
    (a) 121.6 g/eq   (b) 60.8 g/eq   (c) 243.2 g/eq
    (d) The answer depends on the reaction undergone by $Sr(OH)_2$.

_____ 5. What is the normality of a 0.40 M NaOH solution?
    (a) 0.40 N (b) 0.20 N (c) 0.80 N
    (d) The answer depends on the reaction undergone by NaOH.

_____ 6. What is the normality of a 0.40 M $Sr(OH)_2$ solution?
    (a) 0.40 N (b) 0.20 N (c) 0.80 N
    (d) The answer depends on the reaction undergone by $Sr(OH)_2$.

_____ 7. What is the molarity of the salt produced in the reaction of 50 mL of 0.0400 M HBr with 100 mL of 0.0300 M $Ca(OH)_2$?
    (a) 0.0150 M (b) 0.00667 M (c) 0.133 M (d) 0.0670 M (e) 0.075 M

_____ 8. What volume of 0.0750 M $Ba(OH)_2$ solution will react exactly with 125 mL of 0.0350 M HCl?
    (a) 29.2 mL   (b) 42.0 mL   (c) 230 mL   (d) 536 mL   (e) 53.6 mL

_____ 9. Calculate the molarity of a sulfuric acid solution if 40.0 mL of this H_2SO_4 solution reacts with 0.212 g of Na_2CO_3. One mole of Na_2CO_3 = 106 g.
    (a) 0.025 M (b) 0.050 M (c) 0.075 M (d) 0.100 M (e) 0.090 M

_____ 10. A solution of H_2SO_4 contains 9.11 mg of H_2SO_4 per mL. What is the normality of the solution? Assume the H_2SO_4 is to be completely neutralized by a base.
    (a) 0.0482 N (b) 0.0930 N (c) 0.186 N (d) 0.372 N (e) 0.684 N

_____ 11. Calculate the volume of 2.00 M H_2SO_4 required to prepare 200 mL of 0.100 N H_2SO_4. (Assume that the acid is to be completely neutralized.)
    (a) 5.0 mL   (b) 7.5 mL   (c) 10 mL   (d) 12.5 mL   (e) 15 mL

_____ 12. What is the normality of 35.0 mL of a solution of $Ba(OH)_2$ that exactly neutralizes 10.0 mL of 0.150 molar H_2SO_4?
    (a) 0.0857 N (b) 0.150 N (c) 0.0525 N (d) 0.214 N (e) 0.107 N

_____ 13. When we complete and balance the equation

$$SO_3^{2-} + MnO_4^- \rightarrow SO_4^{2-} + Mn^{2+} \qquad \text{(acidic solution)}$$

the coefficients of SO_3^{2-} and Mn^{2+} are, respectively, _____ and _____.
    (a) 5 and 3   (b) 2 and 6   (c) 5 and 2   (d) 6 and 5   (e) 3 and 1

_____ 14. When we complete and balance the equation

$$Cr_2O_7^{2-} + H_2S \rightarrow Cr^{3+} + S \qquad \text{(acidic solution)}$$

the coefficients of $Cr_2O_7^{2-}$ and S are, respectively, _____ and _____.
    (a) 1 and 3   (b) 3 and 2   (c) 2 and 7   (d) 1 and 6   (e) 10 and 4

_____ 15. When we complete and balance the equation

$$NO_3^- + I_2 \rightarrow NO + IO_3^- \quad \text{(acidic solution)}$$

the coefficients in front of I_2 and NO are, respectively, _____ and _____.
(a) 3 and 6   (b) 3 and 10   (c) 10 and 3   (d) 1 and 2   (e) 4 and 7)

Questions 16 and 17 refer to the following reaction:

$$MnO_4^- + 8H^+ + 5Fe^{2+} \rightarrow Mn^{2+} + 5Fe^{3+} + 4H_2O$$

_____ 16. What is the molarity of a solution of $FeSO_4$ if 25.0 mL of it reacts with 38.0 mL of 0.1214 M $KMnO_4$?
(a) 0.185 M (b) 0.399 M (c) 0.0798 M (d) 0.426 M (e) 0.923 M
_____ 17. What volume of 0.100 molar $KMnO_4$ will react with 1.00 g of $FeSO_4$?
(a) 164 mL   (b) 52.6 mL   (c) 26.3 mL   (d) 13.2 mL   (e) 65.8 mL

NOTE: It takes a great deal of practice to learn to use the material in this chapter. You will need much more practice at balancing redox equations and at working with molarity and normality than the questions in this guide provide. You should work many of the text Exercises suggested in the Study Goals.

Answers to Preliminary Test

Do not just look up answers. Think about the reasons for the answers.

Short Answer
1. 1
2. 1/2
3. 1
4. 1
5. 1. The convenience of the use of equivalent weights is seen in Questions 3 through 5, compared with the use of moles in Questions 1 and 2.
6. molarity x mL
7. normality x mL
8. 63.02. Each mole of HNO_3 supplies one mole of H^+ in this acid-base reaction.
9. 63.02. Just as in question 8, each mole of HNO_3 gives up one mole of H^+.
10. 63.02 mg or 0.06302 g
11. 90.04/2 = 45.02
12. 74.10/2 = 37.05
13. 0.15 N. The equivalent weight and the formula weight of HCl are the same, because HCl produces 1 mole H^+ per mole HCl. Thus molarity = normality.
14. 0.45 N. The equivalent weight of H_3PO_4 is only one-third as great as its formula weight. Thus, a given mass of H_3PO_4 would contain three times as many equivalents of H_3PO_4 as moles of H_3PO_4. Normality is always equal to or greater than molarity.
15. 37.5 mL. Compare this problem with Example 11-12.
16. standard

17. titration
18. primary standard
19. H^+; H_2O; OH^-
20. total increase in oxidation number; oxidized; total decrease in oxidation number; reduced
21. gain of electrons; reduced; loss of electrons; oxidized
22. half-reactions

Multiple Choice

1. (e). This is also true for bases. Study the application of the idea of equivalent weight in Section 11-4.
2. (d). This also applies to bases as well as to acids.
3. (a)
4. (b)
5. (a)
6. (c)
7. (b)
8. (a)
9. (b). See Example 11-13
10. (c)
11. (a). You need to convert molarity to normality and to account for the dilution.
12. (a)
13. (c). The balanced equation is

$$5SO_3{}^{2-} + 6H^+ + 2MnO_4{}^- \rightarrow 5SO_4{}^{2-} + 2Mn^{2+} + 3H_2O$$

We must balance H and O by adding H^+ and H_2O, because the reaction takes place in acidic solution. We cannot add OH^- to either side of the equation.

14. (a). The balanced equation (remember that it is in an acidic solution) is

$$Cr_2O_7{}^{2-} + 3H_2S + 8H^+ \rightarrow 2Cr^{3+} + 3S + 7H_2O$$

15. (b). The balanced equation is

$$10NO_3{}^- + 3I_2 + 4H^+ \rightarrow 10NO + 6IO_3{}^- + 2H_2O$$

16. (e). This is the same type of calculation as in Example 11-23.
17. (c)

12 Gases and the Kinetic-Molecular Theory

Chapter Summary

For the previous several chapters, we have been concerned primarily with the bonding, properties, and reactions of specific substances. In this chapter we begin a thorough study of some aspects of the behavior of matter that are more dependent on the physical state of the substances than on their chemical identity. We are first concerned, in this chapter and the next, with pure substances in the three states of matter—gas, liquid, and solid. In Chapter 14 we will study mixtures of substances. The introduction and first section of Chapter 12 point out some of the characteristics that distinguish the gas, liquid, and solid states.

We are first concerned with the observable **macroscopic behavior** of gases, which we then relate to proposed **molecular behavior**. Two examples of this approach are seen in Section 12-1. (1) The observation that gases have low density tells us that in gases the molecules are much further apart than in liquids or solids, and (2) the ability of gases to fill any container in which they are placed and to diffuse and mix tells us that in gases the molecules are in rapid motion. In Sections 12-2 through 12-11, we will deal with measurable properties of gases and the observed relationships among these properties—such quantities as **density, pressure, volume**, amount of gas **(number of moles)**, and **temperature**. Then in Section 12-12 we will interpret these relationships in terms of the kinetic-molecular theory. Most of our study of this chapter is made easier by the observation that all gases behave the same under many conditions. This is referred to as **ideal gas** behavior. In Section 12-14 we will study some conditions for which this ideal behavior is not observed.

We begin our study of the properties pressure (P), volume (V), number of moles (n), and temperature (T) and the ways in which these are related for typical gases. (Of these properties, pressure is the one for which students often have the least prior understanding. Section 12-3 introduces the concept of pressure and some ways of measuring it.) One useful way to interrelate such properties is to keep two of these variables constant and then to see how the other two are related. Many of the relationships in this chapter are valid only under certain conditions, so you *must* keep in mind the limitations of each one. As you study these relationships, you will see the various ways we represent them: (1) **In words**. You should be able to make a verbal statement of each relationship. It is more important that you understand any statement than that you just memorize it. Be certain that you know what is meant when you use words like "proportional," "inversely," etc. (2) **In mathematical relationships**. To use quantitative relationships effectively, we must be able to understand and use the equations that describe them. This includes using the relationships to solve problems. (3) **In graphs**. This pictorial way of representing mathematical relationships can be a great help in understanding, remembering, and applying the relations.

In Section 12-4 we see the relationship between pressure and volume for a fixed amount of gas (constant n) at a fixed temperature (constant T). This relationship is known as **Boyle's Law**. This law can be represented in various mathematical ways, some more useful than others. One use of Boyle's Law is to relate some initial set of conditions of P and V to a final set of conditions of these two variables. One form that we find very useful is

$$P_1V_1 = P_2V_2 \qquad \text{(constant } n, T)$$

Again, remember the set of conditions to which this equation is limited. The relation $P_1V_1 = P_2V_2$ must not be expected to hold if the amount of gas changes (n not constant) or if the temperature changes (T not constant).

Experimentally, we also find that the relationship

$$\frac{V_1}{T_1} = \frac{V_2}{T_2} \qquad \text{(constant } n, P)$$

known as **Charles's Law** (Section 12-5), describes the relation of volume and temperature of a gas, so long as we keep the amount of gas and the pressure of the gas constant. The relation takes this simple form only if the temperature is expressed on the **absolute** or **Kelvin** scale. Be sure that you understand the rationale behind establishing this scale and its relation to other temperature scales. Keep in mind that the symbol T in chemistry always refers to the temperature measured on the Kelvin scale. The establishment of a set of conditions of T and P as **standard temperature and pressure (STP)**, as discussed in Section 12-6, helps us to compare different gases easily.

In Section 12-7, we see how we can combine Boyle's and Charles's Laws into the **combined gas law,**

$$\frac{P_1V_1}{T_1} = \frac{P_2V_2}{T_2} \qquad \text{(constant } n)$$

This equation is convenient to use when several of the quantities P, V, and T change. A common use of this equation is to correct the description of a gas at some arbitrary set of conditions to those that would apply at STP.

We observe that at the same temperature and pressure, equal volumes of all ideal gases contain the same numbers of molecules (Avogadro's Law, Section 12-8). Thus the volume occupied by one mole of the gas at STP (the **standard molar volume**) is the same for all gases. Another statement of Avogadro's Law is

$$\frac{V_1}{n_1} = \frac{V_2}{n_2} \qquad \text{(constant } T, P)$$

This gives the basis for calculations relating volumes and amounts of gas at constant T and P. One mole of each different gas would have a different mass, so different gases have different densities, even at the same conditions.

Each of the gas laws studied up to now applies only at certain conditions and covers only certain relationships. They are useful primarily to determine changes in one quantity if we know changes in another. In Section 12-9, all of these laws are combined into a very useful relation, the **ideal gas equation**:

$$PV = nRT$$

Be sure that you know the significance of the **gas law constant, R.** Its value does not depend on what gas the equation is being applied to, and it can be expressed in different sets of units (with different numerical values), depending on the units used to describe the other quantities in the equation. Because R is a constant, this equation allows us to predict the value of any one of the four quantities P, V, n, and T, given values for the other three. A summary of the ideal gas laws appears on page 453.

Until now, you have seen the molecular weight, or formula weight, obtained only by adding up the atomic weights of the atoms in the formula. An important use of the ideal gas equation is the *experimental* determination of molecular weight by measurement of gas volumes (Section 12-10).

This is the first of several methods you will learn for experimentally determining molecular weights even if the chemical formula is not known. (Watch for other such methods in this and later chapters.) This knowledge lets us find the molecular formulas of unknown substances. This determination can be carried out by measuring P, V, T, and the mass of the sample. We calculate the number of moles of substance present from the ideal gas equation, and then divide the observed mass by the number of moles to obtain the molecular weight. This approach can also be used to find the molecular weights of **volatile** liquids (those that evaporate easily to form gases).

To this point, all that we have described about gases is independent of the identity of the gas. Thus it is not surprising that some aspects of the behavior of gas mixtures (provided they do not react) are also quite simply described. **Dalton's Law of Partial Pressures**, Section 12-11, is such a description, based first on direct observation, but shown to be consistent with the ideal gas equation. In this section we are introduced to another method for describing the composition of a mixture, by using the dimensionless quantity **mole fraction**. One very useful application of Dalton's Law is the calculation associated with the collection of a gas over water, which exerts a pressure by evaporating.

In Sections 12-4 through 12-11, we have been studying the observed properties of gases, with no reference to the properties or even the existence of molecules. We see that at the macroscopic level, all gases behave essentially the same. We now describe the molecular level of behavior. We assume that the behavior of gas molecules is essentially independent of their chemical identity. We then see whether this molecular behavior is consistent with the observed macroscopic properties. The result is the **kinetic-molecular theory**, Section 12-12. Not only does this theory provide us with a molecular-level understanding of such quantities as pressure and temperature, but it can also be used to predict additional types of behavior. One such prediction deals with rates of diffusion and effusion of gases. This is the basis for another method for experimentally determining molecular weights of unknown gaseous compounds. Based on how consistent it is with the observed behavior of gases, we now believe that the kinetic-molecular theory gives a good description of how gas molecules act.

Much of the discussion until now in this chapter has been applicable at high temperature (that is, quite a bit higher than the boiling points of the substances), and at low or moderate pressures. At those conditions, all gases behave alike. However, if the temperature is too low or the pressure is too high, the gas molecules do not behave independently of one another. The gases then do not behave ideally, and this deviation from ideal behavior is different for different gases. We can take two approaches to describe such "**real gases**" (Section 12-14). One is to develop equations that "fit" the observed properties of gases better at high P and low T than do the various ideal equations studied so far. One of these is the **van der Waals equation**. This equation contains two constants, a and b, that are different for every gas to which the equation is to be applied. The second approach, now at the molecular level, is to describe how molecular size and intermolecular attractions can affect molecular properties at low P and high T.

In our study of stoichiometry in Chapter 3, we saw that the fundamental relationship among amounts of reactants and products is in terms of the numbers of moles. At that stage, the only way we had to "measure" the number of moles was in terms of the mass of the substance. Now we have seen that the volume of a gas at specified conditions of temperature and pressure is a measure of the number of moles present. We now use this as a basis to extend our study of stoichiometry. The volume change that accompanies a reaction is proportional to the changes in numbers of moles of gaseous reactants and products. Using the molecular weight, we can extend these ideas to deal with mass-volume relationships in reactions involving gases (Section 12-16).

Study Goals

Remember to answer many of the Exercises suggested at the end of the text chapter. Be systematic in your solutions to numerical problems, writing down all steps and reasoning.

1. Summarize the general physical properties of gases. (Sections 12-1 and 12-2; Exercises 1, 2, and 8)

2. Explain what pressure is, and illustrate how we measure it with (a) an inverted-tube mercury barometer and (b) a mercury manometer connected to a container of gas. Convert pressure measurements to various units. (Section 12-3; Exercises 3 through 7 and 9 through 11)

3. State in words and in mathematical relationships the Laws of (a) Boyle, (b) Charles, (c) Avogadro, and (d) Dalton. (Sections 12-4 through 12-6, 12-8, 12-11, and 12-13; Exercises 12, 13, 22, 29, 30, 36, and 56)

4. Explain the basis of the absolute temperature scale. Convert temperatures from other scales to the Kelvin scale. (Section 12-5; Exercises 19 through 21)

5. Perform calculations using the relationships of Study Goal 3 to relate volume changes, pressure changes, temperature changes, and changes in number of moles of gas. (Sections as listed for Study Goal 3; Exercises 14 through 18, 23 through 28, 35 through 42, and 57 through 67)

6. State the combined gas law in mathematical terms and understand the relationships it expresses. (Section 12-7; Exercise 29, 30, and 34)

7. Given, for a gaseous system, any five out of six variables (P_1, P_2, V_1, V_2, T_1, T_2), be able to solve for the value of the missing sixth variable. (Section 12-7; Exercises 31 through 33)

8. State the ideal gas equation and its assumptions; derive the ideal gas equation from the combined gas laws. Perform calculations relating the variables in the ideal gas law. Given any three of the variables P, V, n, and T for a sample of ideal gas, be able to solve for the value of the other variable. (Section 12-9; Exercises 43 through 48)

9. Be able to calculate the molecular weight of a pure gaseous substance, given the density of the gas at any condition of temperature and pressure. Use this information to help determine the molecular formula of a gaseous substance. (Section 12-10; Exercises 49 through 55)

10. Know the standard molar volume of an ideal gas. Relate the volume of a gas at a given set of temperature and pressure conditions to the number of moles (and grams) of the gas. (Sections 12-8 and 12-9; Exercises 35 and 37 through 42)

11. Given the total pressure and composition of a mixture of gases, be able to solve for the partial pressure of each gas present in the mixture. (Section 12-11; Exercises 57 through 67)

12. Given the vapor pressure of water at various temperatures, be able to correct volumes and pressures of wet gases at specified temperatures and barometric pressures for the vapor pressure of water. (Section 12-11; Exercises 63 and 64)

13. State the assumptions of the kinetic-molecular theory. Account for the laws listed in Study Goal 3 in terms of the kinetic-molecular theory. (Section 12-12; Exercises 81 through 88)

14. Distinguish between real gases and ideal gases. Relate the van der Waals' equation to the ideal gas equation. Relate the van der Waals' constants a and b to deviations of molecular properties of the gases using the assumptions of the kinetic-molecular theory. (Section 12-14; Exercises 89 through 92)

15. Be able to use the van der Waals' equation to carry out calculations relating n, P, V, and T for real gases. Understand both when and why these results differ significantly from those predicted by the ideal gas equation. (Section 12-14; Exercises 93 through 98)

16. Be able to carry out calculations involving (a) volumes of gases involved in chemical reactions, and (b) mass-volume relationships for chemical reactions involving gases. (Section 12-16; Exercises 68 through 80)
17. Use your understanding of this chapter to recognize and answer a variety of types of questions. (Mixed Exercises, 98 through 109)
17. Answer conceptual questions based on this chapter. (Conceptual Exercises 110 through 117)
18. Apply concepts and skills from earlier chapters to the ideas of this chapter. (Building Your Knowledge exercises, 118 through 123)

Some Important Terms in This Chapter

Write the meanings *in your own words*. Check the Key Terms list and the chapter reading. Then rewrite your definitions, still in your own words, to improve them. Study other new terms, and review terms from preceding chapters if necessary.

fluid

equation of state

absolute zero

Boyle's Law

Charles's Law

Avogadro's Law

standard conditions (STP)

standard molar volume

Law of Partial Pressures

mole fraction

ideal gas

ideal gas equation

universal gas constant

partial pressure

Law of Partial Pressures (Dalton's Law)

kinetic-molecular theory

Graham's Law

real gas

van der Waals equation

Preliminary Test

This test will check your understanding of basic concepts and types of calculations. Be sure to answer *many* of the additional textbook Exercises. Those are more challenging than these preliminary test questions, so they will give you additional insight into the meaning and applications of this chapter material. For all Preliminary Test questions in this chapter, assume that the gas is acting ideally, unless you are told otherwise.

True-False

Mark each statement as true (T) or false (F).

_____ 1. Because the noble gases do not easily react to form compounds, they are present in a very high amount in the atmosphere.

_____ 2. All substances are less dense as gases than as liquids.

_____ 3. For any ideal gas, doubling the pressure would always cause its volume to decrease by a factor of two.

_____ 4. For any gas, whether ideal or not, doubling the pressure of a fixed amount of gas at a fixed temperature would cause its volume to decrease by a factor of two.

_____ 5. Doubling the temperature always doubles the volume of a fixed amount of ideal gas at constant pressure.

- -

We have two identical containers. One is filled with hydrogen fluoride gas, HF, and the other is filled with neon gas, Ne. The two containers are at the same temperature and pressure. Questions 6 through 14 all refer to these two samples of gas. The atomic weights to be used for these questions are as follows: H = 1, F = 19, Ne = 20. For Questions 6 through 14, consider the gases to be ideal.

_____ 6. The two gas samples have the same number of molecules.

_____ 7. The two gas samples have the same number of atoms.

_____ 8. The two gas samples have the same total mass of gas.

_____ 9. The two gas samples have the same density.

_____ 10. The molecules in the two gas samples have the same average kinetic energy.

_____ 11. The molecules in the two gas samples have the same average speed.

_____ 12. If the same size hole were opened in each container, the pressure in the neon container would decrease faster than the pressure in the hydrogen fluoride container.

_____ 13. If we were to remove all of the Ne from one box and put it into the other box with the HF, the molecules would mix so as to cause no change in total pressure in the second box.

_____ 14. If the two gases were taken to the same low temperature and the same very high pressure, we would expect HF to behave more ideally than Ne.

Short Answer

1. The pressure of one atmosphere is equal to _____ torr.
2. Charles's Law may be expressed mathematically as _____.

142

3. The volumes of gases, measured at _____, that are involved in chemical reactions are in the ratios of small whole numbers.

4. The statement "Equal volumes of gases at the same conditions of pressure and temperature contain equal numbers of molecules" is a statement of _____ Law.

5. Tripling the absolute temperature (in K) and doubling the pressure of a given volume of an ideal gas would give a new volume of gas. This new volume could be calculated by multiplying the original volume of the gas by a factor of _____.

6. To cause an ideal gas sample to change from a volume of 1.4 L to 4.2 L at constant temperature, the pressure must be changed to _____ times its initial value.

7. To cause an ideal gas sample to change from a volume of 1.4 L to 4.2 L at constant pressure, the absolute temperature must be changed to _____ times its initial value.

8. The zero of absolute temperature is taken as that temperature at which all gases would exhibit _____. Actually, before this temperature is reached, any real gas would form _____ and then _____.

9. If 1.00 L of an ideal gas at 50.0°C is heated at constant pressure to 100°C, the new volume would be _____ L.

10. For one mole of an ideal gas, PV/T = _____. (The answer is to be a number, with correct units.

11. The conditions we refer to as standard temperature and pressure are _____ and _____.

12. A mixture of H_2 and O_2 is in a 6.00 L vessel. The partial pressure of H_2 in this mixture is 500 torr, and the partial pressure of O_2 is 250 torr. The volume that H_2 would occupy alone at these conditions of temperature and total pressure is _____ L.

13. The mole fractions of H_2 and O_2 in the mixture described in Question 12 are _____ and _____, respectively.

14. The vapor pressure of water at 30°C is 31.8 torr. We collect hydrogen over water at 30°C until the total pressure of wet hydrogen is 1 atm. If we dried the hydrogen at the same temperature, the pressure it would exert would be _____ torr.

15. At STP, 1.00 L of a pure gaseous substance has a mass of 4.00 grams. The molecular weight of this substance is _____.

16. At standard temperature and pressure, 14.007 grams of nitrogen gas (considered ideal) occupies _____ L.

17. A sample of a pure gaseous substance occupying 4.10 L at 760 torr and 227°C is found to weigh 6.4 grams. The density of the substance under these conditions is _____.

18. The molecular weight of the substance in Question 17 is _____.

19. If 1.00 L of gas X weighs 44 times as much as 1.00 L of H_2 at the same temperature and pressure, then the molecular weight of the gas X is _____.

20. One mole of chlorine gas weighs _____ grams, contains _____ molecules, contains _____ atoms, and occupies _____ L at 0°C and 1 atm.

21. In the kinetic-molecular description of gas behavior, pressure is a measure of _____.

22. In the kinetic-molecular description of gas behavior, temperature is a measure of _____.

23. At the same _____, the molecules of all samples of ideal gases have the same average kinetic energies.

24. A real gas behaves most nearly like an ideal gas at conditions of _____ and _____.

25. In the van der Waals' equation describing real (nonideal) gases, the *a* term is related to the molecular property of _____.

26. The commercial preparation of ammonia is carried out by reaction of nitrogen and hydrogen:

$$N_2(g) + 3H_2 (g) \rightarrow 2NH_3(g)$$

To form 1000 L of ammonia gas, we would have to supply _____ L of nitrogen and _____ L of hydrogen, with all gases measured at the same temperature and pressure. (Treat the reaction as if it goes to completion.)

Carbon disulfide, CS_2, burns according to the equation

$$CS_2(l) + 3O_2(g) \rightarrow CO_2(g) + 2SO_2(g).$$

The following three questions, 27 through 29, refer to the complete combustion of one mole of CS_2, according to this equation.

27. The volume of O_2, measured at standard temperature and pressure (STP), required for the reaction is _____.

28. The volume of CO_2, measured at STP, that would be produced is _____.

29. The volume of SO_2, measured at STP, that would be produced is _____.

Multiple Choice

_____ 1. Each of the following plots refers to the behavior of a sample of an ideal gas. Which plot is *not* correct?

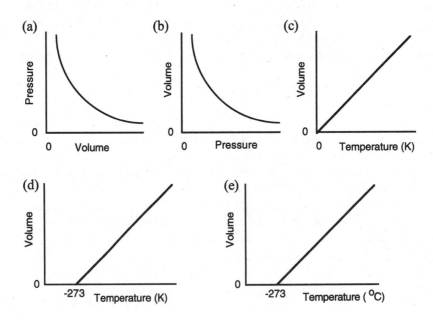

144

_____ 2. A sample of a gas having a volume of 1 L at 25°C and 1 atm pressure is subjected to an increase in pressure and a decrease in temperature. The density of the gas
(a) decreases.   (b) increases.   (c) remains the same.   (d) becomes zero.
(e) either increases or decreases, depending on the sizes of the pressure and temperature changes.

_____ 3. A sample of a gas having a volume of 1 L at 25°C and 1 atm pressure is subjected to an increase in pressure and an increase in temperature. The density of the gas
(a) decreases.   (b) increases.   (c) remains the same.   (d) becomes zero.
(e) either increases or decreases, depending on the sizes of the pressure and temperature changes.

_____ 4. At the same temperature, the molecules of all gases have the same
(a) average masses.   (b) van der Waals' constants a and b.
(c) average kinetic energies.   (d) average speeds.
(e) molecular diameters.

_____ 5. How many molecules of an ideal gas are contained in 8.2 L at –73°C and 0.50 atm?
(a) 0.25   (b) 1.5×10^{23} (c) $–4.1 \times 10^{23}$ (d) 7.5×10^{23} (e) 4.2×10^{-25}

_____ 6. One liter of phosphorus gas reacts with six liters of hydrogen gas to form four liters of phosphine gas. All gas volumes are measured at the same conditions of temperature and pressure. What is the formula of phosphine? (Remember that the formula for phosphorus is P_4.)
(a) P   (b) P_2 (c) PH   (d) PH_2 (e) PH_3

_____ 7. We place equal weights of O_2 and N_2 in separate containers of equal volume at the same temperature. Which one of the following is true?
(a) Both flasks contain the same number of molecules.
(b) The pressure in the nitrogen flask is greater than that in the oxygen flask.
(c) Molecules in the oxygen flask are moving faster on the average than the ones in the nitrogen flask.
(d) The nitrogen molecules have a greater average kinetic energy.
(e) All N_2 molecules have the same energy, and all O_2 molecules have the same energy, but these two energies are not the same.

_____ 8. At 0°C, which of the following gases would have the lowest average molecular speed?
(a) NH_3 (b) CO_2 (c) Ar   (d) N_2 (e) H_2

_____ 9. Which of the following gases would diffuse most rapidly?
(a) CO_2 12+32   (b) O_2 32   (c) CH_4 12+4   (d) He 4   (e) N_2 28

_____ 10. Analysis of a gas gives 80.0% C and 20.0% H by mass. A sample of the gas occupying 2.24 L at 0°C and 1 atm is found to weigh 3.0 grams. Which one of the following substances could this gas be?
(a) methane, CH_4 (b) ethane, C_2H_6 (c) ethylene, C_2H_4
(d) acetylene, C_2H_4 (e) benzene, C_6H_6

_____ 11. An 1120 mL sample of a pure gaseous compound, measured at 0°C and 1 atm, is found to weigh 2.86 grams. What is the molecular weight of the compound?
(a) 2.86   (b) 28.6   (c) 5.72   (d) 57.2   (e) 14.3

_____ 12. What volume of pure oxygen gas, O_2, measured at 546 K and 760 torr, is formed by complete dissociation of 0.200 mole of Ag_2O?

$$2Ag_2O(s) \rightarrow 4Ag(s) + O_2(g)$$

(a) 1.12 L   (b) 2.24 L   (c) 17.92 L   (d) 4.48 L   (e) 8.96 L

_____ 13. How many moles of $KClO_3$ are needed to form 2.8 L of O_2, measured at STP, according to the following equation?

$$2KClO_3(s) \rightarrow 2KCl(s) + 3O_2(g)$$

(a) 1/12 mole   (b) 1/6 mole   (c) 1/3 mole   (d) 1/4 mole   (e) 1/2 mole

_____ 14. What volume of chlorine gas, measured at 0°C and 1 atm pressure, is needed to react completely with 5.61 grams of KOH in the following equation?

$$Cl_2(g) + 2KOH(aq) \rightarrow KCl(aq) + KClO(aq) + H_2O(l)$$

(a) 0.56 L   (b) 1.12 L   (c) 2.24 L   (d) 4.48 L   (e) 22.4 L

_____ 15. One step in the industrial manufacture of nitric acid involves the production of nitric oxide by the oxidation of ammonia:

$$4NH_3(g) + 5O_2(g) \rightarrow 6H_2O(g) + 4NO(g)$$

We wish to produce 100 L of NO gas. How many liters of oxygen gas, measured at the same temperature and pressure, would be required?

(a) 100 L   (b) 125 L   (c) 22.4 L   (d) 100 x 22.4 L

(e) Cannot be calculated unless we know the values of temperature and pressure.

Answers to Preliminary Test

True-False

1. False. See Table 12-2. In fact, these are sometimes referred to as the rare gases.
2. True
3. False. This is only true at constant temperature and for a fixed amount of gas.
4. False. Only for ideal gases. We cannot predict in general what the factor would be for other gases.
5. False. This is true only if the temperature is expressed on the absolute (Kelvin) scale.
6. True
7. False
8. True. But only because the molecular weights of these two particular gases are the same.
9. True. Same comments as for Question 8.
10. True. This would be true even if the molecular weights of the gases were not the same.
11. True. Same comments as for Question 8.
12. False. The two types of gas molecules have the same average speed.
13. False. The pressure would double.
14. False. HF is more polar, so it would behave less ideally than Ne.

Short Answer

1. 760
2. $\dfrac{V_1}{T_1} = \dfrac{V_2}{T_2}$ (constant n, P). This is incorrect without specifying its limitations.
3. the same temperature and pressure

4. Avogadro's
5. 3/2. Recall that V is directly proportional to T and is inversely proportional to P.
6. 1/3
7. 3
8. zero volume; liquids; solids
9. 1.15. Volume is proportional to Kelvin temperature, which changes from 323 K to 373 K. Either multiply old volume by 373/323 or use the mathematical formulation of Charles's Law: $V_1 = 1.00$ L, $T_1 = 323$ K, and $T_2 = 373$ K.
10. $0.0821 \frac{\text{L·atm}}{\text{K}}$. Other values, with appropriate units, could also be correct.
11. 0°C, 1 atm
12. 4.00. Apply Dalton's Law. Because 2/3 of the total pressure of the mixture is due to H_2, it must account for 2/3 of the total number of moles. If we removed the O_2 (1/3 of the total moles) and adjusted to the same total pressure, the moles of H_2 would occupy 2/3 of the original total volume.
13. 2/3, 1/3. You can reason this either from the fraction of total moles or the fraction of total pressure contributed by each gas. Remember that mole fraction is a dimensionless quantity. Notice that the sum of the mole fractions is 1.
14. 728.2. Of the total 760 torr (1 atm) exerted, 31.8 torr is due to water vapor. Only $760 - 31.8 = 728.2$ torr would be due to hydrogen.
15. 89.6. One mole is the amount that would occupy 22.4 L at STP. This would weigh 22.4 x 4.0 grams, or 89.6 grams.
16. 11.2. Remember that nitrogen is diatomic, so 14.007 grams is only 1/2 mole. Each mole occupies 22.4 L at STP.
17. 4.0 g/L
18. 164. Apply $PV = nRT$, with $T = 227 + 273 = 500$ K; $P = 760$ torr $= 1$ atm; $V = 4.1$ L. Solving for n, we find that there is 0.10 mole of gas present. This weighs 16.4 grams, so molecular weight = (16.4 g)/(0.10 mol) = 164 g/mol.
19. 88
20. 71.0; 6.022 x 10^{23} (molecules); 1.204 x 10^{24} (atoms); 22.4
21. collision of molecules with the walls of the container
22. average molecular kinetic energy
23. temperature
24. high T, low P
25. intermolecular attraction. The b term is related to the size of the molecules.
26. 500; 1500. At the same temperature and pressure, the mole ratio (given in the balanced chemical equation) is the same as the gas volume ratio.
27. 67.2 L
28. 22.4 L
29. 44.8 L

Multiple Choice

1. (d)
2. (b)
3. (e)

4. (c)
5. (b). Use the ideal gas equation to determine the number of moles. Remember that there are 6.0×10^{23} molecules/mole (Avogadro's number).
6. (e)
7. (b)
8. (b). All have the same average kinetic energy; CO_2 has the highest molecular weight, so it would be moving most slowly at the same kinetic energy.
9. (d)
10. (b). You can answer this by determining the simplest formula (Chapter 2), CH_2. The gas volume data can be used to find that the molecular weight is 30 g/mol, or twice the simplest formula weight.
11. (d)
12. (d). The 0.100 mole O_2 formed would occupy 0.100 x 22.4 L at 273 K and 1 atm. Corrected to 546 K and 1 atm, it would occupy 4.48 liters. Alternatively, use the ideal gas equation to find the volume of 0.100 mole of O_2 at the desired temperature and pressure.
13. (a). 2.8 L (STP) is 2.8/22.4 mol $= \frac{1}{8}$ mol. According to the stoichiometry of the reaction, this would require $\frac{1}{12}$ mole. Use unit factors.
14. (b)
15. (b)

13 Liquids and Solids

Chapter Summary

In Chapter 13, we will take the same approach to the study of the two **condensed phases** or states of matter, liquids and solids, in the same way that we did for gases in Chapter 12. We will learn about the observed properties of liquids and solids. We will also describe those properties in terms of the behavior of the molecules that make up the liquids and solids, using the kinetic-molecular theory. We will also learn about qualitative and quantitative aspects of changes among these three states of matter. In molecular terms, the most important difference between gases on the one hand and liquids and solids on the other is that in liquids and solids the molecules are so close together that they can interact quite strongly, whereas in gases they are so far apart that their interactions are very slight. We will use the kinetic-molecular theory to understand differences in properties among these three states of matter. These properties were mentioned briefly in Section 12-1 and are listed in Table 13-1.

As you learn about solids and liquids in this chapter, you will see that we have several goals in mind. One is to understand the properties of liquids by describing what the molecules are doing and how they interact. Another is to gain a similar understanding of properties of solids relative to their molecular and ionic interactions. A third aspect of our study deals with changes between phases— liquid $\rightleftharpoons$ gas, solid $\rightleftharpoons$ liquid, and solid $\rightleftharpoons$ gas. In discussing this last aspect, either descriptively or quantitatively, we use our understanding of molecular properties in the individual states. We have concluded from a study of their observed properties (Chapter 12) that in gases the molecules are far apart and have very little attraction for one another. In this chapter, we will see that properties such as high density, surface tension, and viscosity tell us that in liquids the molecules are so close together that their attractions for one another are very strong and are quite important in determining properties. Evaporation and boiling both involve a substance changing from a liquid to a gas. To understand these processes, we consider the amount of energy that must be put in to overcome the attractive forces in the liquid and convert it to a gas.

We can extend the ideas of the kinetic-molecular theory to apply them to liquids and solids (Section 13-1); this provides the basis for the discussion in Chapter 13. The properties of liquids and solids depend on **intermolecular** forces of attraction. These forces may be classified (Section 13-2) as **ion-ion interactions**, **dipole-dipole interactions**, **hydrogen bonding**, and **dispersion forces**. Different types of forces vary considerably in their strengths. They have a great influence on melting and boiling points, heats of fusion, vaporization, and many other physical properties of substances. These effects are summarized in Table 13-6. An understanding of these attractive forces will also aid in understanding properties of solutions, to be discussed in Chapter 14. In the next few sections, we see the relation of observed macroscopic properties of a liquid such as **viscosity** (13-3), **surface tension** (13-4), and **capillary action** (13-5) to the intermolecular attractions.

The next several sections are concerned primarily with a description of changes between liquid and gas. When the kinetic energy of a molecule near the surface of a liquid is high enough to overcome the attraction of its neighbors, it can **evaporate** or **vaporize**; conversely, gas molecules with low enough kinetic energy can **condense** (Section 13-6). If the container is closed, the two processes eventually occur at equal rates, and the **vapor pressure** (Section 13-7) characteristic of the liquid at that temperature is established. The ideas of **dynamic equilibrium** and **LeChatelier's Principle** that are presented here will be central to our understanding of many equilibrium situations, including chemical reactions—pay close attention to these important concepts. Section 13-8 deals with a related phenomenon—**boiling**—and with the use of differences in boiling point from one substance to another to separate components of a liquid mixture—**distillation.**

We now begin a quantitative consideration of some of the energy changes involved in heating substances and in their phase changes. How can we understand the energy change in molecular terms? We recall that increasing the temperature of a substance means that the average kinetic energy of its molecules increases. To raise the temperature of a substance, we must put in energy, often in the form of heat (we speak of "heating the substance"). Likewise, for a liquid to evaporate, the kinetic energy of its molecules must be raised sufficiently to overcome the attractive forces between molecules. Again, this is done by putting in heat. In Section 13-9, we learn about several kinds of calculations with heat transfer involving liquids. Before beginning that section, you should review the units and measurement of heat, Section 1-13. The amount of heat required to raise the temperature of a substance by a certain amount is expressed as the **specific heat** or the **molar heat capacity** of the substance; the quantity of heat needed to cause vaporization of a liquid is the **heat of vaporization,** which is often expressed on a molar basis. The **Clausius-Clapeyron equation** describes how the vapor pressure of a liquid changes with temperature. One of its important uses is to predict how the boiling point of a liquid changes when we change the external pressure.

In the same way, the process of melting (and the reverse process, freezing) involves a change of state, this time between liquid and solid. When these two processes are in equilibrium, the substance is at its **melting point** (Section 13-10). Again, this equilibrium is a dynamic one. In molecular terms, melting involves putting in enough heat energy for the molecules or ions to overcome the strong attractions that hold them in their fixed positions in the solid. Terms analogous to those we have already seen are used to describe the heat transfer involving solids—the **specific heat** and **molar heat capacity** of a solid, as well as its **heat of fusion,** again often on a molar basis (Section 13-11). The corresponding transitions from solid to gas **(sublimation)** and gas to solid **(deposition)** and the related idea of the vapor pressure of solids are discussed in Section 13-12.

Much data regarding phase transitions for any pure substance can be summarized in its **phase diagram,** Section 13-13. To help interpret this diagram for various substances, remember that all points along any horizontal line in the diagram are at the same pressure, while all points along any vertical line are at the same temperature.

Solids can be either **amorphous** (lacking internal regularity) or **crystalline** (Section 13-14). Most solids are crystalline, characterized by a regularly repeating pattern of molecules or ions, somewhat like a three-dimensional wallpaper pattern. Because of the regularity of crystalline solids, we can describe and understand them in a systematic way. Much of our information about the structures of solids has been obtained by the technique of **X-ray diffraction** (Enrichment, page 510). The scattering of X-rays (a very short-wavelength form of light) by the regularly repeating arrays of atoms leads to interference of the rays. Analysis of the directions and intensities of these scattered rays has allowed us to determine the three-dimensional structures of many solids. This method has provided one of our major sources of information about the three-dimensional structures of molecules and polyatomic ions (Chapter 8) that make up many solids.

Section 13-15 discusses the various patterns by which atoms, ions, or molecules are arranged in crystals. The main point here is that even though solids consist of many different kinds of particles, the requirement that they are regularly arranged in a repeating pattern in the crystal means that only a few types of *arrangements* are possible. These are the seven crystal systems, given in Table 13-9.

This lists the restrictions on the shapes of the possible box (**unit cell**) by which the structure repeats. You should realize that this is just an imaginary box that describes the size and shape of the repeating unit; the repeating unit itself consists of atoms, molecules, or ions, arranged in a particular way.

We often classify crystalline solids according to the types of particles making up the arrangement and the type and strength of the bonding or interactions among them. These categories are summarized in Table 13-10 and discussed in detail with examples in Section 13-16.

(1) In **metallic solids,** the metal atoms occupy lattice sites, with valence electrons free to move throughout the array of atoms.

(2) **Ionic solids** consist of extended arrays of ions, with very strong attractions for one another. These strong attractions are due to the arrangement in which each positive ion has negative ions as its nearest neighbors, and vice versa. As seen in Chapters 2 and 7, these substances do not consist of molecules. Knowing the size and shape of the repetitive pattern of such substances can often give us information about ionic sizes, as seen in the various examples of this section.

(3) In **molecular solids,** the fundamental unit of the crystal is a molecule. Each molecule contains strong covalent bonds, but the molecules interact in the solid only weakly, giving these solids their characteristic properties.

(4) **Covalent solids** are those in which the atoms are covalently bonded together over many unit cells. Examples of these are diamond, graphite, quartz, and many minerals.

The **band theory** of metals (Section 13-17) describes the properties of metallic solids in terms of their electronic structures. The "Chemistry in Use" essay in this chapter discusses the important technological applications of solids in semiconductors.

Study Goals

There are many important ideas and skills in this chapter. To help you to organize your study, concentrate on them in the following groups. Answer as many of the text Exercises as you can.

1. Compare and contrast the properties of gases, liquids, and solids. (Introduction)
2. Apply the kinetic-molecular theory to explain the similarities and differences in properties of the three states of matter. (Section 13-1; Exercises 1, 2, 21, 75, and 76).
3. Describe and illustrate each of the following kinds of intermolecular attractions: (a) ion-ion interactions, (b) dipole-dipole interactions, (c) hydrogen bonding, and (d) dispersion forces. (Section 13-2; Exercises 1 through 4, 8, 15 through 18, and 31)
4. Describe and illustrate the following terms concerning liquids: (a) viscosity, (b) surface tension, (c) cohesive and adhesive forces, (d) capillary action, (e) evaporation and condensation, (f) vapor pressure, (g) boiling point, and (h) distillation. (Sections 13-3 through 13-8; Exercises 9, 21 through 23, 27 through 29, and 35 through 42)
5. Understand the significance of the types of intermolecular forces in Study Goal 3 in determining properties listed in Study Goal 4. (Sections 13-2 through 13-8; Exercises 3 through 7, 9 through 18, 30, and 32 through 34)
6. Use the Clausius-Clapeyron equation to relate vapor pressure, temperature, and heat of vaporization. (Section 13-9; Exercises 57 through 64)
7. Describe and illustrate terms concerning solids: (a) melting point, (b) fusion and solidification, and (c) sublimation and deposition. Understand the significance of the types of intermolecular

forces in Study Goal 3 in determining these properties of solids. (Sections 13-10 through 13-12; Exercises 39, 44, and 76 through 83)

8. Be able to write equations for phase transitions. (Sections 13-6, 13-9, 13-10, and 13-12)

9. Be able to use the kinetic-molecular theory to describe the heat flow that accompanies changes of temperature and phase transitions. (Sections 13-7, 13-9, and 13-11; Exercise 45)

10. Given the appropriate heat capacities or heats of fusion, vaporization or sublimation of a substance, be able to relate energy (heat) gain or loss to temperature changes and phase transitions of the substance. (Sections 13-9 and 13-11; Exercises 43 and 46 through 56)

11. Be able to interpret phase diagrams. (Section 13-13; Exercises 65 through 74)

12. Distinguish between amorphous and crystalline solids. Describe a glass as a supercooled liquid, as distinguished from a true solid. (Section 13-14; Exercise 75)

13. Illustrate the ideas underlying X-ray diffraction by crystals. Carry out some X-ray diffraction calculations. (Enrichment section; Exercises 98 through 101)

14. Characterize the seven crystal systems by names and by patterns of axial lengths and interaxial angles. (Section 13-15)

15. Given a picture or description of a unit cell and its contents, determine the numbers of atoms, ions, or molecules per unit cell. (Section 13-15; Exercises 85 and 94)

16. Describe and distinguish between the close-packed crystal structures. Sketch the three kinds of cubic lattices. (Section 13-15; Exercises 84 and 86)

17. Classify solids into one of the following four categories and summarize the distinguishing characteristics of these classes: (a) metallic solids, (b) ionic solids, (c) covalent solids, and (d) molecular solids. (Section 13-16; Exercises 77 through 83)

18. Perform calculations to relate unit cell data to such quantities as atomic and ionic radii and volumes, densities of solids, and the number of particles per unit cell. (Section 13-16; Exercises 86 through 101)

19. Describe metallic bonding in terms of band theory. Use band theory to describe electrical conductors, nonconductors (insulators), and semiconductors. (Section 13-17; Exercises 102 through 104)

20. Recognize and answer types of questions based on the concepts in this chapter. (Mixed exercises, 105 through 117)

21. Answer conceptual questions based on this chapter. (Conceptual Exercises 118 through 121)

22. Apply concepts and skills from earlier chapters to the ideas of this chapter. (Building Your Knowledge exercises, 122 through 128)

Some Important Terms in This Chapter

This chapter has a very large number of new terms, many of which will be used in later chapters. This is *only a partial list*, so be sure you study others as well. Write the meanings *in your own words*. Check the Key Terms list and the chapter reading. Then rewrite your definitions, still in your own words, to improve them. Study other new terms, and review terms from preceding chapters if necessary.

vapor pressure

boiling point

normal boiling point

intramolecular forces

intermolecular forces

ion-ion interactions

dipole-dipole interactions

hydrogen bonding

dispersion forces

crystalline solid

molecular solid

covalent solid

ionic solid

metallic solid

heat of fusion (and molar heat of fusion)

heat of vaporization (and molar heat of vaporization)

Clausius-Clapeyron equation

LeChatelier's Principle

dynamic equilibrium

metallic bonding

band

conduction band

forbidden zone

band gap

Preliminary Test

As in other chapters, this test will check your understanding of basic concepts and types of calculations. Be sure to practice *many* of the additional textbook Exercises indicated with the Study Goals.

Short Answer

1. The much lower compressibility of solids and liquids compared with gases tells us that in these condensed phases the molecules are _____ than in gases.
2. Forces between particles of a substance are referred to as _____ forces, whereas those within molecules are called _____ forces.
3. Forces between molecules within a liquid are called _____ forces.
4. Forces between molecules of a liquid and its container are called _____ forces.
5. Because the forces acting on molecules at the surface of a liquid are not equal in all directions, the liquid exhibits a property known as _____.
6. The movement of a liquid up a tube of small diameter is called _____; this occurs when the _____ exceed the _____.
7. The stronger the intermolecular forces in a liquid, the _____ is its vapor pressure, the _____ is its heat of vaporization, and the _____ is its boiling point.
8. A high viscosity observed for a liquid tells us that the intermolecular forces in the liquid are _____.
9. A method of separating the components of a mixture of liquids with sufficient different boiling points is called _____ .
10. When a liquid is evaporating exactly as fast as its vapor is condensing, a state of _____ is said to exist; the pressure exerted by the gas at this condition is said to be the _____ of the liquid.
11. For a pure substance at any pressure, the temperature at which the vapor pressure is equal to the applied pressure is called the _____ of the liquid. If the pressure is 1 atm, this temperature is termed the _____.
12. The vapor pressure of a liquid at 25°C is 200 torr. If the atmospheric pressure over the liquid is lowered to 200 torr at 25°C, the liquid will _____.
13. Substance A has greater tendency to evaporate at a given temperature than does substance B. We describe substance A as being more _____ than substance B. We also expect that the _____ of A would be higher and that its _____ would be lower.
14. The amount of heat that is required to melt one gram of a substance is called the _____ of the substance. The amount of heat required to melt one mole of the substance would be called the _____ of the substance.
15. For any substance at a given temperature, the molar heat of vaporization must equal the _____.
16. In order to calculate the molar heat capacity of a substance, we would have to _____ its specific heat by its _____.
17. Because of the highly polar nature of an H–F bond, we should expect the molar heat of vaporization of HF to be considerably _____ than that of F_2.
18. Even though both are gases are room temperature, we would expect the temperature at which CH_4 would condense to a liquid to be _____ than that for NH_3.

19. The abnormally high boiling and melting points of water compared with those of H_2S, H_2Se, and H_2Te are due to the presence of _____ in water.

20. Consider the substances $CaCl_2$, PH_3, N_2, H_2, and NH_3. The order of increasing boiling points of these substances would be _____.

21. The two substances CO and N_2 have the same molecular weight, 28. It is observed that the molar heat of fusion of carbon monoxide is higher than that of nitrogen. The reason for this is the _____ of the CO molecules.

22. Hydrogen bonding interactions represent a special, unusually strong form of _____.

23. The temperature at which the solid and liquid phases of a substance are in equilibrium with one another is called the _____ of the substance.

24. The set of conditions, at a particular pressure, at which the solid and gaseous states of a substance can exist in equilibrium is called the _____ of the substance.

25. The set of conditions at which a substance can exist simultaneously as solid, liquid, and gas in equilibrium is called the _____ of the substance.

26. A solid that shows no order (or only limited order) on the molecular scale is called _____.

27. The melting points of ionic crystals are usually _____ than those of molecular crystals.

28. Two examples of substances that exist as covalent crystals are _____ and _____.

29. A method that is used for the study of atomic arrangements in solids is _____.

30. In a single layer of spheres of equal size in a close-packed arrangement, the number of nearest neighbors of each sphere is _____.

31. In a three-dimensional close-packed arrangement of spheres of equal size, each sphere has _____ nearest neighbors.

32. In the NaCl structure, each Cl^- ion is surrounded by _____ nearest neighbors, each of which is a _____.

33. Two types of defects in the structures of crystalline solids that affect the properties of the solids are _____ and _____.

Multiple Choice

_____ 1. A drop of liquid has a spherical shape due to the property of
  (a) surface tension.   (b) capillary action.   (c) viscosity.
  (d) vapor pressure.   (e) close packing.

_____ 2. As a liquid is heated, its vapor pressure
  (a) does not change.   (b) increases.   (c) decreases.   (d) disappears.
  (e) may increase or decrease, depending on the liquid.

_____ 3. When the external pressure is decreased, the boiling point of a liquid
  (a) does not change.   (b) increases.   (c) decreases.   (d) disappears.
  (e) may increase or decrease, depending on the liquid.

_____ 4. If energy, in the form of heat, is added to a pure liquid substance at its boiling point while keeping the pressure constant, the
  (a) temperature will increase.   (b) temperature will decrease.
  (c) temperature will remain constant until all of the liquid has vaporized.
  (d) temperature of the vapor will be greater than that of the liquid.
  (e) temperature of the liquid will be greater than that of the vapor.

_____ 5. As we increase the temperature of a liquid, its properties change. Which of the following would *not* be an expected change in the properties of a liquid as we increase its temperature?
   (a) decrease in viscosity
   (b) decrease in density
   (c) increase in surface tension
   (d) increase in vapor pressure
   (e) increase in tendency to evaporate

_____ 6. The vapor pressure of all liquids
   (a) is the same at 100°C.
   (b) is the same at their freezing points.
   (c) increases with volume of liquid present.
   (d) decreases with the increasing volume of the container.
   (e) increases with temperature.

_____ 7. On a relative basis, the weaker the intermolecular forces in a substance,
   (a) the larger its heat of vaporization.
   (b) the more it deviates from the ideal gas law.
   (c) the greater its vapor pressure.
   (d) the larger its molar heat capacity as a liquid.
   (e) the higher its melting point.

_____ 8. It is found that 600 joules of heat are required to melt 15 grams of a substance. What is the heat of fusion of this substance?
   (a) 600×15 J/g
   (b) $\dfrac{600 \times 15}{1000}$ kJ/g
   (c) $\dfrac{600 \times 1000}{15}$ kJ/g
   (d) $600/15$ J/g

   (e) cannot be answered without knowing the specific heat of the substance.

_____ 9. Substances have properties that are related to their structures. Which of the following statements regarding properties of substance is *not* expected to be correct?
   (a) Molten KBr should be a good conductor of electricity.
   (b) Diamond should have a high melting point.
   (c) Solid sodium should be a good conductor of electricity.
   (d) Solid CO_2 should have a low melting point.
   (e) Solid K_2SO_4 should sublime readily.

_____ 10. Which of the following, in the solid state, would be an example of a molecular crystal?
   (a) carbon dioxide
   (b) diamond
   (c) barium fluoride
   (d) iron
   (e) none of the preceding

_____ 11. Which of the following, in the solid state, would be an example of a covalent crystal?
   (a) carbon dioxide
   (b) diamond
   (c) barium fluoride
   (d) iron
   (e) none of the preceding

_____ 12. Which of the following, in the solid state, would be an example of an ionic crystal?
   (a) carbon dioxide
   (b) diamond
   (c) barium fluoride
   (d) iron
   (e) none of the preceding

_____ 13. Which of the following, in the solid state, would be an example of a metallic crystal?
   (a) carbon dioxide
   (b) diamond
   (c) barium fluoride
   (d) iron
   (e) none of the preceding

_____ 14. The number of nearest neighbors of each atom in the face-centered cubic crystalline arrangement is
   (a) 4.
   (b) 6.
   (c) 8.
   (d) 12.
   (e) 16.

_____ 15. The unit cell constants for a particular crystalline compound were determined by X-ray diffraction to be $a = 12.52$ Å, $b = 5.67$ Å, $c = 12.88$ Å, $\alpha = 90.8°$, $\beta = 98.2°$, $\gamma = 101.3°$. What is its crystal system?
   (a) cubic
   (b) orthorhombic
   (c) tetragonal
   (d) monoclinic
   (e) triclinic

157

_____ 16. A hypothetical metal, atomic weight 60, forms a simple cubic crystals. The nearest distance between centers of adjacent atoms is 2.0 Å. What is the density of the crystalline material?
(a) 12.43 g/cm^3 (b) 2.54 g/cm^3 (c) 62.16 g/cm^3 (d) 124.3 g/cm^3 (e) 1.24 g/cm^3

- -

Questions 17 through 20 refer to the following phase diagram, which is for a hypothetical substance.

_____ 17. Where would you find the conditions necessary for the existence of a solid as a single phase?
(a) region F   (b) region G   (c) region H   (d) along line AE   (e) along line AC

_____ 18. The normal boiling temperature of the substance is indicated in the diagram by which of the following letters?
(a) A   (b) B   (c) C   (d) J   (e) L

_____ 19. In the diagram, the point A is the
(a) triple point.   (b) critical point.   (c) normal freezing point.
(d) normal boiling point.   (e) point of maximum density.

_____ 20. Which of the following represents a pressure below which the substance cannot exist as a liquid?
(a) K   (b) B   (c) L   (d) D   (e) I

- -

_____ 21. Which of the following is the band theory explanation for the conduction of electricity by metals?
(a) The occupied band of molecular orbitals overlaps a vacant conduction band of lower energy.
(b) The forbidden zone becomes polarized.
(c) Addition of electron-rich impurities by doping causes the conduction.
(d) The conduction band is fully occupied.
(e) The occupied band is either partially filled or overlaps a conduction band of higher energy.

Answers to Preliminary Test

Short Answer

1. much closer together
2. intermolecular; intramolecular
3. cohesive
4. adhesive
5. surface tension
6. capillary action; adhesive forces; cohesive forces
7. lower, higher, higher
8. stronger
9. diffusion
10. equilibrium; vapor pressure
11. boiling point; normal boiling point
12. boil
13. volatile; vapor pressure; boiling point
14. heat of fusion; molar heat of fusion
15. molar heat of condensation
16. multiply; formula weight (or molecular weight)
17. higher
18. lower. NH_3 is polar, with strong intermolecular attractions due to hydrogen bonding. Remember that the temperature at which the gas condenses to a liquid is the same at that at which the liquid boils. Thus, you can apply the reasoning of Section 13-2 to questions such as this.
19. hydrogen bonding
20. $H_2 < N_2 < PH_3 < NH_3 < CaCl_2$. Reason as follows: $CaCl_2$ is ionic, so it has the highest boiling point. NH_3 and PH_3 are both polar covalent molecules, but NH_3 is more polar and can undergo hydrogen bonding, so it boils at a higher temperature than PH_3. Both H_2 and N_2 are nonpolar. N_2 is larger and has more electrons, so it boils at a higher temperature than H_2, due to its larger dispersion forces. See Section 13-2.
21. polar nature (or polarity). We have to supply more energy to overcome the dipole–dipole interactions between CO molecules.
22. dipole–dipole interaction
23. melting point. Note that the melting point of a substance changes with pressure much less than the boiling point. Can you see that in the phase diagrams of Figure 13-17?
24. sublimation point
25. triple point
26. amorphous
27. higher
28. Several answers would be correct here. Among these are diamond, graphite, quartz, and many minerals.
29. X-ray diffraction
30. 6
31. 12
32. 6; Na^+ ion
33. vacancies; substitutions of different types of atoms

Multiple Choice

1. (a)
2. (b)
3. (c)
4. (c)
5. (c). In general, at higher temperature, the molecules are less able to attract one another. This *would* cause the results listed in (a), (b), (d), and (e) to occur, along with a *decrease* in surface tension.
6. (e)
7. (c). Weaker forces between molecules would make it easy to evaporate (*lower* heat of vaporization, *higher* vapor pressure), would make it obey the ideal gas law better, would make it easier to move the molecules past one another in the liquid (smaller molar heat capacity), and would make it easier to melt.
8. (d)
9. (e). KBr and K_2SO_4 are ionic solids, diamond is a covalent solid, Na is a metallic solid, and CO_2 is a molecular solid. The characteristics of various types of solids are summarized in Table 13-10.
10. (a)
11. (b)
12. (c)
13. (d)
14. (d)
15. (e)
16. (a). This problem is similar to Examples 13-8 and 13-9, except that this simple cubic cell contains only one metal atom (eight corners, each with $\frac{1}{8}$ atom).
17. (a)
18. (b)
19. (a)
20. (e)
21. (e)

14 Solutions

Chapter Summary

Much of our previous study of substances and their properties has been of pure elements and compounds. But substances usually occur in mixtures, so we must consider how they behave in the presence of other substances. In Chapter 14 we study solutions and their properties. The chapter introduction stresses the importance of solutions in some common processes. It also reminds you of some of the terminology we use to refer to the components of the solution (you should also review Section 3-6). Sections 14-1 through 14-7 describe the ways in which substances interact to form solutions. In general terms, a solution can be in any phase—solid, liquid, or gas. Most of our study in this chapter will concern liquid solutions.

Throughout this chapter, we continue our general theme of describing macroscopic properties as the result of properties and interactions of molecules and ions. In Section 14-1, we introduce the important idea that a process is favored (1) by a **decrease in energy** and (2) by an **increase in randomness or disorder**. In Chapter 15 we will consider these ideas in more detail. To understand the dissolution process, we must think about the various kinds of attractions that occur on the molecular level: solute-solute attractions, solvent-solvent attractions, and solute-solvent attractions. The stronger the solute-solute attractions and the stronger the solvent-solvent attractions, the more each kind of species will tend to self-associate, and the less **soluble** or **miscible** the substances will be. Conversely, very strong solvent-solute attractions can favor the dissolution process. For instance, the strong electrostatic attractions in an ionic solid and the strong attractions due to hydrogen bonding in water can be overcome by the even stronger interactions (**hydration**) between ions and water. As a result, many ionic substances are quite soluble in water. By contrast, even though the molecules of a nonpolar hydrocarbon such as benzene have relatively weak attractions for one another, they are not able to **solvate** ions very well. These strong solute-solute attractions, weak solvent-solvent attractions, and very weak solvent-solute attractions generally make ionic substances quite insoluble in benzene and other hydrocarbons. These ideas will help you to organize your study of the dissolution of solids, liquids, and gases in liquids (Sections 14-2 through 14-4). As you study these sections, pay attention to the factors that help to determine the strengths of the interactions—such features as charge density of ions, hydrogen bonding in liquids, dipole-dipole interactions, relative sizes of solute and solvent particles, and the tendency of some covalent substances to ionize in solution.

The **rate of dissolution** of a solid solute in a solvent depends on such factors as surface area; this is opposed by the **rate of crystallization**, the rate at which dissolved solute particles form crystals of undissolved solid (Section 14-5). When these two rates are equal, a state of **dynamic equilibrium** exists. At that point, the solution is **saturated**. Keep in mind that it is incorrect to describe this state of equilibrium as a situation where "nothing is happening". On the molecular and ionic level, much is happening, but there is *no net change* in the amount of dissolved solute. This is another application of some ideas regarding equilibrium that we first encountered in Chapter 12.

The effects of temperature and pressure on solubility are summarized and related to molecular-level behavior in Sections 14-6 and 14-7. Pressure changes usually have a significant effect only on the solubility of gases in liquids; this relationship is quantitatively summarized in **Henry's Law**. It is important to understand the effects of temperature and pressure on solubility in terms of **LeChatelier's Principle**, which was introduced in Chapter 13.

Section 14-7 discusses two new methods for describing solution concentrations, **molality** and **mole fraction**. You should also review the concentration calculations in Sections 3-6, 3-7, and 12-11. Different methods of expressing concentrations are useful in different applications in chemistry. You should learn all of the methods and practice solving problems associated with each one.

In the next several sections, you will study some solution properties that are closely related to concentration. A number of physical properties of solutions depend on the *number* and not on the *kind* of solute particles. These properties, which apply to solutions in which the *solvent* is volatile and the *solute* is nonvolatile, are called **colligative properties**. The four colligative properties are **vapor pressure lowering, freezing point depression, boiling point elevation**, and **osmotic pressure**. Be sure you understand that vapor pressure, melting (freezing) point, and boiling point are *not* colligative properties—*changes* in these (depression of the first two, elevation of latter) are considered colligative properties.

Why are we interested in the colligative properties of solutions? One reason is to be able to *describe* the behavior, both verbally and mathematically where possible. A second reason is to *understand* this behavior in terms of the interactions between molecules and ions. A third objective of this study is to be able to *use* these properties. Uses discussed in this chapter include separation of mixtures by **fractional distillation** and **reverse osmosis**, alteration of the freezing and boiling points of fluids such as the coolant in automobile radiators, experimental determination of the relative strengths (degrees of ionization) of electrolytes, and methods for the experimental determination of molecular weight. The latter two uses strongly emphasize that the colligative properties depend quantitatively on the particles in solution and not on the kind of particles. Sections 14-9 through 14-15 discuss the colligative properties, their interpretation, and their uses.

When both the solvent and the solute are volatile (i.e., able to evaporate, so vapor pressures are not zero), we can apply Raoult's Law (Section 14-9) to *each* component. The composition of the vapor above a solution is always richer in the more volatile component than the liquid solution is. This composition difference is the basis for the separation of components of a solution by simple distillation (Section 13-8) or by fractional distillation (Section 14-10).

The last portion of the chapter, Sections 14-16 through 14-18, discusses **colloids**. These are mixtures that may be viewed as intermediate between homogeneous and heterogeneous. In a colloid, the dispersed particles are larger than in a true solution, so the mixture is not entirely homogeneous, but they are too small to settle out, so it is not truly heterogeneous either. These dispersed particles may be collections of molecules or ions, or they may be individual large molecules such as proteins. Table 14-4 summarizes the various types of colloids and important examples of each. Colloids are involved in many problems of biological and ecological significance. Colloids are frequently recognized by their ability to scatter light (the **Tyndall effect**, Section 14-16). They consist of finely divided particles with large total surface area, so their properties are understood in terms of surface phenomena, such as **adsorption** (Section 14-17). The interactions that keep the colloidal particles suspended or dispersed depend on the nature of the colloidal particle. Thus, we often classify colloids as **hydrophilic** or **hydrophobic**, depending on whether or not they interact with water so as to remain suspended. Examples of these two important classes of colloids are discussed in Section 14-18, which emphasizes some useful characteristics of colloids.

Study Goals

Write your own outline of the ideas presented in each study goal. Exercises at the end of the text chapter are indicated for your study.

1. Describe and distinguish among (a) solvent, solute, and solution; (b) solution and colloid. (Introduction to chapter; introduction to colloids; Exercises 1, 3, and 91)

2. Give several examples of various kinds of solutions, involving different combinations of solids, liquids, and gases as solvent and solute. (Introduction to chapter; Sections 14-1 through 14-4; Exercise 2)

3. Describe the relative effects on solubility of the following kinds of interactions: (a) solute-solute attractions, (b) solvent-solvent attractions, and (c) solvent-solute attractions. Be able to discuss these effects for solid-liquid, liquid-liquid, and gas-liquid solutions. (Sections 14-1 through 14-4; Exercises 4 through 6 and 13)

4. Describe and illustrate the mechanism of dissolution of ionic solids and polar covalent substances in water. (Section 14-2; Exercises 7 through 12)

5. Distinguish between exothermic and endothermic dissolution processes. Know the effects of exothermicity or endothermicity and of an increase in disorder on the spontaneity of the dissolution process. Describe the effects of exothermicity or endothermicity on the temperature dependence of water solubility of a compound. (Sections 14-1 and 14-6; Exercise 6 and 22)

6. Relate the relative magnitudes of crystal lattice energies of ionic solids to their formulas. (Section 14-2; Exercises 19 through 22)

7. Distinguish among unsaturated, saturated, and supersaturated solutions. (Section 14-5; Exercise 16)

8. Describe the effect on solubility of changing the pressure when the solute is (a) a gas, (b) a liquid, or (c) a solid. (Section 14-7; Exercises 14, 17, and 18)

9. Relate specific amounts of solute dissolved in specific amounts of solvent or solution to concentration expressed as (a) percent by mass, (b) molarity, (c) mole fraction, and (d) molality. (Review Sections 3-5, 3-6, and 12-11; Section 14-8; Exercises 23 through 35)

10. Describe the colligative properties and the factors on which they depend. (Sections 14-9, 14-11, 14-12, and 14-15; Exercises 36 and 81)

11. Use Raoult's Law to describe the vapor pressures of (a) solutions of nonvolatile solutes and (b) solutions in which both components are volatile. (Section 14-9; Exercises 36 through 45)

12. Relate Study Goal 10 to the purification technique of fractional distillation. (Section 14-13; Exercises 41 through 45)

13. Carry out calculations involving freezing point depression, boiling point elevation, and osmotic pressure to relate these physical properties of solutions to the amount of nonvolatile, nonelectrolyte solute dissolved in a given amount of solvent. (Sections 14-11, 14-2, and 14-15; Exercises 46 through 62 and 84 through 90)

14. Use the colligative properties to determine molecular weights of solutes. (Sections 14-13 and 14-15; Exercises 59 through 61 and 90)

15. Describe solutions of strong electrolytes, weak electrolytes, and nonelectrolytes, giving specific examples. (Review Section 4-2; Section 14-14; Exercises 11, 12, and 15)

16. Understand how the properties of electrolyte solutions depend on such factors as ion association and degree of ionization. Be able to determine and interpret the van't Hoff factor, i. (Section 14-14; Exercises 63 through 69)

163

17. Carry out calculations concerned with percent ionization and the van't Hoff factor. (Section 14-14; Exercises 70 through 80, 86, and 89)
18. Describe and illustrate osmosis and reverse osmosis. (Sections 14-15 and 14-16; Exercises 81 through 87)
19. Know the features and properties of colloids. Distinguish among, and give common examples of, the kinds of colloids listed in Table 14-4, such as gels, sols, foams, and so on. (Introduction to colloids; Sections 14-16 and 14-17; Exercises 91 through 93)
20. Describe and distinguish between hydrophilic and hydrophobic colloids. (Section 14-18; Exercise 94)
21. Describe and illustrate the action of emulsifying agents, soaps, and detergents. (Section 14-18; Exercises 95 through 97)
22. Recognize and answer questions about solutions and their properties. (Mixed Exercises 98 through 107)
23. Answer conceptual questions based on this chapter. (Conceptual Exercises 108 through 115)
24. Apply concepts and skills from earlier chapters to the ideas of this chapter. (Building Your Knowledge exercises, 116 through 119)

Some Important Terms in This Chapter

This is another chapter with *many* important new terms, some of which are listed here. You should study many others, too. Write the meanings *in your own words*. Check the Key Terms list and the chapter reading. Then rewrite your definitions to improve them. Study other new terms and review terms from preceding chapters.

solvation

hydration

Henry's Law

saturated solution

mole fraction

molality

164

colligative properties

Raoult's Law

distillation (including fractional)

boiling point elevation

freezing point depression

percent ionization

osmosis

osmotic pressure

van't Hoff factor

colloid

hydrophilic colloids

hydrophobic colloids

coagulation

emulsification

Preliminary Test

As in other chapters, this test will check your understanding of basic concepts and types of calculations. Be sure to practice *many* of the additional textbook Exercises, including those indicated with the Study Goals.

True-False

Mark each statement as true (T) or false (F).

_____ 1. When one substance dissolves in another the result is a situation of greater disorder.
_____ 2. All cases of solids dissolving in water are endothermic.
_____ 3. All cases of gases dissolving in water are exothermic.
_____ 4. It is impossible, under any conditions, to dissolve more solute in a solvent than the amount to saturate it.
_____ 5. Ions are usually dissolved in water without any interaction with the solvent.
_____ 6. In general, the larger the charge and the smaller the size of an ion, the more easily it can be hydrated.
_____ 7. An ionic substance such as KCl is able to dissolve in water because in the solution each K^+ ion can be surrounded closely by several Cl^- ions, while each Cl^- ion has several close K^+ neighbors.
_____ 8. The more negative the lattice energy for a solid substance, the less likely that it is able to dissolve in a specific solvent.
_____ 9. Gases can dissolve in liquids only if they react.
_____ 10. Increasing the pressure usually makes solids more soluble in liquids.
_____ 11. The degree to which an electrolyte is ionized in solution is greater for more concentrated solutions.
_____ 12. For a substance to be very soluble in water, it must be a strong electrolyte.
_____ 13. Nitric acid is a weak electrolyte solution.
_____ 14. Strong electrolytes are those that are mostly ionized in dilute aqueous solutions.
_____ 15. A solution that is described as concentrated would have a higher molarity than a dilute solution.
_____ 16. Before we can answer any question regarding a solution we must decide which component is the solute and which is the solvent.
_____ 17. Any solution that is 20% A dissolved in B has a mole fraction of A less than that of B.
_____ 18. A solution that is described as a strong acid solution would have a higher molarity than a weak acid solution.

_____ 19. For a given solution, it would probably be easier to measure the freezing point depression than the boiling point elevation.

_____ 20. It is not necessary to use measurements such as colligative properties to determine the degree of ionization of $CaCl_2$ in solution, because we can predict from the formula that 1 mole of $CaCl_2$ must dissolve to give 3 moles of solute particles.

_____ 21. All colloid particles are clumps of molecules or ions, just small enough to keep from settling out.

Short Answer

Answer with a word, a phrase, a formula, or a number with units as necessary.

1. A solution is an example of a _____ mixture.
2. When we describe a solution of a little sugar in water, we describe sugar as the _____ and water as the _____.
3. The reverse of the dissolution process is called _____.
4. A solution that contains dissolved solute in equilibrium at a given temperature with excess undissolved solute is said to be _____.
5. The ease of the dissolution process depends on two factors: the change in _____ and the change in _____.
6. A process in which the substances involved decrease in energy is called _____.
7. The three types of interactions that are useful to consider in assessing the change in heat content on dissolution are _____, _____, and _____.
8. The dissolution process is favored by _____ solute-solute attractions, by _____ solvent-solvent attractions, and by _____ solute-solvent attractions.
9. The close clustering of solvent molecules around a solute particle is generally called _____ ; when the solvent is water it is called _____.
10. Classify each of the following substances as soluble or insoluble in water (for review).

(a) KNO_3 _____   (k) $Ca(OH)_2$ _____

(b) $SrCl_2$ _____   (l) $AgNO_3$ _____

(c) SO_2 _____   (m) $Mg(CH_3COO)_2$ _____

(d) $Al(OH)_3$_____   (n) $Zn(ClO_4)_2$ _____

(e) $(NH_4)_2S$ _____   (o) H_3PO_4_____

(f) FeS _____   (p) $Ba_3(PO_4)_2$ _____

(g) $CaCO_3$ _____   (q) MgS_____

(h) HBr _____   (r) $(NH_4)_2CO_3$ _____

(i) $PbCl_2$_____   (s) KOH _____

(j) $NaSCN$ _____   (t) Rb_2SO_4 _____

11. To have a solution of acetone and water that has a mole fraction of acetone equal to 0.20, we must add _____ moles of acetone to 6.0 moles of water.
12. The lowering of the vapor pressure of a solvent due to the presence of dissolved solute is described by _____ Law, which states in words that _____

    _____.
13. The vapor pressure of pure water is always _____ the vapor pressure of an aqueous solution of a nonvolatile solute at the same temperature.

- -

We make up two solutions. Solution A contains 0.10 mole of NaBr dissolved in 1000 grams of water, while solution B contains 0.10 mole of sugar dissolved in 1000 grams of water. The next four questions, 14 through 17, refer to these two solutions.

14. We expect that the vapor pressure of solution A would be _____ that of solution B.
15. We expect that the freezing point of solution A would be _____ that of solution B.
16. We expect that the boiling point of solution A would be _____ that of solution B.
17. If each solution is put in contact through a semipermeable membrane with a sample of pure water, solution A would exhibit _____ osmotic pressure compared to solution B.

- -

At 25°C, the vapor pressure of pure chloroform, $CHCl_3$, is 172.0 torr and that of pure carbon tetrachloride, CCl_4, is 98.3 torr. The formula weights of $CHCl_3$ and CCl_4 are 119.4 and 153.8, respectively. Questions 18 through 20 refer to ideal solutions of these two liquids at 25°C.

18. In a solution containing 40.0 g of $CHCl_3$ and 70.0 g of CCl_4, the mole fraction of $CHCl_3$ is _____ and the mole fraction of CCl_4 is _____.
19. In the solution of Question 18, the vapor pressure of $CHCl_3$ is _____ torr, the vapor pressure of CCl_4 is _____ torr, and the total vapor pressure is _____ torr.
20. In a solution of $CHCl_3$ and CCl_4 with a total vapor pressure of 150.0 torr, the mole fraction of $CHCl_3$ is _____ and the mole fraction of CCl_4 is _____.

- -

21. We prepare a solution by dissolving 0.01 mole of $MgCl_2$ in 1.0 kg water and observe the freezing point. We also calculate the freezing point depression, using $\Delta T_b = K_b m$, with $m = 0.01$ molal. The ratio of observed freezing point depression to calculated freezing point depression would be about _____.
22. The passage of solvent molecules selectively through a membrane is called _____.
23. Liquids diffuse through a membrane from regions of high solvent concentration to regions of low solvent concentration due to a force known as _____.
24. When we distill a mixture of the very volatile substance A and the moderately volatile substance B, the first distillate (liquid obtained by condensing the first vapor to boil off) should be richer in substance _____.
25. Of the four colligative properties of a solution, the one that would give the most easily determined measurement from which to calculate the solute molecular weight would be _____.
26. In order to carry out "reverse osmosis," the pressure that we exert on the solution must

    _____.
27. The Tyndall effect is the name given to the observation that _____

    _____.
28. A fog is an example of a(n) _____.

29. The only combination of phases that cannot lead to formation of a colloid would be _____ dispersed in _____.
30. Colloids consisting of solids or liquids dispersed in a gas medium are called _____.
31. The process by which ions and molecules stick to the surface of a particle is called _____.
32. Colloid particles that readily attract water molecules to their surfaces are called _____.
33. To keep them from coming together to form a separate phase, the particles of a hydrophobic colloid are treated with an agent called a(an) _____.

Multiple Choice

_____ 1. For a gas that does not react chemically with water, the solubility of the gas in water
  (a) increases with increasing gas pressure and decreases with increasing temperature.
  (b) decreases with increasing gas pressure and decreases with increasing temperature.
  (c) increases with increasing gas pressure and increases with increasing temperature.
  (d) decreases with increasing gas pressure and increases with increasing temperature.
  (e) does not change appreciably with change in either gas pressure or temperature.

_____ 2. The formation of tiny bubbles when a beaker of water is mildly heated indicates that
  (a) the boiling point is being approached.
  (b) water is being decomposed into hydrogen and oxygen.
  (c) water is extensively hydrogen bonded.
  (d) air is less soluble in water at higher temperature.
  (e) the lattice energy of water is large.

_____ 3. Hydrogen chloride in the gaseous state is in the form of covalent molecules, yet in aqueous solution it behaves as a strong electrolyte, conducting electricity. The explanation for this solution behavior is that
  (a) HCl forms ionic crystals in the solid state.
  (b) HCl reacts with water to form ions.
  (c) although it is a covalent substance, the molecules of HCl are polar and stay in solution as polar molecules, conducting electricity.
  (d) there is more than one form of gaseous HCl.
  (e) at lower temperatures, HCl becomes ionic.

_____ 4. Which of the following alcohols would be the *least* miscible with water?
  (a) hexanol, $CH_3CH_2CH_2CH_2CH_2CH_2OH$ (b) pentanol, $CH_3CH_2CH_2CH_2CH_2OH$
  (c) propanol, $CH_3CH_2CH_2OH$ (d) ethanol, CH_3CH_2OH (e) methanol, CH_3OH

_____ 5. In four of the following groups, the three substances are either all strong electrolytes, all weak electrolytes, or all nonelectrolytes. Which group has one member that is not like the others in the group?
  (a) H_2SO_4, NaOH, HCl   (b) NH_3, CH_3COOH, HCN   (c) NaCl, $Ca(OH)_2$, HNO_3
  (d) O_2, $CO(NH_2)_2$, CH_3OH (e) $BaCl_2$, $NaNO_3$, HF

_____ 6. Hydrogen chloride is observed to be soluble in benzene, although not to the extent that it is in water. We would expect that
  (a) the HCl in benzene would be ionized, as in water.
  (b) the solution of HCl in benzene would be a good conductor of electricity.
  (c) the HCl molecules would be largely intact in benzene.
  (d) the benzene molecules would solvate the HCl molecules very well.
  (e) the limited solubility of HCl in benzene is due to the very strong interaction between benzene molecules.

_____ 7. Three of the following are colligative properties of solutions. Which one is not a colligative property?
(a) molality   (b) vapor pressure lowering   (c) osmotic pressure
(d) freezing point depression   (e) boiling point elevation

_____ 8. A 1-molar solution has
(a) 1 mole of solute per 1 liter of solvent.
(b) 1 mole of solute per 1 liter of solution.
(c) 1 mole of solute per 1 kilogram of solvent.
(d) 1 mole of solute per 1 kilogram of solution.
(e) 1 mole of solute per 1 mole of solvent.

_____ 9. A 1-molal solution has
(a) 1 mole of solute per 1 liter of solvent.
(b) 1 mole of solute per 1 liter of solution.
(c) 1 mole of solute per 1 kilogram of solvent.
(d) 1 mole of solute per 1 kilogram of solution.
(e) 1 mole of solute per 1 mole of solvent.

_____ 10. We dissolve 7.5 grams of urea (a nonelectrolyte, molecular weight = 60) in 500 grams of water. At what temperature would the solution boil? Additional useful data may be found in Table 14-2.
(a) 100.46°C   (b) 100.13°C   (c) 0.13°C   (d) 99.54°C   (e) 99.87°C

_____ 11. An aqueous solution of compound X is described as "a 10% X solution." Its density is 1.10 g/mL. Which of the following statements about this solution is not true?
(a) 1 liter of the solution would weigh 1.10 kg.
(b) 1 liter of the solution would contain 100 grams of compound X.
(c) 1 kg of the solution would contain 100 grams of compound X.
(d) The molality of the solution would be greater than the molarity.
(e) The boiling point of the solution would be higher than 100°C.

_____ 12. How many mL of a 0.320-molar solution of sucrose in benzene (density = 0.8787 g/mL) are needed to have 0.160 moles of sucrose?
(a) 500 mL   (b) 498 mL   (c) 438 mL   (d) 567 mL   (e) 219 mL

_____ 13. Which of the following aqueous solutions would exhibit the highest boiling point?
(a) 0.1 _m_ urea   (b) 0.1 _m_ NaCl   (c) 0.1 _m_ $CaCl_2$ (d) 0.1 _m_ HCl
(e) All would show the same boiling point.

_____ 14. According to Raoult's Law, the lowering of vapor pressure of the solvent in dilute solutions of a nonvolatile solute is proportional to the
(a) molarity of the solute.   (b) molarity of the solvent.
(c) mole fraction of the solute.   (d) mole fraction of the solvent.
(e) None of the preceding answers is correct.

Answers to Preliminary Test

Do not just look up answers. Be sure you understand the reasons for the answers.

True-False

1. True. The two kinds of molecules are mixed in the solution, so they are less ordered.
2. False. Some are exothermic, including those examples listed in Section 14-2.

3. True. See the discussion in Section 14-4.
4. False. Read about supersaturated solutions, Section 14-5.
5. False. Hydration must occur.
6. True. Remember the definition of charge density and know how it relates to solubility of ionic substances, Section 14-2.
7. False. The solubility of KCl in water is due to the ability of water to surround each ion and thus to *prevent* the kind of interaction described.
8. True. Lattice energy is a measure of the attraction of the ions (or molecules) in the solid. The more negative it is, the more strongly these particles are held in the solid.
9. False. If they do react (e.g., HCl, SO_2, CO_2 to a limited extent) they are often much more soluble, but some nonreacting gases are slightly soluble in water (O_2, N_2).
10. False. Changing the pressure usually has no appreciable effect on the solubilities of either solids or liquids in liquids, but it does greatly alter the solubility of gases in liquids.
11. False. High concentration promotes "clumping" of ions, rather than each ion existing separately.
12. False. Many substances that are quite soluble in water are either weak electrolytes (e.g., HCN, HNO_2, acetic acid, ammonia) or are nonelectrolytes (e.g., sugar, alcohol, urea).
13. False. Remember that nitric acid is one of the common strong acids (Chapters 4 and 10) and is thus essentially completely ionized.
14. True
15. True
16. False. This terminology is a convenience, but sometimes it is neither easy nor necessary to identify one component as solute.
17. False. This would depend on the molecular weights of A and B. Suppose that A had a much lower molecular weight than B, so that the fewer number of grams were a greater number of moles—then the mole fraction A would be greater than that of B.
18. False. It may or may not. Remember that the terms "strong" and "weak" describe the efficiency of ionization of the solute, not its concentration.
19. True. See Table 14-2. For all the solvents listed (in fact, for all solvents) the value of K_f is greater than that for K_b. Thus, a solution of a given concentration would undergo a bigger change in freezing point than in boiling point. Another factor is that freezing points are much less sensitive to external atmospheric pressure.
20. False. The actual degree of ionization depends on concentration—the more dilute the solution, the greater the degree of ionization of a particular solute.
21. False. Some are individual macromolecules, such as proteins or polymers.

Short Answer

1. homogeneous
2. solute; solvent
3. crystallization (or precipitation)
4. saturated
5. energy; disorder
6. exothermic
7. solute-solute attractions; solvent-solvent attractions; solute-solvent attractions
8. weak; weak; strong
9. solvation; hydration

10. (a) soluble   (k) soluble
   (b) soluble   (l) soluble
   (c) soluble   (m) soluble
   (d) insoluble   (n) soluble
   (e) soluble   (o) soluble
   (f) insoluble   (p) insoluble
   (g) insoluble   (q) soluble
   (h) soluble   (r) soluble
   (i) insoluble   (s) soluble
   (j) soluble   (t) soluble

You should *know* and be able to apply the solubility guidelines in Section 4-2.5. You may need to review chemical nomenclature of inorganic compounds, Sections 4-11 and 4-12.

11. 1.5. We want 0.20 of the total number of moles to be acetone, so the total number of moles $(6.0 + x)$ can be calculated (with x = no. of moles of acetone) as $x = 0.2(6.0 + x)$, or $x = 1.5$.

12. Raoult's; the vapor pressure of a solvent above a solution is equal to the mole fraction of the solvent in the solution times the vapor pressure of the pure solvent.

13. higher than

14. lower than. NaBr dissolves to give ions in solutions, resulting (ideally) in 2 moles of solute particles per mole, while sugar gives only 1 mole of solute particles per mole. The *lowering* of the vapor pressure depends on the total concentration of dissolved solute particles.

15. lower than. Freezing point is depressed, again by an amount depending on the total concentration of dissolved solute particles.

16. higher than. Boiling point is elevated, by an amount depending on the total concentration of dissolved solute particles.

17. higher. Osmotic pressure, like all colligative properties, depends on the total concentration of solute particles.

18. 0.424, 0.576. The solution contains 40.0 g/(119.4 g/mol) = 0.335 mol $CHCl_3$ and 70.0 g/(153.8 g/mol) = 0.455 mol CCl_4. Calculate the mole fraction of each component as in Example 14-3.

19. 72.9, 56.6, 129.5. The vapor pressure for each component is equal to its mole fraction in the solution times its vapor pressure when pure. The total vapor pressure is the sum of the vapor pressures of the two components.

20. 0.701, 0.299. This uses the same ideas as Question 19, except that the mole fractions X_{CHCl_3} and X_{CCl_4} are unknown. We can relate these two unknowns, because their sum must equal 1;

$$X_{CHCl_3} + X_{CCl_4} = 1$$

This lets us eliminate one variable and solve.

21. 3. Each mole of $MgCl_2$ dissolves to give 1 mole of Mg^{2+} and 2 moles of Cl^-, or 3 moles total of solute particles per mole of $MgCl_2$ dissolved. This would probably be a pretty good prediction for such a dilute solution.

22. osmosis

23. osmotic pressure

24. A. The more volatile substance evaporates more, so the vapor is richer in the more volatile substance.

25. osmotic pressure. See the discussion in Section 14-15.

26. exceed the osmotic pressure of the solution.

27. colloidal particles scatter light

28. colloid (or aerosol, which is a type of colloid)
29. gases; gases. This is because all gas mixtures are homogeneous.
30. aerosols
31. adsorption. Do you know the difference between adsorption and absorption?
32. hydrophilic
33. emulsifier (or emulsifying agent)

Multiple Choice

1. (a)
2. (d)
3. (b)
4. (a). The three alcohols propanol, ethanol, and methanol are miscible with water in all proportions, but the alcohols with larger nonpolar hydrocarbon ends are much less so. Of the alcohols listed, hexanol is the most hydrocarbon-like, and hence least miscible with water.
5. (e). In (a), all substances are strong electrolytes; in (b), all are weak electrolytes; in (c), all are strong electrolytes; in (d), all are nonelectrolytes. In answer (e), $BaCl_2$ and $NaNO_3$ are strong electrolytes and HF is a weak electrolyte. You may need to review the discussion of strong and weak electrolytes in Section 4-2.
6. (c). Benzene, a nonpolar solvent, cannot solvate very well, so the ionization process does not readily occur, as it does in water.
7. (a)
8. (b)
9. (c)
10. (b). $K_b m = \Delta T_b$. Molality (m) is $\dfrac{7.5 \text{ g}}{0.500 \text{ kg solvent}} \times \dfrac{1 \text{ mole}}{60 \text{ g}} = 0.24$ molal.

    From Table 14-2, K_b is $0.512°C$/molal, so $\Delta T_b = 0.13°C$. Remember that the boiling point is raised from its normal value by this amount.
11. (b). One liter would weigh 1100 grams. Compound X would be 10% or 110 grams, of this total.
12. (a). Molarity directly relates moles of solute to volume of solution. Density is not needed to answer this question.
13. (c). Consider the number of dissolved particles in each solution. Urea is a nonelectrolyte.
14. (c)

173

15 Chemical Thermodynamics

Chapter Summary

At many stages in our study of chemistry, we have been concerned with questions such as these: Which is more stable—situation A or situation B? Will a specific chemical reaction or other process occur? Will the gas expand or contract if we change the pressure? Which is more stable at the stated conditions—solid, liquid or gas? Will the substance dissolve in water? Is the dissolution process exothermic or endothermic? What energy changes are associated with the process of melting or vaporization? Varied as they seem, all of these are versions of the same question, "Is a proposed process spontaneous at a specified set of conditions?" **Chemical thermodynamics** (Chapter 15) is a general approach to answering such questions, based on the study of energy changes that accompany a process. In this chapter, we will also learn how to understand and predict the amount of heat produced or absorbed when a change takes place—the topic called **thermochemistry**.

As we saw briefly in Chapter 14, the spontaneity of a process is favored by two factors: (1) a lowering of the energy (exothermicity) and (2) an increase in the disorder. It may help to recall that rocks roll *downhill* (lower potential energy), not uphill, and that it is easier to *mix* things (scrambling an egg) than to unmix them. In this chapter we study these two factors quantitatively and find out how both factors affect whether a particular process can occur. These ideas, which may seem somewhat abstract at this point, are exemplified in Section 15-1. Be sure that you understand that just because a process is **exothermic** (gives off energy to its surroundings), it is not necessarily spontaneous. Exothermicity favors, but does not guarantee, spontaneity.

Section 15-2 introduces you to some thermodynamic terms, some of which have been used earlier in the text. You will encounter others throughout the chapter. In order to understand the logic of chemical thermodynamics, you must pay close attention to the terms used and interpret them as they were meant to be. In the terminology of thermodynamics, a **process** may be any proposed change in the **system** under observation—a change of state, an expansion or contraction of the sample, the absorption or evolution of heat with or without a change of state, a chemical reaction, and so on.

As we saw above, one factor that helps to make a process spontaneous is a lowering of the energy. In thermodynamics, we express the change in the energy of a sample (the system) as ΔE. If the final energy is greater than the initial energy, ΔE is positive. The total amount of energy in the universe is constant, so in order for the system to increase its energy it must absorb energy from its surroundings.

One very useful comment about notation can be made here. We are using a symbol that will become quite common. Whenever we write Δ(any quantity) we mean the change in that quantity. This is always calculated as (the final value of the quantity) minus (the initial value of the quantity). This is the idea of a **state function**. This important characteristic of a state function will be used throughout the chapter. (Study the bottom blue marginal note on page 595 for an analogy to help you remember this concept.) All of the Δ values we will see, ΔE, ΔH, and so on, can be considered in this way.

Let us clarify another use of terminology that sometimes confuses students. When we say that some variable is "constant" for a process, we mean only that it has the same value at the end of the process as at the beginning -- not that it necessarily had that value throughout the change. For instance, when we talk about a reaction "at constant temperature and pressure", we mean that after the reaction has occurred, we adjust the products to the same temperature and pressure as before the reaction. We do not necessarily mean that the temperature or the pressure were held invariable while the process was occurring.

We use the thermodynamic quantity (or state function), the **enthalpy change** ΔH (Section 15-3) to describe the heat change at constant pressure. A lowering of the enthalpy content of a system, that is, a *negative* value of ΔH, is favorable to spontaneity of the process. The measurements of chemical thermodynamics involve heat changes, so we must have methods for measuring the flow of heat from one body to another. Heat energy can only be given up by one body if it is absorbed by another. We can use the method of **calorimetry** (Section 15-4) to measure the release or absorption of heat by a sample. The type of calorimeter described here, the coffee-cup calorimeter, can be used to measure the heat change at constant pressure, that is, ΔH. Before you study the calculations associated with calorimetry, review specific heat calculations in Section 1-13.

We use a **thermochemical equation** (Section 15-5) to describe the relation between the amounts of reactant and products involved in a reaction and the amount of heat liberated or consumed at constant pressure. Be sure you know how to interpret such equations. Pay special attention to the conventions presented at the bottom of page 599, and know how to apply them. If ΔH is evaluated at 1 atm pressure and **thermodynamic** standard temperature (25°C, 298 K), its value for the reaction is written as ΔH^0 and is referred to as the standard enthalpy change. Be sure that you understand what is meant by the **standard state** of a substance (Section 15-6). Notice also that the so-called standard temperature for thermodynamic studies is different from that used in gas law problems. Section 15-7 introduces the idea of a **standard molar enthalpy of formation** of a substance. This is the enthalpy change for the reaction in which *one mole* of the substance is produced from the elements, all reactants and products being at standard-state conditions. Therefore, the standard molar enthalpy of formation of a substance, ΔH_f^0 is just a measure of how much higher or lower the substance is in enthalpy content than are its constituent elements.

According to **Hess's Law** of heat summation (Section 15-8), we can calculate what would be the ΔH (or ΔH^0) for any proposed reaction. To do this, we can either (1) think of reactions to measure that would total algebraically to the desired reaction, or (2) find tabulations of the ΔH_f^0 values for all substances involved in the reaction. This is a very important idea. We will use the value of ΔH^0 for a proposed reaction to help tell us whether it is spontaneous, and we can now calculate this important quantity without having to carry out the reaction. Another use would be to tell us how much heat we would either have to supply, be prepared to dissipate, or use for doing work when the reaction occurs.

In earlier chapters we have often referred to the stabilities of chemical bonds in covalent substances. Now we have a way of measuring these **bond energies** experimentally, Section 15-9. As you study the section, keep in mind that the bond energy calculation refers only to the step in which the bonds in gaseous molecules break to form isolated gaseous atoms. Usually we cannot measure this directly. We must deduce it from first measuring the enthalpy change associated with the formation of bonds from the elements in their standard states, and then correcting, via Hess's Law, for the conversion of elements from their stable forms (solids, diatomic molecules, etc.) to isolated atoms. Many students find diagrams such as Figure 15-5 helpful in understanding bond energy calculations. Once derived, tabulations of bond energies can help us to understand why reactions that involve the breaking of very stable bonds are hard to carry out. They can also be used to estimate ΔH^o_{rxn} for reactions in which all reactants and products are gases.

Section 15-10 discusses the change in internal energy, ΔE, for a system and its relationship to heat and work. We can understand this important relationship as follows. When we put heat into a system it can (1) increase the internal energy of the system, (2) cause the system to do some work (for example, by expanding against an external force), or result in some combination of (1) and (2). Because energy can be neither created nor destroyed, the total increase in energy of the system (ΔE) must be the sum of the amount of heat absorbed by the system (q) and the work done on the system (w). We can express this as $\Delta E = q + w$, the **First Law of Thermodynamics**. You must remember the conventions for when the values of q, w, and ΔE are positive. The most direct way to measure ΔE is in a bomb calorimeter; in this experiment, the volume is held constant ($\Delta V = 0$) so no work is done and the value of ΔE is given directly by the observed heat flow. The relationship between ΔH and ΔE for a reaction is discussed in Section 15-11. You can think of this as the comparison between the amount of heat absorbed at constant pressure (ΔH) and at constant volume (ΔE); the difference is the work done on the system. For a process involving expansion or contraction without reaction, the relationship

$$\Delta H \;=\; \Delta E \;+\; P\Delta V$$

should be used; when the volume change is due to a chemical reaction, the equation

$$\Delta H \;=\; \Delta E \;+\; (\Delta n)RT$$

is more convenient. As you work with this relationship, remember that Δn is equal to the change in the number of moles of *gaseous substances* in the balanced chemical equation.

So far in this chapter we have studied the first factor that favors spontaneity, the exothermicity of the process. But many endothermic reactions are also spontaneous, at least under some conditions. A second favorable factor, an increase in the disorder of the system, is introduced in a general way in Section 15-12. The next two sections deal in more detail with the second favorable factor, an increase in disorder. After some general remarks about the more disordered situation being more favorable (the **Second Law of Thermodynamics**, Section 15-13), we introduce a thermodynamic measure of disorder called **entropy** (Section 15-14). Thus, an increase in entropy of the system, ΔS positive, favors a process being spontaneous. The discussion of melting/freezing that accompanies Table 15-4 will help you to clarify these ideas. The introduction of a scale of absolute entropies (the **Third Law of Thermodynamics**) allows us to use a relationship analogous to Hess's Law to calculate entropy changes for any proposed process.

Now, in the final sections of the chapter, we are ready to combine the two factors that affect spontaneity into a single measure, the change in Gibbs free energy, ΔG. As seen in Section 15-15, the equation,

$$\Delta G \;=\; \Delta H \;-\; T\Delta S \qquad \text{(constant } T, P)$$

allows us to formulate simple rules for predicting whether a reaction will be spontaneous. A negative value of ΔG means that the reaction is **product-favored** (spontaneous) and *can* occur, whereas a positive value of ΔG means that the reaction is **reactant-favored** (not spontaneous) under the conditions specified. (Realize, however, that just because a reaction is spontaneous does not guarantee that it will occur at a measurable rate.)

Just as for other thermodynamic quantities, we usually carry out the calculation of free energy change with respect to special conditions. One important set of conditions is with all reactants and products at 1 atm pressure or at concentrations of 1 M. We designate the free energy change at these special conditions as

$$\Delta G^0 = \Delta H^0 - T\Delta S^0$$

The presence of T in this equation tells us that *some* reactions that are impossible at some temperatures can be made spontaneous at other temperatures (Section 15-16). If the ΔH^0 aspect is favorable (–) and the $T\Delta S^0$ aspect is unfavorable (–), lower temperatures make the unfavorable term less significant and the reaction can become product favored (spontaneous). Conversely, if the ΔH^0 aspect is unfavorable (+) and the $T\Delta S^0$ aspect is favorable (+), higher temperatures make the favorable term more significant and the reaction can become product-favored. We can calculate from this relation the temperature at which the reaction would become spontaneous, if it could be spontaneous at any temperature. It is very important that you understand the discussion accompanying Table 15-7 in the text. Be sure that you can calculate ΔG^0 either (1) from known or experimentally determined values of ΔH_f^0 and S^0 or (2) from known ΔG_f^0 values. Be sure you remember that the latter method can be used *only* for the temperature for which the ΔG_f^0 values are tabulated—25°C for the values in Appendix K. In the former method, be careful of the units. We will see many uses of ΔG^0 in subsequent chapters, including its relation to equilibrium constants (Chapter 19) and electrochemistry (Chapter 21).

There is much detail and considerable subtlety in this chapter. You must proceed carefully through the chapter, being sure that you master the terminology and the calculations of one section before proceeding. Do not be content with just seeing how to do the problems; because they involve only addition, subtraction, and multiplication of numbers, they are easy to calculate. The usual difficulty with chemical thermodynamics at first encounter is in seeing what these calculations mean. Remember that we are dealing with energy changes, calculating the predicted energy changes that will accompany reactions, and trying to use these energy changes to predict whether or not a reaction can go or to understand why it does or does not go at certain conditions. And do not just memorize—*think*!

Study Goals

It may help you to write out a brief summary of the ideas of each study goal, summarizing the applicable parts of your class notes and text readings. Be sure to answer many of the suggested Exercises related to the Study Goals.

1. Be familiar with the calculations and units used to describe heat transfer. (Review Section 1-13; review Exercises 53 and 58 of Chapter 1)
2. Understand the basic terminology of thermodynamics as presented and used in this chapter (be able to give examples of each): (a) system, (b) surroundings, (c) state of a system, (d) state function, (e), standard conditions, (f) standard state, (g) endothermic process, (h) exothermic process, (i) spontaneous process, and (j) nonspontaneous process. (Section 15-2, unnumbered paragraphs on page 620, Section 15-12; Exercises 1 through 9 and 22)
3. Know what is meant by the enthalpy change of a reaction. Know what is meant by the standard enthalpy change of a reaction. (Sections 15-3 and 15-6; Exercises 10 and 23)
4. Be familiar with the measurements and calculations of calorimetry, either at constant pressure or at constant volume. Be able to carry out the calculations, using experimental calorimeter data, to

determine enthalpy or energy changes for reactions and processes. (Section 15-4; Exercises 57 through 69)

5. Know how to write and interpret thermochemical equations. (Section 15-5; Exercises 12 through 27)

6. Know what is meant by the standard molar enthalpy of formation of a substance. (Section 15-7; Exercises 26 and 27)

7. Understand the meaning of Hess's law and why it works. Know how to carry out the calculations of Hess's Law to determine enthalpy changes for specified reactions. (Section 15-8; Exercises 9 and 28 through 41)

8. Give the enthalpy changes for appropriate reactions or physical changes, be able to calculate average bond energies. Given appropriate bond energies, be able to calculate enthalpy changes for reactions. (Section 15-9; Exercises 44 through 56)

9. Understand the First Law of Thermodynamics, both in words and in mathematical form. Carry out calculations relating ΔE, q, and w. (Section 15-10; Exercises 70 through 79)

10. Explain the relationship between energy change and enthalpy change. Explain why the use of ΔH is often preferred. (Section 15-11)

11. State the Second Law of Thermodynamics and summarize its implications with respect to reaction spontaneity. (Section 15-13; Exercises 80 and 83)

12. Understand the interpretation of entropy of a system in terms of microscopic disorder within that system. Be able to predict the sign of ΔS for various kinds of chemical and physical changes. (Section 15-14; Exercises 84 through 90)

13. Use tabulated values of absolute entropies of reactants and products to calculate the entropy change for a reaction. (Section 15-14; Exercises 91 and 92)

14. Explain how ΔH and $T\Delta S$ are related to spontaneity of a reaction. (Sections 15-12 and 15-14; Exercises 93 and 94)

15. Explain what is meant by the Gibbs free energy change for a process. Be able to relate it, both mathematically and descriptively, to enthalpy and entropy changes. (Section 15-15; Exercises 93 and 94)

16. Use the standard Gibbs free energy change for a reaction as an indicator of spontaneity. Be able to calculate the standard Gibbs free energy change for a reaction (a) from standard Gibbs free energies of formation of reactants and products, and (b) from standard molar enthalpies of formation and standard entropies of reactants and products. (Section 15-15; Exercises 95 through 101)

17. Determine the temperature range of spontaneity of a reaction from tabulated thermodynamic data. Make general statements regarding the spontaneity of a reaction at relatively low and high temperatures on the basis of the signs of ΔH^0 and ΔS^0. (Section 15-16; Exercises 102 through 112)

18. Be able to recognize and solve a variety of types of problems concerning the material of this chapter. (Mixed Exercises 113 through 118)

19. Answer conceptual questions based on this chapter. (Conceptual Exercises 119 through 123)

20. Apply concepts and skills from earlier chapters to the ideas of this chapter. (Building Your Knowledge exercises, 124 through 130)

> **NOTE:** After you believe you have mastered the material of this chapter, you should work many of the exercises in the text. This will give you valuable experience in recognizing problems of various kinds as well as applying the concepts and skills you have learned in the chapter.

Some Important Terms in This Chapter

Write the meanings *in your own words*. Check the Key Terms list and the chapter reading. Then rewrite your definitions, still in your own words, to improve them. Study other new terms, and review terms from preceding chapters if necessary.

kinetic energy

potential energy

thermodynamics

thermochemistry

First Law of Thermodynamics

system

surroundings

state function

enthalpy

enthalpy change

standard enthalpy change

standard molar enthalpy of formation

Hess's Law of heat summation

bond energy

spontaneity

entropy

absolute entropy

free energy

free energy change

standard free energy change

Preliminary Test

True-False

Mark each statement as true (T) or false (F).

_____ 1. For a body to have energy it must do some work.
_____ 2. Potential energy can be changed into kinetic energy.
_____ 3. Kinetic energy can be changed into potential energy.

_____ 4. Chemical energy is a form of kinetic energy.

_____ 5. The heat capacity of a body is the same as its specific heat.

_____ 6. When we mix water at 100°C with water at 20°C, the final water temperature averages out to 60°C.

_____ 7. When we put a 100-g piece of metal at 100°C into 100 g of water at 20°C, the final temperature is 60°C.

_____ 8. The reverse of an exothermic process must be endothermic.

_____ 9. The term ΔH means $H_{final} - H_{initial}$.

_____ 10. The term ΔT means $T_{final} - T_{initial}$.

_____ 11. A process in which the system absorbs heat from its surroundings is called endothermic.

_____ 12. For a process in which the system absorbs heat from its surroundings, the system has a negative ΔH value.

_____ 13. The process described by the thermochemical equation

$$H_2(g) + I_2(g) \rightarrow 2HI(g) + \text{heat}$$

is described as exothermic.

_____ 14. The process described by the thermochemical equation

$$H_2(g) + I_2(g) \rightarrow 2HI(g) + \text{heat}$$

has a negative value of ΔH.

_____ 15. The thermochemical equation

$$C(\text{graphite}) + \tfrac{1}{2}O_2(g) \rightarrow CO(g) \qquad \Delta H = -110.5 \text{ kJ/mol}$$

refers to the reaction of one atom of C with one-half molecule of O_2 to form one molecule of CO.

_____ 16. If the temperature of a reaction mixture in a coffee-cup calorimeter increases as the reaction proceeds, the reaction is endothermic, because the mixture is absorbing heat from the surrounding air.

_____ 17. Hess's Law is true because enthalpy is a state function.

_____ 18. The standard molar enthalpy of formation of a substance is the amount of heat one mole of the substance contains at standard conditions.

_____ 19. The enthalpy change at standard conditions for the reaction

$$H_2(g) + I_2(g) \rightarrow 2HI(g)$$

is referred to as the standard molar enthalpy of formation of HI(g).

_____ 20. The enthalpy change at standard conditions for the reaction

$$\tfrac{1}{2}H_2(g) + \tfrac{1}{2}I_2(g) \rightarrow HI(g)$$

is referred to as the standard molar enthalpy of formation of HI(g).

_____ 21. The enthalpy change at standard conditions for the reaction

$$CO(g) + \tfrac{1}{2}O_2(g) \rightarrow CO_2(g)$$

is referred to as the standard molar enthalpy of formation of CO_2(g).

_____ 22. According to the First Law of Thermodynamics, the only way we can increase the internal energy of a system is to heat it.

_____ 23. The advantage of a bomb calorimeter is that the volume is held constant, so that the heat absorbed by the system is equal to the change in internal energy of the system.

_____ 24. A reaction that is exothermic must be spontaneous.
_____ 25. A reaction that is spontaneous must be exothermic.
_____ 26. The reverse of an ordering reaction is a disordering reaction.
_____ 27. The reverse of a spontaneous reaction is a nonspontaneous reaction.
_____ 28. When we say that ΔE is a state function, we mean that the energy change of the system does not depend on its initial or final state.
_____ 29. The enthalpy change, ΔH, of a system is a state function.
_____ 30. The heat, q, absorbed by a system is a state function.
_____ 31. The work, w, done on a system is a state function.
_____ 32. The entropy change, ΔS, of a system is a state function.
_____ 33. The Gibbs free energy change, ΔG, of a system is a state function.
_____ 34. The bond energy of any binary compound (consisting of only two elements) is equal to its ΔH_f^0 value divided by the number of bonds formed per molecule.
_____ 35. The enthalpy change associated with the reaction in which 2 moles of H atoms are formed from 1 mole of H_2 is called the bond energy of the H–H bond (all substances at standard conditions).
_____ 36. The enthalpy change associated with the reaction in which 1/2 mole of H_2 and 1/2 mole of Cl_2 are formed from 1 mole of HCl is called the bond energy of the H–Cl bond (all substances at standard conditions).
_____ 37. The enthalpy change associated with the reaction in which 1 mole of H atoms and 1 mole of Cl atoms are formed from 1 mole of HCl is called the bond energy of the H–Cl bond (all substances at standard conditions).
_____ 38. The enthalpy change of a gas phase reaction is the energy required to break all the bonds in reactant molecules minus the energy required to break all the bonds in product molecules.
_____ 39. For the vaporization of 1 mole of water at 1 atm at its normal boiling point, $\Delta E = \Delta H$.

The next five questions, 40 through 44, refer to the following standard reaction:

$$SiH_4(g) + 2O_2(g) \rightarrow SiO_2(s) + 2H_2O(l) \qquad \Delta H_{rxn}^0 = -1516 \text{ kJ/mol}$$

_____ 40. The reaction is exothermic.
_____ 41. When this reaction is carried out at constant pressure with 1 mole of SiH_4 and sufficient oxygen, the reaction mixture gives off 1,516,000 joules of heat to its surroundings.
_____ 42. When this reaction is carried out at constant pressure with 2 moles of SiH_4 and sufficient oxygen, the reaction mixture gives off 1,516,000 joules of heat to its surroundings.
_____ 43. When this reaction is carried out at constant pressure, the reaction mixture does work against its surroundings.
_____ 44. When this reaction is carried out at constant volume at room temperature with 1 mole of SiH_4 and sufficient oxygen, the reaction mixture gives off 1,516,000 joules of heat to its surroundings.

_____ 45. The entropy of a substance is greater as a liquid than as a solid.
_____ 46. The entropy of a substance is greater as a gas than as a liquid.
_____ 47. All reactions become more spontaneous at higher temperatures.

Short Answer

1. The sample or portion of the universe that we wish to study or describe in thermodynamics is called the _____.

2. Three examples of state functions of a gaseous sample are _____, _____, and _____.

3. In calculations with the First Law of Thermodynamics, we use the convention that heat absorbed by the system from its surroundings is given a _____ sign.

4. In calculations with the First Law of Thermodynamics, we use the convention that work done on the system by its surroundings is given a _____ sign.

5. If a gas sample absorbs 1,000 joules of heat from its surroundings and does 400 joules of work by expansion against its surroundings, its internal energy must _____ by _____ joules.

6. A reaction in which the total chemical potential energies of the products is greater than the total chemical potential energies of the reactants is said to be _____.

7. Measurements in calorimetry are based on equating heat lost by the sample to heat gained by the surrounding. This is an application of the Law of _____.

8. A pure substance is said to be in its standard state at conditions of _____ and _____.

9. For a change that takes place at constant volume, the heat gained or lost by the system is a measure of _____ for the system.

10. For a change that takes place at constant pressure, the heat gained or lost by the system is a measure of _____ for the system.

11. A statements of Hess's Law in words is that the enthalpy change for a reaction is the same whether it occurs _____ or _____.

12. The enthalpy change associated with the formation of 1 mole of any substance at standard conditions from its elements, also at standard conditions, is called the _____ _____ of the substance.

13. The enthalpy change associated with the reaction in which 1 mole of HBr(g) at standard conditions is formed from the appropriate numbers of moles of hydrogen gas and bromine liquid, all at standard conditions, is called the _____ of HBr.

14. The enthalpy change associated with the reaction in which 1 mole of HBr(g) at standard conditions breaks apart to form the appropriate numbers of moles of hydrogen atoms and bromine atoms is called the _____ of HBr.

- -

The following equations, referred to by letter, are to be used in the next five questions, 15 through 19. Consider all reactants and products to be at 1 atm, 298 K.

A. $C(graphite) + O_2(g) \rightarrow CO_2(g)$

B. $C(g) + O_2(g) \rightarrow CO_2(g)$

C. $C(graphite) + 2O(g) \rightarrow CO_2(g)$

D. $C(g) + 2O(g) \rightarrow CO_2(g)$

E. $CO(g) + \frac{1}{2}O_2(g) \rightarrow CO_2(g)$

F. $C(graphite) + \frac{1}{2}O_2(g) \rightarrow CO(g)$

G. $C(g) + \frac{1}{2}O_2(g) \rightarrow CO(g)$

H. $C(graphite) + O(g) \rightarrow CO(g)$

I. $C(g) + O(g) \rightarrow CO(g)$

15. The enthalpy change for Reaction _____ is referred to as the standard molar enthalpy of formation of carbon monoxide.

16. The enthalpy change for the reverse of reaction _____ is referred to as the bond energy of carbon monoxide.

17. The enthalpy change for Reaction _____ is referred to as the standard molar enthalpy of formation of carbon dioxide.

18. If we subtract the ΔH^o value for Reaction C from that for Reaction A, the result would be equal to the _____ in O_2.

19. The standard enthalpy change for Reaction _____ could be called the standard molar enthalpy of combustion of carbon monoxide.

- -

20. When we reverse a chemical equation, we must multiply its ΔH by _____.

21. The symbol for enthalpy is _____.

22. The symbol for enthalpy change for a process at any conditions is _____.

23. The symbol for enthalpy change for a process in which all substances are at standard conditions is _____.

24. The symbol for standard molar enthalpy of formation of a substance is _____.

25. The energy holding the atoms, molecules, or ions in a crystal in their regular arrangement is called the _____.

26. A reaction in which the total heat content of the products is lower than the total heat content of the reactants is said to be _____.

27. A reaction in which the system absorbs heat from its surroundings is called _____.

28. A reaction in which the system absorbs heat from its surroundings has a value of ΔH that is _____.

29. The standard enthalpy change for the reaction

$$\frac{1}{2}H_2(g) + \frac{1}{2}F_2(g) \rightarrow HF(g)$$

is called the _____ of HF(g).

30. The reaction for which the standard enthalpy change represents the standard molar enthalpy of formation of $NH_3(g)$ is _____.

- -

For each of the changes described in Question 31 through 40, indicate whether the entropy of the system would increase (I), decrease (D), or remain unchanged (U). You should also know which ones would be described as having ΔS = +, 0, or −.

31. A liquid is frozen to make a solid. _____

32. A new deck of cards is carried home from the store, its wrapper intact. _____

33. A new deck of cards is shuffled. _____

34. A shuffled deck of cards is rearranged according to suits. _____

35. Salt is dissolved in water. _____
36. Two moles of hydrogen atoms combine to form one mole of hydrogen molecules: $2H(g) \rightarrow H_2(g)$. _____
37. Calcium carbonate decomposes on heating: $CaCO_3(s) \rightarrow CaO(s) + CO_2(g)$. _____
38. Nitrogen gas and hydrogen gas combine to form ammonia gas:
   $N_2(g) + 3H_2(g) \rightarrow 2NH_3(g)$. _____
39. A sample of ammonia gas, $NH_3(g)$, is carried from New York to London. _____
40. Dry ice, $CO_2(s)$, sublimes to form $CO_2(g)$. _____

- -

41. Complete the relationship $\Delta G = \Delta H$ _____.
42. A reaction in which ΔS is positive tends to become more spontaneous as the temperature _____.

Multiple Choice

_____ 1. If a system absorbs heat and also does work on its surroundings, its energy
   (a) must increase.   (b) must decrease.   (c) must not change
   (d) may either increase or decrease, depending on the relative amounts of heat absorbed and work done.

_____ 2. A calorimeter is
   (a) a dieting aid.
   (b) a device used to measure transfer of heat energy.
   (c) equal to the molar enthalpy of reaction.
   (d) an indicator of spontaneity.
   (e) only useful in measuring exothermic reactions.

_____ 3. To change 1 mole of ice to 1 mole of liquid water at 0°C, 334 calories of heat are needed. For this process,
   (a) $q = 0$.   (b) $\Delta E = 0$.   (c) $\Delta H = 334$ joules
   (d) $\Delta H = -334$ joules.   (e) $w = 334$ joules.

_____ 4. All spontaneous reactions
   (a) give off heat.   (b) are fast.   (c) have $\Delta G < 0$.
   (d) have $\Delta H < 0$.   (e) have $\Delta S > 0$.

_____ 5. Given the following standard heats of reaction:
   $S(s) + H_2(g) \rightarrow H_2S(g)$ $\Delta H^0 = -20.6$ kJ/mol
   $H_2(g) \rightarrow 2H(g)$ $\Delta H^0 = 436.0$ kJ/mol
   $S(s) \rightarrow S(g)$ $\Delta H^0 = 278.8$ kJ/mol
   The bond energy for the H–S single bond is
   (a) 736 kJ/mol.   (b) 507 kJ/mol.   (c) 368 kJ/mol
   (d) 239 kJ/mol.   (e) 173 kJ/mol.

_____ 6. For a reaction in which gases are neither produced nor consumed, ΔH is _____ ΔE .
   (a) the same as   (b) less than   (c) greater than   (d) unrelated to

_____ 7. For a reaction in which more moles of gas are produced than are consumed, ΔH is _____ ΔE .
   (a) the same as   (b) less than   (c) greater than   (d) unrelated to

_____ 8. A reaction occurring under certain conditions has an enthalpy change of $\Delta H = -50$ kJ/mol. This means that 50 kJ of heat will be liberated when one mole of reaction occurs
   (a) at constant temperature and volume.
   (b) at constant temperature and pressure.
   (c) at constant volume and pressure.
   (d) at constant temperature, volume, and pressure.
   (e) under any conditions.

_____ 9. Is an exothermic process spontaneous?
   (a) Yes, always   (b) No, never   (c) Sometimes

_____ 10. Given the following reactions and their associated enthalpy changes:

$CH_4(g) \rightarrow C(g) + 4H(g)$	$\Delta H^0 = 1663$ kJ/mol
$O_2(g) \rightarrow 2O(g)$	$\Delta H^0 = 498$ kJ/mol
$H_2O(g) \rightarrow 2H(g) + O(g)$	$\Delta H^0 = 927$ kJ/mol
$CO_2(g) \rightarrow C(g) + 2O(g)$	$\Delta H^0 = 1608$ kJ/mol

Use these values to calculate ΔH^0 for the combustion (burning) of 1 mole of CH_4 to form $H_2O(g)$ and $CO_2(g)$ as described by the reaction

$$CH_4(g) + 2O_2(g) \rightarrow 2H_2O(g) + CO_2(g)$$

(a) –372 kJ/mol (b) 4696 kJ/mol (c) –807 kJ/mol (d) 6121 kJ/mol (e) 124 kJ/mol

_____ 11. The value of ΔH^0 for the following reaction, as written, is –2220 kJ/mol. How much heat is produced when 11.0 g of propane gas, C_3H_8, is burned in a constant pressure system?

$$C_3H_8(g) + 5O_2(g) \rightarrow 3CO_2(g) + 4H_2O(l)$$

(a) 555 kJ (b) 2220 kJ (c) 50.5 kJ (d) 24420 kJ (e) 25.96 kJ

_____ 12. When a liquid boils, which of the following must be true?
   (a) $\Delta S < 0$ (b) $\Delta S > 0$ (c) $\Delta T < 0$ (d) $\Delta T > 0$

_____ 13. Which one of the following statements is correct?
   (a) The standard molar enthalpy of formation of an element is usually positive.
   (b) ΔH can never equal ΔE.
   (c) q and w are state functions.
   (d) $q + w$ is a state function.
   (e) The standard entropy S^0 of an element is zero.

_____ 14. For some reaction carried out at constant atmospheric pressure and at constant temperature of 25°C, it is found that $\Delta H^0 = -38.468$ kJ/mol and $\Delta S^0 = +51.4$ J/mol·K. What is the value of ΔG^0 for this reaction at these conditions?
   (a) –53.785 kJ/mol (b) –84.454 kJ/mol (c) 53.785 kJ/mol (d) 84.454 kJ/mol

_____ 15. Is the reaction referred to in Question 14 spontaneous at the conditions given?
   (a) Yes (b) No (c) Insufficient information is given to answer.

_____ 16. For the process $H_2O(l) \rightarrow H_2O(g)$ at 1 atm and 100°C, the change in enthalpy is
   (a) positive.   (b) negative.   (c) zero.

_____ 17. For the process referred to in Question 16, the change in entropy is
   (a) positive.   (b) negative.   (c) zero.

_____ 18. For the process referred to in Question 16, the change in Gibbs free energy is
   (a) positive.   (b) negative.   (c) zero.

_____ 19. A reaction for which $\Delta H = +$ and $\Delta S = -$
   (a) is spontaneous at all temperatures.
   (b) is not spontaneous at any temperature.
   (c) could become spontaneous at high temperature.
   (d) could become spontaneous at low temperature.
_____ 20. A reaction for which $\Delta H = -$ and $\Delta S = -$
   (a) is spontaneous at all temperatures.
   (b) is not spontaneous at any temperature.
   (c) could become spontaneous at high temperature.
   (d) could become spontaneous at low temperature.
_____ 21. A reaction for which $\Delta H = +$ and $\Delta S = +$
   (a) is spontaneous at all temperatures.
   (b) is not spontaneous at any temperature.
   (c) could become spontaneous at high temperature.
   (d) could become spontaneous at low temperature.

NOTE: This Preliminary Test has not provided you with enough practice at thermodynamic calculations. Be sure that you work *many* problems from the end of Chapter 15 in the text.

Answers to Preliminary Test

Do not just look up answers. Think about the reasons for the answers.

True-False

For those that are false, it is especially important to think carefully *why* they are false.

1. False. Energy is the *capacity* to do work or transfer heat; it is not necessary that work actually be done or heat actually be transferred.
2. True. Think about a rock, whose higher potential energy at the top of the hill can be changed (partly) into kinetic energy as it rolls down the hill.
3. True. Suppose you throw a ball into the air. As it goes up its kinetic energy decreases; that kinetic energy is transformed into increasing potential energy. After it has reached the top of its travel, the reverse begins to happen.
4. False. It is a form of potential energy.
5. False. The heat capacity of a body is the amount of heat required to raise the temperature of that body (whatever its mass) by 1 degree Celsius. The specific heat refers specifically to one gram of the substance of which body is composed. In the terminology of Chapter 1, heat capacity is an extensive property, whereas specific heat is an intensive property. However, we often cite the *molar heat capacity* of a substance, which is the heat capacity of one mole of that substance.
6. False. Common sense would tell you, for instance, that putting 1 teaspoon of 100ºC water into a gallon of 40ºC water cannot make the final temperature equal to 60ºC! The result depends on the relative masses of water. Review the heat transfer calculations in Section 1-13.
7. False. The temperature change depends on the specific heat of each substance as well as the mass of each. Review the heat transfer calculations in Section 1-13.
8. True

9. True
10. True
11. True
12. False. If it absorbs heat from the surrounding, its enthalpy is increasing, so that $H_{final} > H_{initial}$. Then $\Delta H = H_{final} - H_{initial} > 0$.
13. True
14. True
15. False. Thermochemical equations always refer explicitly to the numbers of *moles* of substances involved. Be sure you are clear about the distinction between *molecules* and *moles* (Chapter 2).
16. False. The styrofoam in this kind of calorimeter serves as in insulator. Any increase in temperature must come from heat liberated by the reacting substances, so the reaction is exothermic.
17. True. See the discussion in Section 15-8.
18. False. It is the heat content (enthalpy) of 1 mole of the substance at its standard state *compared to* those of the corresponding elements in their standard states. These latter values are conventionally taken to be zero, but this does not mean that they "have no heat content."
19. False. The value must refer to the formation of *1 mole* of the substance.
20. False. The reaction must involve all reactants and products *at their standard states*. I_2 is a solid at its standard state.
21. False. All reactants must be *elements* in their standard states.
22. False. $\Delta E = q + w$. Doing work on the system can also increase its entropy.
23. True
24. False. Changes in disorder of the system also affect spontaneity. Think of the example of water freezing (an exothermic process); it is certainly not spontaneous at 10°C, although it is at −10°C.
25. False. For example, the melting of water is spontaneous at 10°C, and this is an endothermic process (absorbs energy from the surroundings).
26. True
27. True
28. False. The change in a state function *does* depend on the initial and final states. For instance, think about (1) the heating of a substance at constant volume from the initial state of 25°C to a final state of 50°C, compared to (2) heating of a substance at constant volume from the initial state of 25°C to a final state of 75°C. It is apparent that ΔE is different for these two changes. You may need to review Section 15-2 where the concept of the thermodynamic state of a system is discussed.
29. True
30. False
31. False
32. True
33. True
34. False. The bond energy is ΔH for the reaction in which the *only* thing that happens is the breaking of 1 mole of bonds to form isolated atoms. The value of definition of ΔH_f^0 involves the elements in their standard states, and for many elements the standard state is not isolated gaseous atoms. Many elements are diatomic, or solid, or liquid, etc. in their standard states.
35. True. This reaction *does* involve only the breaking of 1 mole of H–H bonds to form isolated H atoms.

36. False. This reaction also involves the energy of forming 1 mole of H–H bonds and that of forming 1 mole of Cl–Cl bonds. See the answer to Question 34 above.

37. True

38. True. Even though reactions do not usually proceed by first breaking all bonds in reactants and then forming products from the resultant atoms, we can calculate enthalpy changes as if they did. This is a major advantage of thermodynamic calculations—the answer does not depend on the pathway used to get from initial to final states.

39. False. Gas is formed, so Δn is not equal to zero. In other words, the system does work against its surroundings, so ΔE cannot be equal to ΔH.

40. True. Negative values of ΔH^0 mean that the reaction is exothermic (enthalpy of products is less than that of reactants).

41. True. The ΔH^0 value is, among other things, a measure of the heat transferred at constant pressure.

42. False. Remember that the ΔH^0 value is for the specific amount of material referred to in the equation as written. The burning of 2 moles of SiH_4 gives off twice as much heat as the 1 mole that is referred to in the reaction, or 3,032,000 joules.

43. False. The number of moles of gas decreases so at constant pressure the volume would decrease. Thus, the surroundings (atmosphere) does work on this system.

44. False. At constant volume, the amount of heat given off is equal to ΔE. Because Δn for this reaction is –3, ΔE is algebraically greater than ΔH for the reaction. Study Section 15-11, which relates ΔE to ΔH.

45. True. Remember that entropy is a measure of disorder. In the solid state, the particles are more ordered, and thus at lower entropy.

46. True

47. False. Study Section 15-16, with special emphasis on the discussion of Table 15-7 and Examples 15-18 through 15-20.

Short Answer

1. system
2. In this chapter, we have seen many possible answers to this question. Acceptable answers include pressure, volume, temperature, energy, enthalpy, entropy, and Gibbs free energy.
3. positive
4. positive
5. increase; 600
6. endothermic
7. Conservation of Energy
8. 1 atmosphere; 25°C
9. ΔE
10. ΔH
11. by one step; by a series of steps
12. standard molar enthalpy of formation
13. standard molar enthalpy of formation
14. bond energy

15. F. Notice that in thermodynamics we often write the reactions with fractional coefficients. This just means that we are dealing with some fractional numbers of moles. Do not think about a fraction of a molecule!

16. I

17. A. All substances must be in their standard states. For carbon this is solid graphite and for oxygen this is the diatomic gas.

18. bond energy of the oxygen-oxygen bond

19. E. Combustion is a reaction with oxygen, O_2; in this reaction, one mole of carbon monoxide, CO, reacts with O_2.

20. -1

21. H

22. ΔH

23. ΔH^0

24. ΔH_f^0

25. crystal lattice energy

26. exothermic

27. endothermic

28. positive

29. standard molar enthalpy of formation

30. $\frac{1}{2}N_2(g) + \frac{3}{2}H_2(g) \rightarrow NH_3(g)$. Be sure that you write the reaction so that 1 mole of $NH_3(g)$ is formed. You must also always explicitly include the physical state of each substance; in this case, they are all gases.

See the discussion in Section 15-14 as you think about the answers to questions 31 through 40. An increase in entropy is denoted by positive ΔS, a decrease by negative ΔS, and no change by $\Delta S = 0$.

31. D

32. U

33. I

34. D

35. I

36. D

37. I

38. D

39. U

40. I

41. $-T\Delta S$

42. increases. The term in ΔG that includes entropy change is $-T\Delta S$. The more negative this term, the more likely the reaction is to be spontaneous. This term is more important at high temperatures.

Multiple Choice

1. (d)

2. (b)

3. (c). The ice must *absorb* this 334 joules, so $\Delta H = +334$ joules.

4. (c)

5. (c). Compare this to the value of 347 kJ/mol given in Table 15-2; remember that the tabulated values represent *average* bond energies in a number of compounds.

6. (a). Recall that $\Delta H = \Delta E + (\Delta n)RT$, as shown in Section 15-11. Remember that Δn is the change in number of moles of gaseous substances shown in the balanced reaction. In the case mentioned here, $\Delta n = 0$, so $\Delta H = \Delta E$.

7. (c) Now $\Delta n > 0$, so $\Delta H > \Delta E$.

8. (b) See Section 15-5.

9. (c). As seen in this chapter, exothermic processes *may* be spontaneous, but they are not necessarily so. The factor of change in order/disorder (entropy change) also affects spontaneity.

10. (c). Combine the equations, as shown in Examples 15-7 and 15-8, so that they add up to the desired equation. You need to *reverse* the last two equations (while changing the signs of their ΔH values) and also *double* the second and third equations (while doubling their values). If you got answer (a), you forgot to double the appropriate equations and their ΔH^0 values. If you got answer (d), you forgot to reverse the appropriate equations and change the signs of ΔH^0. If you got answer (b), you forgot both the doubling and the reversing. If you got answer (e), you forgot to double the H_2O equation. If you think about it, you would realize that the *positive* answers cannot be correct, because this is a burning reaction. You know that burning liberates heat, so the reaction must be exothermic and have a negative ΔH^0 value.

11. (a) The value of 2220 kJ is for the combustion of *1 mole* of propane. In the problem, we have 11.0 g, or (11.0 g)/(44.0 g/mol) = 0.25 mol.

12. (b)

13. (d)

14. (a)

15. (a)

16. (a). We must put heat into a liquid to boil it—vaporization is an endothermic process.

17. (a). A gas is in a higher state of disorder than a liquid at the same temperature.

18. (c). The boiling point of water at 1 atm is 100°C so the process is at equilibrium. This means that $\Delta G = 0$.

19. (b). This is class 4 of Table 15-7. See the discussion in Section 15-16.

20. (d). This is class 2 of Table 15-7.

21. (c). This is class 3 of Table 15-7.

16 Chemical Kinetics

Chapter Summary

In the preceding chapter, we studied the topic of chemical thermodynamics, which is concerned with the question of *whether* a reaction can go. In Chapter 16, we study an equally important question, that of *how fast* a reaction goes. The subject of **chemical kinetics** deals with the rates of chemical reactions, the factors affecting these rates, and the information that a study of reaction rates can give about the detailed set of steps by which the reaction proceeds. Thus, a study of chemical kinetics will serve several purposes, from the practical one of allowing us to speed up or slow down a reaction to the fundamental one of learning about reactions on the molecular level.

In Section 16-1 we learn how to describe quantitatively the rate of a reaction. We could describe the rate of travel of an automobile by the expression (change in location)/(change in time). The greater the change in location in a given length of time, the higher is the rate of travel. In just the same way, we describe the rate of a reaction by the expression (change in concentration of reactant or product)/(change in time). Several points should be noted here: (1) As you have already seen in an earlier chapter, the symbol Δ means "change in"; (2) the notation [X] means **molar** concentration of the substance X whose formula appears in the brackets; (3) the term for the change in each reactant or product is divided by its stoichiometric coefficient; (4) this change term is multiplied by -1 for reactants, to account for their decrease in concentration. Study very carefully the discussion in Section 16-1 of the $ICl + H_2$ reaction. It will introduce you to several important ideas such as the **determination** of the rate from a plot of concentration vs. time, the concept of **initial rate**, and the realization that **rates change with time** (as reactants are consumed, the reaction slows).

We are then introduced to the factors that affect reaction rates: **nature of the reactants**, **concentrations of reactants**, **temperature**, and **presence of a catalyst**. Section 16-2 describes the influence of the nature of the reactants—how finely divided solid reactants are, whether reactants are in a well-mixed solution, and so on.

The presence of more substance means more can react in a given time, so it is not surprising that concentration of reactants plays a very important part in determining reaction rates. The central idea of Section 16-3 is the **rate-law expression**, which relates the rate of reaction to the concentrations of reactants. Keep in mind that the values of the **specific rate constant** (k), and of the **orders** with respect to the reactants *must* be determined experimentally. They cannot necessarily be related to the coefficients in the balanced overall chemical equation. The several examples in this section will clarify the deduction of rate-law expressions from experimental data. Problems 13 through 31 at the end of the textbook chapter will give you some necessary practice at this type of data analysis.

In Section 16-4, the rate-law expression is reformulated into the **integrated rate equation**. Here the emphasis is on the **time** required for the reaction to proceed by a specified amount. This is sometimes expressed in terms of the **half-life** of reactants. The integrated rate equation is different for reactions with different order rate laws. We will study only the most straightforward cases, those of zero-, first-, and second-order reactions. Again, careful attention to the examples will aid you in understanding the central idea of this mathematically complex section. Table 16-2 summarizes important concentration relationships from Sections 16-3 through 16-4. You will find the optional Enrichment section useful as an illustration of the use of calculus in chemistry.

The integrated rate equation also provides us a method for finding reaction orders from graphical and other analyses of concentration-*vs*-time data (Enrichment section: Using the Integrated Rate Equation to Determine Reaction Order). Once again, the examples will help you to understand the details of this method.

The **collision theory** of reaction rates, Section 16-5, takes the viewpoint that for molecules to react they must collide (1) with sufficient energy to react—that is, to break bonds and form new ones—and (2) with the proper orientation for a reaction to occur. This idea is extended by the **transition-state theory**, Section 16-6, which says that reactants pass through a short-lived, high-energy state known as the **transition state** when they react. The reactants must increase their energy by an amount called the **activation energy (E_a)** to reach the transition state. Thus, the activation energy can be viewed as a barrier to the reaction. If the colliding molecules do not have this much energy, they cannot react; if they do have sufficient energy, the appropriate bonds can break, new bonds can then form, and the reaction (for those particular colliding molecules) can proceed. Some energy is then released as the new molecules are formed from the transition state. **Potential energy diagrams** such as those in Figure 16-10 help us to understand the relation between the activation energies for the forward and reverse reactions and the value of ΔE for the reaction.

One of the most important applications of the study of chemical kinetics is the information this study gives us about the detailed series of elementary steps by which reactions occur, called the **reaction mechanism**. Section 16-7 relates the reaction mechanism and the experimentally determined rate law, introducing the important ideas of **reaction intermediates** and the **rate-determining step**. This interpretation of kinetics makes use of both the collision theory and the transition-state theory presented earlier in the chapter.

We observe that any reaction proceeds faster if we raise the temperature. Section 16-8 relates this observation to the increase in the fraction of molecules with sufficient energy to react (recall **activation energy** and review the kinetic-molecular theory, Section 12-12). This idea is expressed mathematically in the **Arrhenius equation**. Of course, instead of increasing the number of molecules that can "leap the energy barrier" in order to speed up the reaction, we could lower the barrier, thereby lowering E_a. This is the function of **catalysts**, discussed in Section 16-9. The catalyst provides a lower energy pathway by which the reaction can occur, causing the reaction to speed up. It is important to realize that catalysts cannot cause thermodynamically impossible reactions to occur; they can only alter the rates of reactions that are spontaneous. The discussion in this section of several **homogeneous** and **heterogeneous** catalysts will illustrate how catalysis can occur.

The ideas of chemical kinetics from this chapter will be one basis for our study in Chapter 17 of chemical equilibrium.

Study Goals

1. Distinguish between chemical kinetics and chemical thermodynamics. Know what each subject does and does not cover. (Chapter introduction; Exercise 4)
2. Summarize the factors that affect reaction rates. (Sections 16-3, 16-4, 16-6, and 16-7 through 16-9; Exercise 1)
3. For a given reaction, express reaction rate in terms of changes of concentrations of reactants and products per unit time. (Section 16-3; Exercises 3, 5, and 6 through 10)

4. Describe how reaction rates are determined. Know what must be measured and how the data are analyzed. (Section 16-3; Exercise 3)

5. Identify all the terms in an expression such as Rate = $k[A]^x[B]^y$ and describe the significance of each. Be familiar with the terminology of order of a reaction. (Section 16-3; Exercises 3, 5, 6, and 13 through 16)

6. Given experimental data for a particular reaction at a specified temperature, consisting of initial rates corresponding to different combinations of initial concentrations of reactants, determine the rate law for the reaction. (Section 16-3; Exercises 17, 18, and 21 through 27)

7. Given the rate-law expression for a specific reaction at a certain temperature, calculate the initial rate of reaction at the same temperature for any initial concentrations of reactants. Use changes in concentrations to predict changes in initial rates. (Section 16-3; Exercises 19, 20, 27, 28, and 31)

8. Know the relationships for various orders of reaction that are summarized in Table 16-2. (Section 16-4)

9. Perform various calculations with the integrated rate equations for zero-, first-, and second-order reactions relating rate constants, half-lives, initial concentrations, and concentrations of reactants remaining at some later time. (Section 16-4; Exercises 32 through 42 and 46)

10. Analyze concentration-*vs*-time data to determine reaction order. (Enrichment, pages 612-614; Exercises 43 through 45)

11. Describe simple one-step reactions in terms of (a) collision theory and (b) transition state theory. (Sections 16-5 and 16-6; Exercises 47 and 56)

12. Describe activation energy and illustrate it graphically for exothermic and endothermic reactions. Be able to relate E_a for forward and reverse reactions to ΔE_{rxn}. (Section 16-6; Exercise 41)

13. Understand how rate-law expressions are related to reaction mechanisms. (Section 16-7; Exercise 63)

14. Be able to derive the rate law based on a proposed mechanism. Given the observed rate-law expression for a reaction, (a) recognize whether a proposed mechanism could be the correct one, or (b) postulate mechanisms consistent with the data, specifying the slow and fast steps. (Section 16-7; Exercises 64 through 72)

15. Understand, in terms of the distribution of energies of the reactant molecules, how the reaction rate depends on temperature. (Section 16-8; Exercises 47 and 50)

16. Be able to use the Arrhenius equation (1) to find the rate constant from collision frequencies, A, and activation energies, E_a, and (2) to relate rate constants at two different temperatures. Know how to represent and interpret this equation graphically. (Section 16-8; Exercises 29, 30, and 51 through 62)

17. Understand, in terms of potential energy diagrams, how a reaction rate is altered by the presence of a catalyst. Give examples of homogeneous and heterogeneous catalysts. (Section 16-9; Exercises 48 and 49)

18. Be able to recognize and solve a variety of types of problems based on the principles of this chapter. (Mixed Exercises, 74 through 81)

19. Answer conceptual questions based on this chapter. (Conceptual Exercises 82 through 86)

20. Apply concepts and skills from earlier chapters to the ideas of this chapter. (Building Your Knowledge exercises, 87 through 92)

SOME IMPORTANT TERMS IN THIS CHAPTER

Write the meanings *in your own words*. Check the Key Terms list and the chapter reading. Then rewrite your definitions, still in your own words, to improve them. Study other new terms and review terms from preceding chapters if necessary.

rate-law expression

specific rate constant

integrated rate equation

half-life

collision theory

transition state theory

activation energy

Arrhenius equation

reaction mechanism

rate-determining step

catalysis

homogeneous catalyst

heterogeneous catalyst

Preliminary Test

As in other chapters, this test will check your understanding of basic concepts and types of calculations. Be sure to practice *many* of the additional textbook Exercises, including those indicated with the Study Goals.

True-False

Mark each statement as true (T) or false (F).

_____ 1. All reactions taking place at the same temperature occur at the same rate.

_____ 2. All exothermic reactions taking place at the same temperature occur at the same rate.

_____ 3. All reactions taking place at the same temperature with the same value of ΔH occur at the same rate.

_____ 4. All reactions taking place at the same temperature with the same activation energy occur at the same rate.

_____ 5. Reactions with lower (more negative) values of ΔG are more spontaneous, so they usually proceed at a higher rate.

_____ 6. The transition state is a short-lived, high-energy state, intermediate between reactants and products in the reaction.

_____ 7. According to the collision theory, whenever molecules collide with one another, they react.

_____ 8. Once the reactants have absorbed the necessary activation energy to reach the transition state, they are stuck and cannot react further.

_____ 9. Once the reactants have absorbed the necessary activation energy to reach the transition state, they must then go ahead to form products.

_____ 10. The activation energy E_a is usually the same as ΔE for the reaction.

_____ 11. A reaction in which the activation energy of the forward reaction is greater than the activation energy of the reverse reaction is endothermic.

_____ 12. The exponents in the rate law must match the coefficients in the balanced chemical equation for the reaction.

_____ 13. The exponents in the rate law may match the coefficients in the balanced chemical equation for the reaction.

_____ 14. If the exponents in the rate law match the coefficients in the balanced chemical equation, then we know that the reaction takes place in one step.

_____ 15. The rate law for a reaction can be correctly predicted from the balanced chemical equation.
_____ 16. The value of k for a given reaction is constant at all conditions.
_____ 17. The value of k for a given reaction is constant at a given temperature.
_____ 18. If we know the mathematical form of the rate-law expression for a reaction, and we have also determined the value of the specific rate constant at some temperature, we can predict the rate at which the reaction would occur at that temperature for any specified reactant concentrations.
_____ 19. As a rule of thumb, every 10°C increase in temperature roughly doubles the rate of many reactions.
_____ 20. A catalyst works by increasing the average energy of reacting molecules.
_____ 21. A catalytic converter on an automobile is an example of a heterogeneous catalyst.
_____ 22. An effective catalyst increases the rate of the forward reaction but does not affect the rate of the reverse reaction.
_____ 23. A homogeneous catalyst is composed of a single compound, whereas a heterogeneous catalyst is a mixture of compounds.
_____ 24. A catalyst can make a thermodynamically nonspontaneous reaction spontaneous.
_____ 25. If we change the initial concentration of the reactant in a first-order reaction, the half-life changes.
_____ 26. If we change the initial concentration of the reactant in a second-order reaction, the half-life changes.

Short Answer

1. The symbol [X] means _____.
2. In the rate-law expression, the quantity represented by the symbol k is called the

   _____.
3. For a particular reaction, the rate law is determined to be Rate = $k[A]^2[B]$. The reaction is said to be _____ order in A, _____ order in B, and _____ order overall.
4. The rate of a reaction may be expressed either in terms of the _____ in reactant concentration with time or the _____ in product concentration with time.
5. In a reaction for which A + B → products is the rate-determining step, doubling the concentration of A and doubling the concentration of B would cause the rate to _____ by a factor of _____.
6. The amount of energy that reactants must gain above their ground states in order to react is called the _____ of the forward reaction.
7. Four factors that can influence the rate of a chemical reaction are _____,
   _____, _____, and _____.
8. When heated, magnesium metal reacts with the oxygen in air to form magnesium oxide. Suppose we have a cube of magnesium metal, 1 cm on edge, and measure the initial rate at which it forms MgO. Then we cut the magnesium into cubes 1 mm on edge. We would expect that at the same temperature the initial rate of MgO formation would _____ by a factor of _____.
9. The principal reason that a reaction speeds up at elevated temperatures is an increase in the

   _____.
10. A substance that alters the rate of a chemical reaction but is not itself consumed or produced in the reaction is called a _____.

11. The sequence of individual steps by which a reaction occurs is called the _____ of the reaction.
12. The slowest step in a reaction sequence is called the _____.
13. According to the collision theory, when a reaction requires the collision between two reactant molecules for the rate-determining step to proceed, that step is termed _____ and the reaction is called _____ order.
14. The time required for half of a reactant to disappear in a reaction is called the _____ of the reactant.
15. The rate equation $\frac{1}{[A]} - \frac{1}{[A]_o} = akt$ describes the dependence of concentration on time for a reaction that is _____ order; the rate law for this reaction would be _____.
16. For a reaction that is first order in A, the equation that relates concentration and time is

    _____.

- -

The next five questions, 17 through 21, refer to the reaction

$$2A + 2B \rightarrow C + 2D$$

in which all reactants and products are gases. The reaction is observed to follow the rate law   Rate = $k\,[A][B]^2$. The reaction is also known to be strongly exothermic. In each question, we make *only* the one change in the reaction conditions indicated. For each, predict the effect on (a) the initial rate of the forward reaction, (b) the value of the rate constant, k, and (c) the activation energy, E_a, for the forward reaction. The possible answers are: increase (I), decrease (D), or remain unchanged (U).

Change	Initial Rate of Forward Reaction	k	E_a
17. The concentration of A is doubled.	____	____	____
18. The concentration of D is doubled.	____	____	____
19. A catalyst is introduced.	____	____	____
20. The temperature is decreased.	____	____	____
21. The volume of the container is doubled.	____	____	____

- -

22. For a particular reaction, the rate law is determined to be Rate = $k[A]^2[B]$. We first carry out the reaction with some starting concentrations of A and B. Then we repeat it, after doubling the concentrations of A and B. The rate will change by the factor (new rate)/(old rate) = _____ (a number).
23. For a particular reaction, the rate law is determined to be Rate = $k[A]^2[B]$. We first carry out the reaction with some starting concentrations of A and B. Then we repeat it, after halving the concentration of A and doubling the concentration of B. The rate will change by the factor (new rate)/(old rate) = _____ (a number).
24. The half-life of a certain first-order reaction is 2.5 hours. After 10 hours, the fraction of initial reactant remaining is _____.

Multiple Choice

_____ 1. Which of the following reactions is expected to be the slowest at a given temperature?
(a) $Ag^+(aq) + Cl^-(aq) \rightarrow AgCl(s)$
(b) $H^+(aq) + OH^-(aq) \rightarrow H_2O(l)$
(c) $CH_4(g) + 2O_2(g) \rightarrow CO_2(g) + 2H_2O(g)$
(d) $Pb^{2+}(aq) + CrO_4^{2-}(aq) \rightarrow PbCrO_4(s)$
(e) $H^+(aq) + CN^-(aq) \rightarrow HCN(aq)$

_____ 2. One of the reactions used commercially to produce hydrogen gas is

$$H_2O(g) + CO(g) \rightarrow H_2(g) + CO_2(g)$$

Which of the following is a proper expression for the rate of this reaction?

(a) $-\dfrac{\Delta[CO_2]}{\Delta t}$ (b) $-\dfrac{\Delta[H_2]}{\Delta t}$ (c) k (d) $\dfrac{\Delta[CO]}{\Delta t}$ (e) $-\dfrac{\Delta[H_2O]}{\Delta t}$

- -

The following potential energy diagram is to be used in answering Questions 3 through 6.

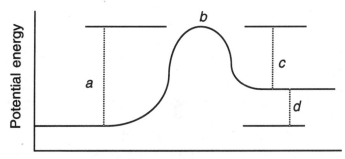

_____ 3. The reaction "reactants → products" carried out at constant volume and temperature
(a) is exothermic.   (b) is endothermic.   (c) takes place without energy change.
(d) is impossible.   (e) may either be exothermic or endothermic.

_____ 4. The energy difference represented by dashed line d is the
(a) energy content of products.   (b) activation energy for the forward reaction.
(c) energy content of reactants.   (d) activation energy for the reverse reaction.
(e) energy change of the reaction.

_____ 5. The region designated by b in the diagram is
(a) the energy of the mixture when half of the reactants have been converted to products.
(b) the energy of the transition state.
(c) the number of moles of transition state that must be formed.
(d) the energy of the forward reaction.
(e) the energy of the reverse reaction.

_____ 6. The energy difference represented by dashed line c is the
(a) reaction energy of the forward reaction.   (b) activation energy of the forward reaction.
(c) reaction energy of the reverse reaction.   (d) activation energy of the reverse reaction.
(e) energy content of the reactants.

7. A reaction has an activation energy of 40 kJ and an overall energy change of reaction of -100 kJ. In each of the following potential energy diagrams the horizontal axis is the reaction coordinate and the vertical axis is potential energy in kJ. Which potential energy diagram best describes this reaction?

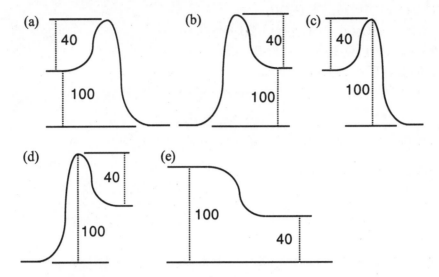

(a) 40   100

(b) 40   100

(c) 40   100

(d) 40   100

(e) 100   40

8. Four of the following changes can affect the forward rate of a chemical reaction. Which one cannot affect this rate?
   (a) raising the temperature   (b) adding a catalyst   (c) removing some of the reactants
   (d) removing of some of the products   (e) grinding a solid reactant into smaller pieces
9. For a given reaction the rate law is
   (a) a constant of proportionality between reaction rate and the concentrations of reactants.
   (b) the sum of the powers to which reactant concentrations appear.
   (c) an equation in which reaction rate is equal to a mathematical expression involving concentrations of reactants.
   (d) the additional energy that the reactants must obtain to be able to react.
   (e) 55 miles per hour.
10. A catalyst
   (a) is used up in a chemical reaction.
   (b) changes the value of ΔG^0 of the reaction.
   (c) is always a solid.
   (d) does not take part in the reaction in any way.
   (e) changes the activation energy of the reaction.
11. The units of the rate constant for a second order reaction can be
   (a) $M^{-1}s^{-1}$ (b) M (c) $M\,s^{-1}$ (d) s^{-1} (e) M^2s^{-1}

- -

In Questions 12 through 14, we refer to the gaseous reaction A + B → C, for which the reaction rate is experimentally observed to follow the relationship Rate = k [A]^2[B].

_____ 12. The overall order of the reaction is
    (a) first.   (b) second.   (c) third.   (d) zero.
    (e) Cannot be answered without a detailed knowledge of the reaction mechanism
_____ 13. If the concentration of A is tripled and the concentration of B is doubled, the reaction rate would be increased by a factor of ____.
    (a) 6   (b) 9   (c) 12   (d) 18   (e) 36
_____ 14. Which of the following would change the value of the specific rate constant k?
    (a) decreasing the temperature   (b) changing the concentration of A
    (c) changing the concentration of B   (d) changing the concentration of C
    (e) letting the reaction go on for a long time

- -

Rate data have been determined at a particular temperature for the overall reaction

$$2NO + 2H_2 \rightarrow N_2 + H_2O$$

in which all reactants and products are gases. These data are shown in the table:

Trial Run	Initial [NO]	Initial [H$_2$]	Initial Rate (Ms^{-1})
1	0.10	0.20	0.0150
2	0.10	0.30	0.0225
3	0.20	0.20	0.0600

Questions 15 through 18 refer to this reaction and these data.

_____ 15. The rate law is
    (a) Rate = k [NO]^2[H$_2$]^2 (b) Rate = k [NO][H$_2$]^2
    (c) Rate = k [NO]^2[H$_2$]   (d) Rate = k [NO][H$_2$]
    (e) None of the preceding answers is correct.
_____ 16. The value of the specific rate constant at this temperature is
    (a) 0.75 $M^{-1}s^{-1}$ (b) 7.5 $M^{-2}s^{-1}$ (c) 3.0 x 10$^{-3}M^{-2}s^{-1}$
    (d) 3.0 x 10^{-4} $M^{-1}s^{-1}$ (e) 375 $M^{-2}s^{-1}$
_____ 17. What would be the initial rate with the molar concentration of NO = 0.30 M and the molar concentration of H$_2$ = 0.10M ?
    (a) 0.0675 M s^{-1} (b) 0.225 M s^{-1} (c) 0.0225 M s^{-1}
    (d) 0.0400 M s^{-1} (e) 0.1000 M^{-1}
_____ 18. Does the reaction take place in one step?
    (a) Yes   (b) No
    (c) Probably yes, but we need more information to be certain.
    (d) Probably no, but we need more information to be certain.

- -

_____ 19. An instructor dismisses class early. Which of the following factors is most important in determining the rate of students exiting from the room (students/min)?
  (a) realizing that the instructor really meant it
  (b) gathering up books and notes
  (c) walking from the seats to the aisle
  (d) walking down the aisle to the door
  (e) crawling through a 1-ft by 2-ft hole in the locked door

_____ 20. Consider the following proposed mechanism.

$$2A \rightarrow C + I \qquad \text{(specific rate constant} = k_1)$$

$$I + B \rightarrow C + D \qquad \text{(specific rate constant} = k_2)$$

If this mechanism for the overall reaction were correct, and if k_1 is much less than k_2, then the observed rate law would be
  (a) Rate $= k_1 k_2 [A]^2 [I][B]$ (b) Rate $= k_2 [I][B]$ (c) Rate $= k_1 [A]^2$
  (d) Rate $= k_1 [A]^2 k_2 [C][D]$ (e) Rate $= k_1 [A]$

_____ 21. A reaction mechanism will usually be
  (a) the only possible explanation for the reaction.
  (b) difficult to verify experimentally.
  (c) proven experimentally to be the only possible mechanism.
  (d) obvious from a consideration of the balanced chemical equation.
  (e) obvious from a consideration of the reaction rate data.

_____ 22. According to the Arrhenius equation for the rate constant of a chemical reaction, the higher the activation energy, the _____ at a given temperature.
  (a) slower a reaction is   (b) more exothermic a reaction is
  (c) more endothermic a reaction is   (d) more atoms are involved in a reaction
  (e) faster a reaction is

_____ 23. The first-order rate constant for the conversion of A $\rightarrow$ B is 0.0551 min^{-1}. How many grams of A would be left after 1 hr if the reaction started with 80 g of A at $t = 0$?
  (a) 2.89 g   (b) 4.42 g   (c) 10.0 g   (d) 52.2 g
  (e) cannot be determined from the data given.

_____ 24. The decomposition of dimethylether at 504°C is first-order with a half-life of 1570 seconds. What fraction of an initial amount of dimethylether remains after 4710 seconds?
  (a) $\frac{1}{3}$ (b) $\frac{1}{6}$ (c) $\frac{1}{8}$ (d) $\frac{1}{16}$ (e) $\frac{1}{32}$

Answers to Preliminary Test

Do not just look up answers. Think about the reasons for the answers.

True-False

1. False
2. False
3. False
4. False

5. False. One of the most important ideas of this chapter is that the rate of the reaction depends on the availability of a low-energy pathway from reactants to products. Spontaneity of a reaction does not determine its rate. Think about Study Goal 1.

6. True

7. False. According to this theory, collision is necessary but not sufficient for reaction. Be sure that you understand what is required for a collision to be effective (Section 16-5).

8. False. They either react to give products or go back to form reactants again, giving up the appropriate amount of energy.

9. False. See the answer to Question 8.

10. False

11. True. Draw a potential energy diagram to help you think about this.

12. False

13. True

14. False. This important idea is discussed in Section 16-7. What this tells us is that it *may* take place in one step, but not necessarily. If the exponents do *not* match the coefficients in the balanced equation, then we know that the reaction definitely *does not* take place in one step.

15. False. This important point can cause much confusion. The reaction may take place in a series of steps with some fast and some slow, so the overall rate is generally impossible to predict from the balanced equation. Study Section 16-7.

16. False. It can be changed either by changing the temperature or by altering the activation energy by introduction of a catalyst.

17. False. It can still be changed by introduction of a catalyst.

18. True

19. True. Of course, the actual change depends on the activation energy—this amount of increase is just typical of many reactions with activation energies near 50 kJ/mol. A calculation of this type was done in Section 16-8.

20. False. It lowers (or, in general, alters) the activation energy, so that at a given temperature, more molecules have enough energy to react.

21. True

22. False. The action of a catalyst is to change the energy of the activated complex (transition state). This changes the values of E_a for both the forward and the reverse reactions, so the rates of both reactions are altered.

23. False. The terms homogeneous and heterogeneous refer to the number of phases of the reaction mixture. When the catalyst is in a different physical state than the reaction mixture, it is called a heterogeneous catalyst.

24. False

25. False. For a first-order reaction, the time required for half of the initial substance to react is the same regardless of amount of starting material.

26. True. For a second-order reaction, the half-life is different depending how much material we start with. Watch out for this in calculations or reasoning involving half-lives.

Short Answer

1. the molar concentration of substance X

2. specific rate constant. Sometimes it is just referred to as the rate constant.

3. second; first; third

4. decrease; increase
5. increase; 4. If this is the rate-determining step, we would observe that Rate = $k[A][B]$.
6. activation energy
7. concentrations of reactants; temperature; nature of reactants; presence of a catalyst
8. increase; 10. The rate would depend on the surface area of magnesium exposed to air. In the 1 cm cube, the total area is 6 cm^2 (6 faces with 1 cm^2 each). After we cut it up, the total area is 60 cm^2. Can you figure this out?
9. fraction of collisions that have energy equal to or exceeding E_a. A secondary factor is the increased number of collisions, but this is less important than the increase in the fraction of high-energy molecules. Can you see this in the Arrhenius equation, Section 16-8?
10. catalyst
11. mechanism
12. rate-determining step
13. bimolecular; second
14. half-life. You may see this referred to in some books as the "half-time."
15. second; rate = $k[A]^2$. Study Section 16-4.
16. $\ln \dfrac{[A]_0}{[A]} = akt$
17. I, U, U. Changing concentration does *not* alter k. What does?
18. U, U, U. Changing concentration of a product does not change the rate of the *forward* reaction.
19. I, I, D. Generally, when we talk about a catalyst we are assuming that it is a catalyst that enhances the rate. If it were an inhibitory catalyst we would have specified it so.
20. D, D, U. Be sure that you know how a temperature change alters the rate.
21. D, U, U. Think about what doubling the volume does to concentrations.
22. 8. Doubling [A] would make the rate $2^2 = 4$ times as fast; doubling [B] would account for an additional factor of 2, so the total effect would be 4 $\times$ 2 = 8.
23. $\frac{1}{2}$. That is, it would go only $\frac{1}{2}$ as fast. Halving [A] would make the rate $(\frac{1}{2})^2 = \frac{1}{4}$ as fast; doubling [B] would account for an additional factor of 2, for a total change of $\frac{1}{2}$.
24. 1/16. Ten hours is four half-life periods. This is a first-order reaction, so the half-life means that $\frac{1}{2}$ of *any* starting amount remains after 2.5 hours. Thus, after 2.5 hours, $\frac{1}{2}$ of the original remains. After the next 2.5 hours, $\frac{1}{2}$ of the amount at the beginning of *that* 2.5 hour period, or $\frac{1}{4}$ of the original, remains. Reasoning in this fashion for four half-lives, we obtain the answer. Be sure you know how to answer a similar question (a) if the time is not an integer number of half-lives, (b) if k instead of half-life is given, or (c) for a second-order reaction. Study the Examples in Section 16-4 carefully.

Multiple Choice

1. (c). This reaction requires that covalent bonds be broken. The others involve only ionic reactants, either forming covalent bonds or forming ionic solids. Reactions that involve the breaking of covalent bonds are usually slower. Can you tie this in with what you know about activation energies?
2. (e)
3. (b). The energy of products is higher than that of reactants.
4. (e)
5. (b)
6. (d)

7. (a)
8. (d)
9. (c)
10. (e)
11. (a). Write down a second-order rate law and think about units.
12. (c)
13. (d)
14. (a)
15. (c). The rationale for this is like that in Examples 16-3 and 16-4 in the text.
16. (b). Put the observed rate and the initial concentrations from any one of the three trial runs into the rate law and solve for k. Be careful of the units. Check to see that you get the same answer regardless of which set of data you use. This can be a check that you deduced the correct rate law as well as a check on your arithmetic.
17. (a). Put these molar concentrations into the rate law, together with the value for k that was obtained in Question 16.
18. (b). The exponents do not match the stoichiometric coefficients, so we know that the reaction *cannot* take place in one step.
19. (e). The slow step determines the rate of overall process.
20. (c). Because k_1 is much less than k_2, the first step is much slower. Thus, the rate of the first step determines the rate of the entire reaction. Can you write the equation for the overall reaction? Note that I is an intermediate—the reaction is

$$2A + B \rightarrow 2C + D.$$

21. (b). Read the discussion in Section 16-7.
22. (a)
23. (a). The calculation is similar to that for N_2O_5 in Example 16-6. Proceed as follows:

$$\ln \frac{[A]_o}{[A]} = akt \text{ where } a = 1, k \text{ is } 0.0551 \text{ min}^{-1}, \text{ and } t = 60 \text{ min}$$

Then $\ln \frac{[A]_o}{[A]} = 3.32$; taking antilogarithms, $\frac{[A]_o}{[A]} = 27.7$.

We cannot calculate either molar concentration of A because we do not know its formula weight or the volume of the container. However, the volume does not change, so the volumes will cancel in the ratio, as will molecular weight:

$$\frac{[A]_o}{[A]} = \frac{\dfrac{mass_o}{MW_A \cdot volume}}{\dfrac{mass}{MW_A \cdot volume}} = \frac{mass_o}{mass}$$

Thus, $\frac{mass_o}{mass} = 27.7$ or $mass = \frac{mass_o}{27.7}$. Then $mass = \frac{80.0}{27.7} = 2.89$ g.

24. (c). 4710 seconds is $4710/1570 = $ three half-lives. This is a first-order reaction, so each half-life period consumes $\frac{1}{2}$ of the amount present at the beginning of *that* period.

17 Chemical Equilibrium

Chapter Summary

Most chemical reactions do not go to completion; rather they attain a state of **chemical equilibrium** that consists of some mixture of reactants and products. Some important questions regarding this equilibrium state are these: How can we describe the equilibrium mixture in terms of concentrations? What factors determine the equilibrium, and how can we use these factors to enhance the occurrence of some desirable reaction? How can we predict what amounts of products will have been formed and what amounts of reactants will remain when equilibrium has been reached? In Chapter 17, we bring together many ideas from our earlier studies to get a general answer to these and other questions about chemical equilibrium. We will first describe from the standpoint of chemical kinetics (Chapter 16) how chemical equilibrium is attained. Then we will use this description to carry out calculations of equilibrium concentrations of reactants and products and to understand the factors that influence equilibrium. Finally we shall see the relation of our equilibrium discussion to chemical thermodynamics. This general discussion of chemical equilibrium ties together many important ideas that we have been developing over the past several chapters. It will also be the basis for study of several specific types of reactions in the next three chapters.

Some basic ideas of equilibrium are presented in Section 17-1. Central to this discussion are three points you must keep in mind: (1) The main conceptual point is that equilibrium is a dynamic situation (sound familiar?) in which two processes, the forward reaction and the reverse reaction, are proceeding at equal rates. (2) We know how rates depend on concentrations (Chapter 16), so we can see that the condition $\text{rate}_{\text{forward}} = \text{rate}_{\text{reverse}}$ can be satisfied for many different combinations of concentrations. (3) The *changes* in concentrations of all reactants and products due to reaction must always be in the ratio of the stoichiometric coefficients, even though the concentrations themselves may not be. Please learn the general layout of analyzing equilibrium problems presented in this section. For each substance,

initial concentration + change in concentration due to reaction = equilibrium concentration

This approach will also apply to equilibrium-concentration problems in the next several chapters.

Section 17-2 defines one of the most important quantities in your study of chemistry, the **equilibrium constant, K**. The algebraic expressions for this constant, unlike those for the rate law expressions, can always be written down directly from the balanced chemical equation. Keep in mind that the concentrations referred to in this expression for the equilibrium constant are the concentrations of reactants and products *when the reaction has reached equilibrium*. Of course, you could put any other values of the concentrations into such an expression and carry out the arithmetic, but this would give the equilibrium constant value *only* if the concentrations described were equilibrium ones. (See the discussion of reaction quotient, later in this chapter). We also see in this section how the *numerical value* of the equilibrium constant for any given reaction can be calculated from a measurement of the concentrations at equilibrium. Some further points should be noticed about equilibrium constants: (1) The symbol for equilibrium constant expressed terms of molar concentrations is K_c. (2) A value of K_c for a particular reaction is valid only at a given temperature. We shall see later in the chapter how this changes with temperature. In Section 17-3 we see how the form and value of K_c vary with the exact form of the balanced equation for a given reaction. We could use any of these forms to solve any particular problem and would always get the same answer for concentration, so long as we are careful to know which representation of the chemical equation the value of K_c corresponds to.

The **reaction quotient, Q,** presented in Section 17-4, is a very useful aid to analyzing whether a reacting system is at equilibrium. You see that it has the same *algebraic form* in terms of concentrations as does K_c, but here the concentrations are *not necessarily* equilibrium concentrations. They might be, but only if those concentrations represent equilibrium. This means that we can use the reaction quotient with any specified combination of concentrations as a criterion (1) of whether or not the reaction is at equilibrium and (2) if not, how it must adjust to get there. The blue box on page 652 summarizes these interpretations.

In Section 17-5, we get to the heart of the matter—the **calculation of concentrations from equilibrium constants.** In this most important kind of calculation, we typically know what the concentrations of all species are at the beginning of the reaction, as well as the value of K_c, and we are asked to calculate the concentrations of some or all of the substances when equilibrium has been reached. Be sure that you see how this differs from the stoichiometric calculations you learned in Chapter 3. In the calculations in this section, the reactants may or may not be present initially in stoichiometric amounts. In stoichiometry problems, you were finding out the maximum amount of product that could be formed *if* the reaction went to completion; now we are recognizing that few reactions actually go to completion.

The strategy of analyzing these problems as shown in Section 17-5 will be very useful. You are sometimes required to solve a quadratic equation in problems of this kind. Pay close attention to the comments in the text regarding which of the two mathematically correct roots of the quadratic equation is the one to be used for the problem at hand. Do not look for a rule to tell you whether to take the + or the − sign in applying the quadratic formula, as this will depend on how you set up the problem in the first place. There is only one good rule—*think* about the physical significance of the answers. The criterion that will always work is that no concentration can be negative—we could not have less than none of a substance! To put it another way, more of a substance cannot react than is present to begin with. (Of course, a *change* in concentration can be negative, meaning that the substance is consumed in the reaction.) Whereas chemical equilibrium problems all use essentially the same reasoning, they can appear in many different disguises and with varied wordings. You must work many of these problems to become comfortable with them. The advice early in Chapter 1 of this guide to be systematic as you interpret and work through problems is never more important than in learning to do equilibrium calculations!

The various factors that may affect chemical equilibria are assessed descriptively in Section 17-6. We think about a reaction at balance (equilibrium), and then throw it out of balance and use LeChatelier's Principle to tell us how it could regain equilibrium. The four "stresses" that are considered are as follows:

(1) **Changes in concentration.** The reaction quotient is useful in analyzing the response to this stress. At equilibrium, Q is equal to K. Changing one or more of the concentrations, by adding or removing a reactant or product, alters the value of Q (but *not* K!). We then reason whether Q must increase or decrease to get back to the equilibrium value of K. Because product concentrations appear in the numerator and reactant concentrations in the denominator, this tells you which way the reaction must proceed to re-establish equilibrium. It is much safer (and surely more satisfying) to learn to analyze the situation yourself than simply to memorize the directions of shift summarized in the text.

(2) **Changes in volume and pressure.** These changes can be interpreted simply in terms of the resulting changes in concentrations and then analyzed as just outlined. Solids and liquids are nearly incompressible, so pressure changes are important only if one or more reactant or product is a gas. Notice again that these changes do not change the value of K, though the value of Q may be altered.

(3) **Changes in temperature.** The direction of such a change can be reasoned out by thinking of heat as a product of an exothermic reaction, or as a reactant in an endothermic reaction, and then applying LeChatelier's Principle. But it is important to see that a change in temperature, like *no other change,* alters the numerical value of K. The change in K caused by changing T will be calculated later in the chapter.

(4) **Introduction of a catalyst.** We saw in Chapter 16 that introduction of a catalyst can change reaction rates dramatically. However, you must remember that it *does not* affect the position of an equilibrium. You should understand how this can be true in terms of the action of a catalyst in lowering the activation energies of both the forward and the reverse reactions. Because not all reactions reach equilibrium quickly, we might desire to introduce a catalyst to help it reach equilibrium in a useful time.

A thoughtful study of the discussion of the Haber process (Section 17-7) for the industrial synthesis of ammonia will help you understand these aspects of chemical equilibrium.

These qualitative considerations of Section 17-6 regarding stresses applied to a system at equilibrium are studied quantitatively in Section 17-8. Study the examples carefully, not just so you can learn to do these kinds of problems, but for the understanding that you can gain of the nature of chemical equilibrium and LeChatelier's Principle. In all equilibrium problems, it is helpful to use the relationship

equilibrium concentration = initial concentration + change in concentration due to reaction

This approach will also be very helpful to you in learning to work problems that involve disturbance from equilibrium. Just think of imposing a change on the equilibrium concentrations to produce the "new initial concentrations." These are then changed by the response of the reaction to the stress as discussed above, to reach the "new equilibrium concentrations." When you have finished working a problem of this type, reason whether the answer is consistent with the changes expected from the qualitative approach you learned in Section 17-6.

The next three sections deal with other ways to represent the equilibrium constant that are sometimes more useful. In Sections 17-9 and 17-10, we see that the equilibrium condition can also be described in terms of the partial pressures of gases, an approach that may be more useful for reactions that take place in the gas phase. The relation between the **pressure equilibrium constant, K_P,** and the **concentration equilibrium constant, K_c,** is discussed in Section 17-10. Remember that Δn has the same meaning here that it had in relating ΔE to ΔH in Chapter 15, the change in number of moles of gaseous substances *in the balanced equation.* Section 17-11 shows how the equilibrium constant is formulated and interpreted in the case of **heterogeneous reactions,** that is, when at least one phase is a pure solid or a pure liquid.

In our study of chemical thermodynamics, we saw that ΔG could be used as a criterion for spontaneity of a reaction. A negative value of ΔG means that the reaction is product-favored (spontaneous in the forward direction) and a positive value means that it is reactant-favored (spontaneous in the reverse direction). This implies that when ΔG is neither negative nor positive, but zero, the reaction is spontaneous in neither direction, i.e., that it is at equilibrium. As we see in Section 17-12, this means that we can relate the **value of ΔG_{xn}^0,** which corresponds to unit concentrations of reactants and products, to the **value of the equilibrium constant, K.** Review (Section 15-15) two methods for the calculation of ΔG_{rxn}^0 from tabulated values. The relation between ΔG^0 and K has two practical uses: (1) If a value of a desired equilibrium constant cannot be found in tables, then more readily available tables of ΔG_f^0 (or ΔH_f^0 and S^0) values can be used to obtain ΔG_{rxn}^0. Then the numerical value of K can be found, from which that the equilibrium concentrations can be calculated. (2) Experimental determination of equilibrium constant values can sometimes be used to evaluate ΔG_f^0 and other thermodynamic quantities for substances that are not otherwise available.

The final section of the chapter, Section 17-13, returns to a quantitative consideration of the change of K with temperature, first described in general terms in Section 17-6. The key relationship of this section is the **van't Hoff equation.** This equation has two practical uses: (1) If we know the value of K at one temperature, T_1, and also know the standard molar enthalpy of reaction, ΔH_{rxn}^0 (as from tabulated values of ΔH_f^0), we can calculate the value of K at another desired temperature, T_2. This is demonstrated in Example 17-17. (2) From experimental measurements of K at different temperatures, we can calculate ΔH_{rxn}^0. This latter application is useful for reactions that involve substances for which ΔH_f^0 values are not tabulated or for reactions that may be difficult to measure with a calorimeter. This might be due, for instance, to scarcity of material, very small amounts of heat given off, or unfavorably slow reaction rates.

It is important to carry out the calculations of these last two sections using the proper form of K. For our calculations, we must remember that when the reaction takes place in the gas phase, the K referred to in Sections 17-12 and 17-13 is the *pressure* equilibrium constant, K_P; when the reaction is in liquid solution, K refers to the *concentration* equilibrium constant, K_c.

At first encounter, the topic of chemical equilibrium is often difficult for students. But it is one of the most important parts of a study of chemistry, and understanding it well is worth your patient, systematic, and diligent study. In the next chapters, we will see the application of these concepts of chemical equilibrium to particular types of processes such as reactions of acids and bases, hydrolysis reactions, and the dissolution and precipitation of slightly soluble ionic substances.

Study Goals

1. Describe the concept of chemical equilibrium as a dynamic equilibrium. Be able to use real or hypothetical reactions in your explanation. Relate chemical equilibrium to forward and reverse reaction rates. (Section 17-1; Exercises 1, 3, 4, 6 through 8, and 10)
2. Write the equilibrium constant expression for any balanced chemical equation, either for a homogeneous or a heterogeneous reaction. (Sections 17-2, 17-3, and 17-9 through 17-11; Exercises 2, 9, 14 through 21, 30, and 31)
3. Be able to reason qualitatively (i.e., without calculations) from information about the magnitude of K. (Section 17-2; Exercises 5, 11, 12, 13, 22, 23, 35, and 36)
4. Given the equilibrium concentrations or the equilibrium partial pressures, evaluate the equilibrium constant, K_c or K_P, for a reaction at a particular temperature. (Sections 17-2, 17-9, and 17-10; Exercises 124 through 26, 28 through 29, and 32 through 34)
5. Distinguish between the reaction quotient, Q, and the equilibrium constant, K. Be able to use Q to assess whether or not a system is at equilibrium, and if not, how it must proceed to approach equilibrium. (Section 17-4; Exercises 37 through 43)

6. Given initial concentrations of all species in a reaction and the value of K_c for this reaction, calculate equilibrium concentrations of all species. (Section 17-5; Exercises 27 and 44 through 52)

7. Use LeChatelier's Principle to determine the direction in which a system at equilibrium will shift (if it does) when changes of the following kinds are made: (a) change in pressure, (b) change in volume, (c) change in temperature, (d) change in amounts of reactants or products present, (e) addition of a catalyst. (Sections 17-5 and 17-6; Exercises 53 through 63)

8. Given concentrations or partial pressures of all species in a system at equilibrium, to which a stress is then applied, determine the concentrations or partial pressures of all species after equilibrium is re-established. (Section 17-8; Exercises 64 through 68)

9. Carry out calculations as in Study Goals 4 and 6, but involving partial pressures and K_P. Be able to convert between K_c and K_P for gas phase reactions. (Section 17-10; Exercises 69 through 79)

10. Given the equilibrium constant for a reaction at a particular temperature, calculate the standard free-energy change, ΔG^0_{rxn}, at that temperature, and vice versa. (Section 17-12; Exercises 81 through 86 and 88)

11. Use the van't Hoff equation to relate the standard enthalpy change (ΔH^0) and the equilibrium constant values at two different temperatures. (Section 17-13; Exercises 80, 87, and 89)

12. Be able to recognize and solve a variety of types of problems based on the principles of this chapter. (Mixed Exercises, 90 through 97)

13. Answer conceptual questions based on this chapter. (Conceptual Exercises 98 through 102)

14. Apply concepts and skills from earlier chapters to the ideas of this chapter. (Building Your Knowledge exercises, 103 through 107)

Some Important Terms in This Chapter

Write the meanings *in your own words*. Check the Key Terms list and the chapter reading. Then rewrite your definitions, still in your own words, to improve them. Study other new terms, and review terms from preceding chapters if necessary.

reversible reaction

chemical equilibrium

dynamic equilibrium

equilibrium constant

reaction quotient

homogeneous equilibrium

heterogeneous equilibrium

LeChatelier's Principle

van't Hoff equation

Preliminary Test

This test will check your understanding of basic concepts and types of calculations. Most students require more practice than they anticipate on the calculations and reasoning of chemical equilibrium. Be sure to practice *many* of the additional text Exercises, including those indicated with the Study Goals.

True-False

Mark each statement as true (T) or false (F).

_____ 1. In any equilibrium mixture, the amounts of substance present must be in ratios given by the stoichiometric coefficients in the balanced chemical equation.
_____ 2. Generally the larger the numerical value of K_c, the more nearly to completion the reaction goes before reaching equilibrium.
_____ 3. A living organism is in a state of chemical equilibrium.
_____ 4. In writing the expression for K, we omit species present as pure solids or pure liquids, but include all substances present in solution or as gases.
_____ 5. When we need to solve a quadratic equation in a chemical equilibrium problem, we always use the positive sign before the square root expression in the quadratic formula.
_____ 6. Because the reaction quotient has the same algebraic form in the concentrations as does the equilibrium constant, Q must always equal K_c.

_____ 7. For the reaction A + B $\rightleftharpoons$ C, for which ΔG^0 = –44.5 kJ, there will be only C left after a long enough period of time if stoichiometric amounts of A and B are allowed to react.

- -

The next eleven questions, 8 through 18, refer to the endothermic equilibrium reaction

$$CoCO_3(s) \rightleftharpoons CoO(s) + CO_2(g)$$

_____ 8. At equilibrium conditions, the forward and the reverse reactions have stopped.
_____ 9. The same equilibrium condition would result if we started with *only* pure $CoCO_3(s)$ in a closed container or if we started with *only* pure $CoO(s)$ in a closed container.
_____ 10. Equal concentrations of $CoO(s)$ and $CO_2(g)$ result from the decomposition of a given amount of $CoCO_3(s)$.
_____ 11. Equal numbers of moles of $CoO(s)$ and $CO_2(g)$ result from the decomposition of a given amount of $CoCO_3(s)$.
_____ 12. Decreasing the volume of the closed system at constant temperature results in more $CoCO_3(s)$ being formed.
_____ 13. Introducing 1.00 atm additional pressure of $CO_2(g)$ to the system at equilibrium in a closed container results in more $CoCO_3(s)$ being formed.
_____ 14. Introducing 1.00 atm additional pressure of $N_2(g)$ to the system at equilibrium in a closed container results in more $CoCO_3(s)$ being formed.
_____ 15. Increasing the temperature of the reaction mixture, but holding the pressure constant by allowing the container to expand or contract, results in more $CoCO_3(s)$ being formed.
_____ 16. The equilibrium constant expression is $K_c = [CO_2]$.
_____ 17. The reaction is a heterogeneous reaction.
_____ 18. If we increase the temperature, the value of the equilibrium constant increases.

Short Answer

1. LeChatelier's Principle may be stated as follows: _____
   _____ .

2. If an equation for a reaction is multiplied by a number n, the original value of K_c is altered by _____ .

3. Suppose we know the numerical value of K_c for a reaction (call this numerical value x). If we reverse the direction in which the equation is written, the numerical value of the equilibrium constant for this reversed equation will be _____ .

4. The equilibrium constant expression, K_c, for the reaction

$$NO_2(g) + SO_2(g) \rightleftharpoons NO(g) + SO_3(g)$$

   is _____ .

5. The pressure equilibrium constant expression, K_P, for the reaction in Question 4 is _____ .

6. The equilibrium constant expression, K_c, for the reaction

$$4FeS_2(s) + 11O_2(g) \rightleftharpoons 2Fe_2O_3(s) + 8SO_2(g)$$

   is _____ .

7. The pressure equilibrium constant expression, K_P, for the reaction in Question 6 is

_____.

8. In qualitative analysis for metal ions, sulfide ions are often produced by the reaction

$$H_2S(aq) + 2H_2O(l) \rightleftharpoons 2H_3O^+(aq) + S^{2-}(aq).$$

The addition of _____ would effectively decrease the S^{2-} concentration in the equilibrium mixture.

9. The slightly soluble substance barium fluoride dissolves only slightly in water according to the reaction

$$BaF_2(s) \rightleftharpoons Ba^{2+}(aq) + 2F^-(aq).$$

The equilibrium constant expression for this reaction is _____.

- -

In the Mond process for refining nickel, the gaseous substance nickel tetracarbonyl, $Ni(CO)_4$, is heated to cause it to decompose into solid nickel and gaseous carbon monoxide, CO. Questions 10 through 13 refer to this reaction.

10. The expression for the equilibrium constant, K_c, for this reaction is _____.
11. The expression for the pressure equilibrium constant, K_P, for this reaction is

_____.

12. The value for Δn for this reaction is _____.
13. The numerical value of K_P for this reaction at room temperature is _____ (larger than, smaller than, the same as) the numerical value for K_c.

- -

Questions 14 through 18 refer to the reaction

$$2NO(g) + 2H_2(g) \rightleftharpoons N_2(g) + 2H_2O(g)$$

This reaction is strongly exothermic. In each question, we allow the reaction to attain equilibrium; then we make *only* the one change in the conditions indicated and allow the reaction to re-establish equilibrium. For each, predict the effect on (a) the amount of N_2 present compared with the original equilibrium conditions and (b) the numerical value of K. The possible answers are: increase (I), decrease (D), or remain unchanged (U).

Change	Amount of N_2	Value of K
14. More NO is added to the reaction mixture.	_____	_____
15. A catalyst is introduced.	_____	_____
16. Some $H_2O(g)$ is removed from the reaction mixture by introducing a drying agent.	_____	_____
17. The temperature is decreased.	_____	_____
18. The volume of the reaction container is doubled.	_____	_____

19. The thermodynamic quantity that is most closely related to the value of the equilibrium constant is _____.

20. The thermodynamic quantity that is most closely related to the change in value of the equilibrium constant is _____.

Multiple Choice

_____ 1. When a reaction reaches equilibrium,
  (a) the rate of the forward reaction and the rate of the reverse reaction are both zero.
  (b) all reaction stops.
  (c) the rate of the forward reaction is zero.
  (d) the rate of the reverse reaction is zero.
  (e) the forward and reverse reaction rates are equal.

_____ 2. In which of the following cases will the least time be required to arrive at equilibrium?
  (a) K_c is a very small number.
  (b) K_c is a very large number.
  (c) K_c is approximately 1.
  (d) Cannot tell without knowing the value of K_c.
  (e) Cannot tell because the time to arrive at equilibrium does not depend on K_c.

_____ 3. Which of the following statements is *not* true?
  (a) A system that is disturbed from an equilibrium condition reacts in a manner to restore equilibrium.
  (b) Equilibrium in molecular systems is static, all molecular interactions having stopped.
  (c) The nature of an equilibrium mixture is the same regardless of the direction from which it was reached.
  (d) A system moves spontaneously toward a state of equilibrium.
  (e) The equilibrium constant usually depends on temperature.

_____ 4. In the reaction

$$A + B \rightleftharpoons C + D$$

for which k_f and k_r are the forward and reverse specific rate constants, respectively, which of the following is correct at equilibrium?
  (a) $k_f = k_r$ (b) $k_f > k_r$ (c) $k_f < k_r$ (d) $rate_f = rate_r$ (e) $K_c = \frac{[A][B]}{[C][D]}$

_____ 5. For the reaction

$$2A + B \rightleftharpoons C + 2D \qquad \text{at } 35^oC$$

the value of k_f is 3.0×10^{-3} $M^{-1}s^{-1}$ and the value of k_r is 1.5×10^{-2} $M^{-1}s^{-1}$. Calculate the value of K_c for this reaction.
  (a) 2.0   (b) 0.5   (c) 0.2   (d) 5.0   (e) 4.5×10^{-5}

_____ 6. Solid ammonium carbamate, $NH_4CO_2NH_2$, dissociates completely into ammonia and carbon dioxide. When pure ammonium carbamate is put into an evacuated container and allowed to come to equilibrium with the gaseous products, the total pressure at 45^oC is found to be 0.6 atm. What is the value of K_P for this reaction at 45^oC?
  (a) 6.0×10^{-2} (b) 4.0×10^{-3} (c) 8.0×10^{-2}
  (d) 3.2×10^{-2} (e) 2.16×10^{-1}

214

_____ 7. The equilibrium constant for the gaseous reaction

$$HCHO \rightleftharpoons H_2 + CO$$

has the numerical value 0.50 at 600°C. A mixture of HCHO, H_2, and CO is introduced into a flask at 600°C. After a short time, analysis of a small sample of the reaction mixture shows the molar concentrations to be [HCHO] = 1.5, [H_2] = 0.5, and [CO] = 1.0. Which of the following statements about this reaction mixture is true?
(a) The reaction mixture is at equilibrium.
(b) The reaction mixture is not at equilibrium, but no further reaction will occur.
(c) The reaction mixture is not at equilibrium, but will move toward equilibrium by forming more HCHO.
(d) The reaction mixture is not at equilibrium, with will move toward equilibrium by using up more HCHO.

_____ 8. The numerical value of the concentration equilibrium constant for the gaseous reaction

$$2SO_2 + O_2 \rightleftharpoons 2SO_3$$

is 0.5 at some temperature. When a reaction mixture is brought to equilibrium at that temperature, [O_2] is found to be 0.5 M and [SO_3] is found to 5.0 M. What is the equilibrium concentration of SO_2 in this mixture, in mol/L?
(a) 0.5   (b) 1.0   (c) 2.0   (d) 5.0   (e) 10.0

_____ 9. The equilibrium constant for the gaseous reaction

$$CO + H_2O \rightleftharpoons CO_2 + H_2$$

is 4.0 at a certain temperature. A reaction is carried out at this temperature starting with 2.0 mol/L of CO and 2.0 mol/L of H_2O. What will be the equilibrium concentration of H_2, in mol/L?
(a) 2.0   (b) 0.75   (c) 1.33   (d) 0.67   (e) 1.5

_____ 10. The introduction of a catalyst into a reaction mixture at equilibrium would
(a) cause a change in the amount of heat absorbed or evolved.
(b) change the value of ΔG^0 for the reaction.
(c) change the value of the equilibrium constant.
(d) change the relative amounts of reactants and products present at equilibrium.
(e) cause none of the changes listed above.

_____ 11. Suppose the reaction mixture

$$N_2(g) + 3H_2(g) \rightleftharpoons 2NH_3(g)$$

is at equilibrium at a given temperature and pressure. The pressure is then increased at constant temperature by compressing the reaction mixture. The mixture is then allowed to re-establish equilibrium. At the new equilibrium
(a) there is more ammonia present than there was originally.
(b) there is less ammonia present than there was originally.
(c) there is the same amount of ammonia present as there was originally.
(d) the nitrogen is used up completely.
(e) the amount of ammonia present may be either greater or smaller than it was originally, depending on the value of K.

215

_____ 12. Suppose the reaction mixture

$$H_2(g) + Cl_2(g) \rightleftharpoons 2HCl(g)$$

is at equilibrium at a given temperature and pressure. The pressure is then increased at constant temperature by compressing the reaction mixture. The mixture is then allowed to re-establish equilibrium. At the new equilibrium
(a) there is more hydrogen chloride present than there was originally.
(b) there is less hydrogen chloride present than there was originally.
(c) there is the same amount of hydrogen chloride present as there was originally.
(d) the hydrogen and chlorine are completely used up.
(e) the amount of hydrogen chloride present may be either greater or smaller than it was originally, depending on the value of K.

_____ 13. An acetic acid solution is allowed to come to equilibrium:

$$CH_3COOH + H_2O \rightleftharpoons H_3O^+ + CH_3COO^-$$

If some silver ion, Ag^+, is then added to the solution, solid silver acetate, $AgCH_3COO$, is formed. The resulting amount of undissociated acetic acid, CH_3COOH, in solution would be
(a) unchanged from that in the original solution.
(b) higher than that in the original solution.
(c) lower than that in the original solution.
(d) zero.

_____ 14. The reaction

$$CaCO_3(s) + 2H_3O^+(aq) \rightleftharpoons Ca^{2+}(aq) + 2H_2O(l) + CO_2(g)$$

may be forced to completion (that is, forced to consume all of the $CaCO_3$ present) by which of the following methods?
(a) Removing some H_3O^+ from the reaction mixture by neutralizing it with base.
(b) Adding more Ca^{2+} to the solution.
(c) Removing CO_2 gas as it is formed.
(d) Adding CO_2 gas to the reaction mixture.
(e) None of the preceding would have any effect on the amount of $CaCO_3$ consumed.

_____ 15. For the heterogeneous reaction

$$2MnO_2(s) \rightleftharpoons 2MnO(s) + O_2(g)$$

the equilibrium constant expression for K_c is which of the following?

(a) $K = \dfrac{[MnO]^2[O_2]}{[MnO_2]^2}$

(b) $K = [O_2]$

(c) $K = 1/[O_2]$

(d) $K = k[MnO_2]^2$

(e) $K = \dfrac{[MnO_2]^2}{[MnO]^2[O_2]}$

_____ 16. Given the equilibrium reaction

$$MgSO_4(s) \rightleftharpoons MgO(s) + SO_3(g)$$

which of the following statements is true?
(a) If the container were opened to the atmosphere, eventually only $MgSO_4(s)$ would remain.
(b) Decreasing the volume of the closed container at constant temperature results in more $MgO(s)$ being formed.
(c) The same equilibrium condition would result if we started with *only* pure $MgSO_4(s)$ in a closed container as if we started with *only* pure $MgO(s)$ in a closed container.
(d) At equilibrium the forward and the reverse reactions have stopped.
(e) If pure $MgSO_4(s)$ were placed in a closed container filled with $N_2(g)$ at 1 atm, the total pressure would increase due to $SO_3(g)$ being formed.

- -

The following three questions, 17 through 19, refer to the exothermic reaction:

$$2SO_2(g) + O_2(g) \rightleftharpoons 2SO_3(g)$$

_____ 17. Suppose we let the reaction come to equilibrium. Then we decrease the temperature of the reaction mixture. What will be the effect on the net amount of $SO_3(g)$ present?
(a) It increases.　　　　(b) It decreases.　　　　(c) It does not change.
(d) The question cannot be answered without knowing the value of K_c or K_P.
(e) The question cannot be answered without knowing the value of ΔH^0 .

_____ 18. Suppose we let the reaction come to equilibrium. Then we decrease the total pressure by increasing the volume of the container. What will be the effect on the net amount of $SO_3(g)$ present?
(a) It increases.　　　　(b) It decreases.　　　　(c) It does not change.
(d) The question cannot be answered without knowing the value of K_c or K_P.
(e) The question cannot be answered without knowing the value of ΔH^0.

_____ 19. Suppose we let the reaction come to equilibrium. Then we decrease the temperature and decrease the total pressure simultaneously. What will be the effect on the net amount of $SO_3(g)$ present?
(a) It increases.　　　　(b) It decreases.　　　　(c) It does not change.
(d) It may either increase, decrease, or remain the same, depending on the sizes of the temperature and pressure changes.
(e) It may either increase, decrease, or remain the same, depending on the numerical value of the equilibrium constant.

- -

_____ 20. At 25°C, the value of K_c for a particular reaction is 30. At 75°C, the value of K_c for this same reaction is 40. Which one of the following statements about this reaction is true?
(a) The reaction is at equilibrium.
(b) LeChatelier's Principle does not apply to this reaction.
(c) The reaction is endothermic.
(d) The reaction is exothermic.
(e) Increasing the pressure would decrease the value of K_c.

Answers to Preliminary Test

Do not just look up answers. Think about the reasons for the answers.

True-False

1. False
2. True
3. False. All reactions in a living organism are obviously not at equilibrium at any given time.
4. True
5. False. The only sure way to know is to find out which root leads to physically reasonable results. Usually this means that the incorrect root would predict negative concentrations for one or more substances in the reaction.
6. False. This is true only at equilibrium. Study Section 17-4 and the uses of the reaction quotient outlined in Section 17-6.
7. False
8. False
9. False. Starting with only pure $CoO(s)$, equilibrium could never be reached. Some $CO_2(g)$ is necessary to allow the reverse reaction to proceed at all.
10. False. We do not refer to "concentrations" of pure solids. The activity of a pure liquid or solid is one.
11. True
12. True
13. True
14. False. Because $N_2(g)$ is neither a reactant nor a product, its presence does nothing to disturb the equilibrium.
15. False
16. True. Remember that solids do not appear in the equilibrium constant expression.
17. True
18. True. Because the reaction is endothermic (consumes heat), the forward reaction would be favored at higher temperatures. This would form more products and consume reactants, so the value of K would increase.

Short Answer

1. if a change of conditions is applied to a system at equilibrium, the system will respond so as to reduce the disturbance to reach a new equilibrium state. (Of course it is much better for you to know this and be able to state it in your own words than just to memorize a set statement.)
2. raising it to the nth power. You do not need to remember this rule—if you understand how to write an equilibrium constant from the chemical equation, you can easily figure it out.
3. $1/x$. Same comments as for Question 2.
4. $\dfrac{[NO][SO_3]}{[NO_2][SO_2]}$
5. $\dfrac{P_{NO}\, P_{SO_3}}{P_{NO_2}\, P_{SO_2}}$

6. $\dfrac{[SO_2]^8}{[O_2]^{11}}$

7. $\dfrac{P_{SO_2}{}^8}{P_{O_2}{}^{11}}$

8. H_3O^+. This is an example of a weak acid equilibrium. This will be covered in more detail in Chapter 18, but you can answer this with your general knowledge about equilibrium from Chapter 17.

9. $[Ba^{2+}][F^-]^2$. Equilibria involving slightly soluble salts will be covered in more detail in Chapter 20. This is clearly just a case of heterogeneous equilibrium, which you have learned about in Chapter 17.

10. $K = \dfrac{[CO]^4}{[Ni(CO)_4]}$. From the description given for the reaction, you should be able to write the chemical equation as

$$Ni(CO)_4(g) \rightleftharpoons Ni(s) + 4CO(g)$$

11. $K_P = \dfrac{P_{CO}{}^4}{P_{Ni(CO)_4}}$

12. 3. Remember that this is the number of moles of gas products minus the number of moles of gas reactants in the balanced equation.

13. larger than. Remember (Section 17-10) that $K_P = K_c (RT)^{\Delta n}$. From the preceding question, Δn is positive.

14. I, U

15. U, U

16. I, U

17. I, I. Only a temperature change can alter the value of K.

18. D, U. The only way to change the value of K is to change the temperature, so K does not change. Doubling the volume of the container has the effect of multiplying all gas concentrations (or all gas pressures) by a factor of $\frac{1}{2}$. Because there are more moles of gas on the left than on the right, the denominator of Q decreases more than the numerator. Thus, Q is now greater than K. The reaction shifts to re-establish equilibrium by proceeding to the left, making more moles of gas, and decreasing the amount of N_2 present.

19. ΔG^0

20. ΔH^0

Multiple Choice

1. (e)
2. (e)
3. (b)
4. (d). As equilibrium is reached, the rates change, but the values of k_f and k_r do not. Do not get the idea that $k_f = k_r$ at equilibrium; rather, the concentrations are such that (d) is true. Answer (e) must be inverted to be correct.
5. (c). From Section 17-2, $K_c = k_f/k_r$.

6. (d). You should be able to figure out that the reaction is

$$NH_4CO_2NH_2(s) \rightleftharpoons 2NH_3(g) + CO_2(g)$$

If the total pressure is 0.6 atm, the partial pressure of NH_3 is 0.4 atm and that of CO_2 is 0.2 atm. You should be able to find these from the stoichiometry of the reaction. Then put them into the expression $K_P = P_{NH_3}^2 P_{CO_2}$.

7. (d). Putting the given values of concentrations into the expression for Q, you find that $Q = 0.333$. This is less than K, so the reaction must proceed to make more products, raising the value of the numerator and lowering the denominator. This uses up HCHO.

8. (e). Put the known value of concentrations into the expression for K_c and then solve for $[SO_2]$.

9. (c). This calculation is like Examples 17-5 through 17-7.

10. (e)

11. (a). There are more moles of gas on the left than on the right of the balanced equation. Thus an increase in pressure would disturb the equilibrium. This stress could be relieved by the reaction proceeding more to the right.

12. (c). In this case, there are equal numbers of moles of gas on the two sides of the balanced equation. Thus, an increase in pressure would not disturb the equilibrium.

13. (c). This combines ideas that will be covered quantitatively in Chapters 18 and 19 (weak acid equilibrium) and in Chapter 20 (slightly soluble salt equilibrium). It can be answered easily from the general principles of equilibrium presented in Chapter 12.

14. (c)

15. (b)

16. (e)

17. (a)

18. (b)

19. (d)

20. (c). You should be able to figure this out by the ideas of Section 17-6. You should also be able to reason it out, without calculation, from the van't Hoff equation, Section 17-13.

18 Ionic Equilibria I: Acids and Bases

Chapter Summary

Many important reactions take place in water solutions—in the laboratory, in living organisms, and in industrial chemistry. Now we apply what we learned about chemical equilibrium in Chapter 17 to solutions of strong and weak electrolytes (Section 4-2), with special emphasis on weak acids and bases (Section 4-2 parts 2 and 4, Chapter 10). Some review of the appropriate sections of the chapters just mentioned would be helpful before you begin your detailed study of Chapter 18.

The chapter begins with a review of the terminology of **strong** and **weak electrolytes** (Section 18-1). Until now you have seen this topic only in a descriptive way. Here we use it to calculate the concentrations of each ionic species present in solution when a strong electrolyte (a strong acid, a strong soluble base, or a soluble ionic salt) dissolves in water. The key idea here is that each reaction is considered to go *virtually to completion*, so we can calculate concentrations based solely on interpreting the stoichiometric coefficients in the balanced chemical equation. We do not apply the ideas of chemical equilibrium to solutions of strong electrolytes. But we will see in this chapter that when one of the ions is the conjugate acid of a weak base or the conjugate base of a weak acid, we still must take equilibrium into account.

Application of chemical equilibrium does become important, however, when we consider the **autoionization of water** to form H_3O^+ and OH^- (Section 18-2). This reaction proceeds only slightly, so we describe it in terms of an equilibrium constant, $K_w = [H_3O^+][OH^-] = 1.0 \times 10^{-14}$ at 25°C. A very important point here is that this equilibrium constant expression *must be satisfied in all aqueous solutions,* no matter what the source of H_3O^+ or OH^- and no matter what other ions may be present. This allows us to relate $[H_3O^+]$ and $[OH^-]$ in any aqueous solution, because their product must be equal to K_w. You must become very familiar with the terminology at the end of Section 18-2, by which we refer to a solution as acidic, basic, or neutral, depending on whether $[H_3O^+]$ is greater than, less than, or equal to $[OH^-]$, respectively. Such solutions can contain Brønsted-Lowry or Lewis acids and bases, but it is an overall characteristic of the solution that we are describing here.

Because of the inconvenience of the great range of numbers necessary to describe $[H_3O^+]$ or $[OH^-]$, chemists use the **pH** and **pOH scales** (Section 18-3). On these scales, (1) a ten-fold change in $[H_3O^+]$ or $[OH^-]$ is described by a change of 1 pH or pOH unit, (2) increasing $[H_3O^+]$ corresponds to decreasing pH, and (3) increasing $[OH^-]$ corresponds to decreasing pOH. The terminology of pH is used widely in science and engineering, so study this section carefully.

There are only a few common strong acids and bases, but many solutions involve **weak acids** and **weak bases.** These substances are involved in dissociation or ionization reactions that reach equilibrium without going to completion. So we describe the ionization reactions of these substances in terms of equilibrium constants (Section 18-4). This is the most important section of this chapter—be sure you understand it well. These equilibrium constants, known as **ionization constants**, are denoted as K_a for weak acids and K_b for weak bases. Of course, we are really just applying the same kind of equilibrium constant calculations that you learned in Chapter 17. The *very small* values typical of K_a and K_b often help us to solve problems more easily. For instance, we can often neglect very small values when added to or subtracted from larger numbers. This is discussed in the Problem-Solving Tip entitled "Simplifying Quadratic Equations" in Section 18-4; be sure that you understand how and when you can make this approximation. In addition to learning to work problems to solve for concentrations, you should (1) have a feel for relative strengths of acids or bases from looking at K_a and K_b values, (2) be familiar with the **percent ionization** representation, and (3) see how the weak acid-strong conjugate base relationships from Chapter 10 apply here. The pH description is used throughout these calculations.

We can also use equilibrium constants and expressions to describe **polyprotic weak acids** (Section 18-5), those that ionize stepwise (with varying strengths), each step yielding H_3O^+. Notice that successive steps are always less complete, so that $K_{a2} < K_{a1}$, and so on. There are no *common* examples of weak bases that ionize in a stepwise fashion, although a few inorganic bases and many weak organic bases do so.

In the remainder of this chapter we continue our study of chemical equilibrium. We now describe equilibria in **hydrolysis reactions** (Section 18-6), in which water or one of its ions is a reactant. Several sections deal with hydrolysis reactions involving salts of Brønsted-Lowry acids or bases, each of which may be weak or strong. As background to this material, you must be familiar with the Brønsted-Lowry terminology, especially with regard to **conjugate acid-base pairs** and their relative strengths, Section 10-4. At the end of the chapter, reactions in which water interacts with metal ions will be considered.

So far in this chapter, we have dealt with solutions to which an acid or base, either strong or weak, had been added. In Sections 18-7 through 18-10, we are concerned with solutions to which only a salt has been added, but no free acid or base. In pure water $[H_3O^+] = [OH^-]$. In aqueous solutions, reactions can occur in which this $[H_3O^+]/[OH^-]$ balance is upset; thus, the resulting solutions could be acidic (excess H_3O^+), basic (excess OH^-), or neutral. Recall (Section 10-6) that salts can be viewed as composed of the cation of a base and the anion of an acid. Acids and bases can each be classified as either strong or weak, so four classes of salts can be identified; the hydrolysis reactions of these four classes of salts are summarized in the following table.

Salt of			Nature of	
Base	*Acid*	**Section**	**Solution**	**Comments**
strong	strong	18-7	neutral	The conjugate base of the strong acid is too weak to react with H_3O^+; the conjugate acid of the strong base is too weak to react with OH^-. The $[H_3O^+]/[OH^-]$ balance remains undisturbed, so the solution remains neutral.
strong	weak	18-8	basic	The anion of the weak acid is strong base so it reacts with H_3O^+ (hydrolyzes), leaving excess OH^-. The water equilibrium shifts, making more H_3O^+ and OH^- in equal amounts, so OH^- is still in excess, and the solution is basic. The equilibrium constant for this hydrolysis reaction is $K_b = K_w/K_a$, where K_a is the ionization constant for the parent weak acid.

(continued)

222

| Salt of | | | Nature of | |
Base	Acid	Section	Solution	Comments
weak	strong	18-9	acidic	The conjugate acid of the weak base is strong so it reacts with OH^- (hydrolyzes), leaving excess H_3O^+. The water equilibrium shifts, making more H_3O^+ and OH^- in equal amounts, so H_3O^+ is still in excess and the solution is acidic. The equilibrium constant for this hydrolysis reaction is $K_a = K_w/K_b$, where K_b is the ionization constant for the parent weak acid.
weak	weak	18-10	acidic, basic, or neutral	Both ions hydrolyze; the result depends on relative strengths of acid and base.

Be sure you can calculate the solution concentrations that result from hydrolysis of salts of strong base–weak acid or weak base–strong acid. These results can be expressed either in terms of H_3O^+, OH^-, pH, pOH, or % hydrolysis. The relation $K_aK_b = K_w$ is very useful in equilibrium constant calculations involving conjugate acid-base pairs.

Another type of hydrolysis reaction occurs in solutions that contain **small, highly charged metal ions** (Section 18-11). Water molecules coordinate with these ions; the resulting strong interaction of the metal ion with the oxygen atom of water weakens the O–H bond. This causes the hydrated complex ion to lose H^+ to a solvent water molecule; the hydrated ion thus acts as a Brønsted-Lowry acid. Thus, the solution contains an excess of H_3O^+ over OH^-, and is acidic. As with other reactions that occur only to a limited extent, the principles of chemical equilibrium can be applied. The equilibrium constant that is used for this type of hydrolysis reaction is given the symbol K_a, but it is called a **hydrolysis constant.** Do not get so involved with the calculation of concentrations that you forget to see which metal ions hydrolyze readily. Pay attention to how the ease of hydrolysis of metal ions is correlated with position in the periodic table, with charge density, and with oxidation state of the metal.

Study Goals

There is large amount of apparently diverse material in this chapter. This material is tied together in terms of acid-base behavior and equilibrium considerations. The following Study Goals will help you to organize the material. Be systematic in your study. Write your own summary of each section, with these goals in mind. Also be systematic in answering Exercises from the text.

1. Review the fundamental ideas of acids and bases from earlier chapters. Memorize the lists of the common strong acids (Table 4-5) and the common strong soluble bases (Table 4-7). Classify common acids as strong or weak; classify common bases as strong soluble, insoluble, or weak. (Review Sections 4-2, 10-4, 10-7, and 10-11; Exercise 1 through 3)
2. Understand the Brønsted-Lowry acid-base theory. Be able to identify Brønsted-Lowry acids and Brønsted-Lowry bases in reactions. Understand and apply the Brønsted-Lowry terminology of conjugate acid-base pairs and their relative strengths. (Review Section 10-4)

3. Be able to calculate the concentrations of constituent ions in solutions of strong electrolytes (whether they are strong acids, strong bases, or soluble salts). (Section 18-1; Exercises 4 through 9)

4. Be familiar with the autoionization of water, including the chemical equation and the significance and uses of its equilibrium constant, K_w. (Section 18-2; Exercises 10 through 16, 34, 35, and 59)

5. Be familiar with the pH and pOH scales. Given any one of the following for any aqueous solution, calculate the other three: hydronium ion concentration, hydroxide ion concentration, pH, or pOH. (Section 18-3; Exercises 17 through 29)

6. Know how to write the equations for the ionization of any weak acid or weak base. Be able to write and interpret the expression for the corresponding equilibrium constant. (Section 18-4; Exercises 30 through 33, 36, and 37)

7. Be able to perform equilibrium calculations for solutions of weak acids and weak bases, to relate initial concentrations, value of ionization constants, and equilibrium concentrations of all species present. Given any one of these quantities, calculate the others. (Section 18-4; Exercises 38 through 58)

8. Be able to apply Study Goal 7 to solutions of polyprotic weak acids. (Section 18-5; Exercises 60 through 65)

9. Describe solvolysis and hydrolysis reactions. Review how the strengths of conjugate acids and bases are related to one another. (Review Section 10-7; Section 18-6; Exercises 66 through 70)

10. Classify salts into one of the following categories: (a) salts of strong soluble bases and strong acids, (b) salts of strong soluble bases and weak acids, (c) salts of weak bases and strong acids, or (d) salts of weak bases and weak acids. (Sections 18-7 through 18-10; Exercise 71)

11. Write the chemical equation and the equilibrium constant expression describing any possible hydrolysis of salts listed in Study Goal 10. Tell whether each resulting solution will be acidic, basic, or neutral. Evaluate the hydrolysis constant in terms of the parent weak acid or base. (Sections 18-7 through 18-10; Exercises 72 through 81, 85 through 89, 91 through 93

12. Given the initial concentration of a salt and the ionization constant for the weak acid or base to which it is related, calculate the concentrations of all species in the solution and the solution's pH, pOH, and % hydrolysis. (Sections 18-7 through 18-10; Exercises 82 through 84 and 90)

13. Write equations for and describe the hydrolysis of small, highly charged metal ions. (Section 18-11; Exercises 94 and 95)

14. Given the hydrolysis constant of a small, highly charged metal ion, calculate the concentrations of all species in a solution of a given concentration of a salt of the cation. (Section 18-11; Exercises 96 and 97)

15. Recognize and solve problems of various types from this chapter. (Mixed Exercises, 98 through 106)

16. Answer conceptual questions based on this chapter. (Conceptual Exercises 107 through 111)

17. Apply concepts and skills from earlier chapters to the ideas of this chapter. (Building Your Knowledge exercises, 112 through 115)

Some Important Terms in This Chapter

Many of the important terms in this chapter have been encountered before, especially in Chapters 4 and 10. Write the meanings of the following terms *in your own words*. Check the Key Terms list and the chapter reading. Then rewrite your definitions, still in your own words, to improve them. Study other new terms, and review as needed.

Brønsted-Lowry acid

Brønsted-Lowry base

conjugate acid-base pair

ionization

autoionization

ionization constant

ion product for water

pH

pOH

monoprotic acid

polyprotic acid

solvolysis

hydrolysis

hydrolysis constant

conjugate acid-base pair

Preliminary Test

This test will check your understanding of basic concepts and types of calculations. Be sure to practice *many* of the additional textbook Exercises, including those indicated with the Study Goals.

Short Answer

1. Strong electrolytes _____ electricity better than weak electrolytes in aqueous solution, because they are more completely _____.
2. The seven common strong acids are _____, _____, _____, _____, _____, _____, and _____. (Give name and formula for each.)
3. The eight strong soluble bases are _____, _____, _____, _____, _____, and _____. (Give name and formula for each.)
4. The five acids listed in Table 18-4 of the text, arranged according to increasing acid strength, are _____.
5. In Table 18-4 of the text, there are five bases listed (excluding water). Arranged according to increasing strength, these bases are _____.
6. The common weak bases discussed in this chapter are the compound _____ and several of its organic derivatives, called _____.
7. The ionization constant for a weak base is represented by the symbol _____.
8. The symbol for the ionization constant of a weak acid is _____.
9. The weaker the acid, the _____ the percent ionization of the acid in solution.
10. The definition of pH is _____.

11. In a solution of pH 9.0, $[H_3O^+]$ = _____.
12. In a solution of pH 9.3, $[H_3O^+]$ = _____.
13. In any aqueous solution, as $[H_3O^+]$ increases, $[OH^-]$ _____.
14. In any aqueous solution at 25°C, _____ = 10^{-14} and _____ = 14.
15. In 0.10 M HCl, $[H_3O^+]$ = _____ M.
16. In 0.10 M HCl, $[OH^-]$ = _____ M.
17. The pH of a solution in which $[H_3O^+]$ = 1.0×10^{-5} is _____.
18. The pH of a solution in which $[H_3O^+]$ = 2.0×10^{-5} is _____.
19. In a solution with pH = 3.0, pOH = _____.
20. In the solution of Question 19, $[H_3O^+]$ = _____ M and $[OH^-]$ = _____ M.
21. The solution of Question 19 would be described as _____ (acidic, basic, neutral).
22. By definition, an acidic solution is one in which $[H_3O^+]$ ____ $[OH^-]$, a basic solution is one in which $[H_3O^+]$ ____ $[OH^-]$, and a neutral solution is one in which $[H_3O^+]$ ____ $[OH^-]$.
23. If the concentration of a monoprotic strong acid is x molar, then the concentration of hydronium ions, $[H_3O^+]$, will be equal to _____ M.
24. In a solution of RbOH that is x molar, $[Rb^+]$ = _____ M and $[OH^-]$ = _____ M.
25. In a solution of Ba(OH)$_2$ that is x molar, $[Ba^{2+}]$ = _____ M and $[OH^-]$ = _____ M.
26. In a 0.10 M solution of any monoprotic strong acid, pH = 1.0; in a 0.10 molar solution of any monoprotic weak acid, pH is _____ (greater than, less than, equal to) 1.0.
27. In a solution containing only a weak monoprotic acid HA, $[H_3O^+]$ is _____ (greater than, less than, equal to) $[A^-]$; if the solution is not very dilute, the concentration of nonionized HA is approximately equal to the _____ of the solution.
28. H_3PO_4 is a _____ acid than $H_2PO_4^-$, and $H_2PO_4^-$ is a _____ acid than HPO_4^{2-}.
29. For all diprotic weak acids, K_{a1} ____ K_{a2}.
30. The reaction of a substance with the solvent in which it is dissolved is called _____; when the solvent is water, these are called _____ reactions.
31. The stronger an acid, the _____ is its conjugate base.
32. There are two types of salts that lead to neutral solutions when dissolved in water. These are (1) salts of _____ acids and strong bases and (2) salts of _____ acids and _____ bases, but only in the latter case if _____ are the same.
33. One reaction whose equilibrium is always disturbed when a hydrolyzable salt is dissolved in water is _____.
34. The hydrolysis reaction that occurs when ammonium chloride is dissolved in water is _____ .
35. The hydrolysis reaction that occurs when potassium fluoride is dissolved in water is _____ .
36. HX is a weak acid. When a salt such as NaX is dissolved in water, the net hydrolysis reaction can be represented as _____.
37. When the salt of a strong base and a strong acid is dissolved in water, the resulting solution is _____. An example would be _____.
38. When the salt of a strong base and a weak acid is dissolved in water, the resulting solution is _____. An example would be _____.
39. When the salt of a weak base and a strong acid is dissolved in water, the resulting solution is _____. An example would be _____.
40. An example of a salt of a weak base and a weak acid is _____.

41. Whether the solution of a salt of a weak acid and a weak base is acidic, basic, or neutral depends on the value of _____ and on the value of
_____ .

42. The value of K_a for benzoic acid, C_6H_5COOH, is 6.3×10^{-5}. When a salt such as sodium benzoate is dissolved in water, the reaction that occurs is _____.

43. The value of K_b for methylamine, $(CH_3)NH_2$, is 5.0×10^{-4}. When we dissolve methyl-ammonium chloride in water, the reaction _____ occurs, leading to a solution that is _____. The value of the equilibrium constant for this reaction is _____.

44. When a salt such as methylammonium benzoate is dissolved in water, the two hydrolysis reactions that occur are _____ and _____. The resulting solution would be _____ because _____
_____.

45. The _____ of the ionization constant for an acid and the hydrolysis constant for its conjugate base is equal to K_w.

46. When a salt of a small, highly charged metal ion such as Al^{3+} is dissolved in water, the resulting solution is _____.

47. Comparing isoelectronic cations in the same period of the periodic table, the ease of extent of hydrolysis increases as the _____ increases.

- -

In questions 48 through 70, classify each of the following as a salt of a *strong*, *weak*, or *insoluble* base and a *strong* or *weak* acid. Tell whether an aqueous solution of each would be *acidic*, *basic*, or *neutral*. It should usually not be necessary to refer to tables of ionization constants, but you may find them helpful in a few cases. Write the equation for any hydrolysis reaction

| Salt | Salt of | | Nature of | Hydrolysis Reaction |
	Base	Acid	Solution	
48. KCH_3COO	____	____	____	_____
49. RbI	____	____	____	_____
50. $CaBr_2$	____	____	____	_____
51. NH_4Cl	____	____	____	_____
52. $(CH_3)_2NH_2Cl$	____	____	____	_____
53. K_3AsO_4	____	____	____	_____
54. $Ba(NO_2)_2$	____	____	____	_____
55. $Ba(NO_3)_2$	____	____	____	_____
56. NaF	____	____	____	_____
57. NaCl	____	____	____	_____
58. NaBr	____	____	____	_____
59. $NaClO_4$	____	____	____	_____

60. Na_2CO_3 ____ ____ ____ _____

61. NH_4ClO_4 ____ ____ ____ _____

62. KCN ____ ____ ____ _____

63. NH_4Br ____ ____ ____ _____

64. $AlCl_3$ ____ ____ ____ _____

65. $Fe(NO_3)_3$ ____ ____ ____ _____

66. $BiCl_3$ ____ ____ ____ _____

67. K_2S ____ ____ ____ _____

68. $BeCl_2$ ____ ____ ____ _____

69. NH_4CN ____ ____ ____ _____

70. NH_4OCN ____ ____ ____ _____

Multiple Choice

____ 1. Which of the following is a weak electrolyte in aqueous solution?
     (a) $HClO_4$ (b) NaOH   (c) NaCl   (d) HNO_2 (e) $Ca(NO_3)_2$

____ 2. Ammonia dissolves in water by the reaction

$$NH_3(g) + H_2O(l) \rightleftharpoons NH_4^+(aq) + OH^-(aq)$$

We would expect the resulting solution to be
     (a) acidic.   (b) neutral.   (c) basic.

____ 3. A solution in which the pH is 12.5 would be described as
     (a) very basic.   (b) slightly basic.   (c) neutral.   (d) slightly acidic.   (e) very acidic.

____ 4. We add enough base to a solution to cause the pH to increase from 7.5 to 8.5. This means that
     (a) $[OH^-]$ increases by a factor of 10.   (b) $[H_3O^+]$ increases by a factor of 10.
     (c) $[OH^-]$ increases by 1 M.   (d) $[H_3O^+]$ increases by 1 M.
     (e) $[OH^-]$ increases by a factor of 8.5/7.5.

____ 5. Calculate the pH of a 0.04 M HNO_3 solution.
     (a) 2.4   (b) 12.5   (c) 1.4   (d) 1.6   (e) 4.0×10^{-2}

____ 6. What would be the $[OH^-]$ concentration in the solution of Question 5?
     (a) $2.5 \times 10^{-10} M$ (b) $12.6 M$ (c) $1.4 M$ (d) $2.5 \times 10^{-13} M$ (e) $0 M$

____ 7. For a given acid HA, the value of K_a
     (a) will change with pH.   (b) cannot be less than 10^{-7}.
     (c) cannot be greater than 10^{-7}.   (d) does not change with temperature.
     (e) None of the preceding answers is correct.

Hypobromous acid, HBrO, is a weak acid with an acid dissociation constant of 2.5×10^{-9}. Questions 8 through 10 refer to a 0.010 M HBrO solution.

_____ 8. Calculate the value of $[H_3O^+]$ in this solution.
   (a) $5.0 \times 10^{-6} M$ (b) $5.0 \times 10^{-5} M$ (c) $2.5 \times 10^{-7} M$
   (d) $2.5 \times 10^{-11} M$ (e) $5.0 \times 10^{-7} M$

_____ 9. What is the pH of this solution?
   (a) 3.5   (b) 4.7   (c) 5.3   (d) 5.7   (e) 6.3

_____ 10. What is the value of $[OH^-]$ in this solution?
   (a) $2.0 \times 10^{-9} M$ (b) $2.0 \times 10^{-8} M$ (c) $1.0 \times 10^{-7} M$
   (d) $5.0 \times 10^{-6} M$ (e) $5.0 \times 10^{-5} M$

_____ 11. In a sample of pure water, only one of the following statements is *always* true at all conditions of temperature and pressure. Which one is *always* true?
   (a) $[H_3O^+] = 1.0 \times 10^{-7} M$ (b) $[OH^-] = 1.0 \times 10^{-7} M$
   (c) pH $= 7.0$ (d) pOH $= 7.0$ (e) $[H_3O^+] = [OH^-]$

_____ 12. Which of the following expressions correctly describes the relationship between pH and pOH in any aqueous solution at 25°C?
   (a) $(pH)(pOH) = 10^{-14}$ (b) $pH/pOH = 10^{-14}$ (c) $pOH/pH = 10^{-14}$
   (d) pH $+$ pOH $= 14$ (e) pH $-$ pOH $= 14$

_____ 13. Which of the following expressions correctly describes the relationship between $[H_3O^+]$ and $[OH^-]$ in any aqueous solution at 25°C?
   (a) $[H_3O^+][OH^-] = 10^{-14}$ (b) $[H_3O^+]/[OH^-] = 10^{-14}$
   (c) $[OH^-]/[H_3O^+] = 10^{-14}$ (d) $[H_3O^+] + [OH^-] = 14$
   (e) $[H_3O^+] - [OH^-] = 14$

_____ 14. In water solution, HCN is only very slightly ionized according to

$$HCN + H_2O \rightleftharpoons H_3O^+ + CN^-$$

This observation shows that CN^- is a _____ than H_2O.
   (a) stronger acid   (b) stronger base   (c) weaker acid   (d) weaker base

_____ 15. When solid NaCN is added to water,
   (a) the pH remains at 7.
   (b) the pH becomes greater than 7 because of hydrolysis of Na^+.
   (c) the pH becomes less than 7 because of hydrolysis of Na^+.
   (d) the pH becomes greater than 7 because of hydrolysis of CN^-.
   (e) the pH becomes less than 7 because of hydrolysis of CN^-.

_____ 16. When the salt NH_4CH_3COO is dissolved in water, the resulting solution is
   (a) acidic.   (b) basic.   (c) neutral.

_____ 17. When the salt NH_4NO_2 is dissolved in water, the resulting solution is
   (a) acidic.   (b) basic.   (c) neutral.

_____ 18. Each of the following salts is completely ionized when dissolved in water. Which of these salts will produce a basic solution when dissolved in water?
   (a) sodium chloride   (b) sodium acetate   (c) ammonium chloride
   (d) calcium nitrate   (e) rubidium perchlorate

230

_____ 19. Which of the following salts will produce an acidic solution when dissolved in water?
       (a) sodium chloride   (b) sodium acetate   (c) ammonium chloride
       (d) calcium nitrate   (e) rubidium perchlorate

_____ 20. The value of K_a for acetic acid is 1.8×10^{-5}. Find the value of K for

$$CH_3COO^- + H_2O \rightleftharpoons CH_3COOH + OH^-$$

    (a) 1.8×10^{-5} (b) 1.8×10^{-10} (c) 1.0×10^{-14} (d) 5.6×10^{-10} (e) 5.6×10^{-5}

Be sure to work many of the text exercises involving weak acid equilibria and salt hydrolysis.

Answers to Preliminary Test

Do not just look up answers. Think about the reasons for the answers.

Short Answer

1. conduct; ionized (dissociated)
2. hydrochloric, HCl; hydrobromic, HBr; hydroiodic, HI; nitric, HNO_3; perchloric, $HClO_4$; chloric, $HClO_3$; sulfuric, H_2SO_4. It may help to remember that three of these are binary acids of halogens. Remember that HF is a weak acid.
3. lithium hydroxide, LiOH; sodium hydroxide, NaOH; potassium hydroxide, KOH; rubidium hydroxide, RbOH; cesium hydroxide, CsOH; calcium hydroxide, $Ca(OH)_2$; strontium hydroxide, $Sr(OH)_2$; barium hydroxide, $Ba(OH)_2$.
4. HCN < HClO < CH_3COOH < HNO_2 < HF. Smaller values of K_a mean weaker acids.
5. F^- < NO_2^- < CH_3COO^- < ClO^- < CN^-. Remember that the stronger the acid, the weaker its conjugate base.
6. ammonia; amines
7. K_b
8. K_a
9. smaller
10. $-\log [H_3O^+]$
11. 10^{-9}
12. 5.0×10^{-10}
13. decreases
14. $[H_3O^+][OH^-]$; pH + pOH
15. 0.10. Remember that HCl is a strong acid and is completely ionized.
16. 1.0×10^{-13}. $[H_3O^+] = 0.10$ (from preceding question) and $[OH^-] = K_w/[H_3O^+]$.
17. 5.00
18. 4.70
19. 11.0
20. 10^{-3}; 10^{-11}
21. acidic
22. >; <; =
23. x. Strong acids are completely dissociated, so x moles of a monoprotic strong acid would yield x moles of H_3O^+ on dissociation.

24. x; x. RbOH is a strong soluble base.
25. x; $2x$. Ba(OH)$_2$ is a strong soluble base that gives 2 moles OH$^-$ for each mole dissolved.
26. greater than. In a $0.10\ M$ weak acid solution, [H$_3$O$^+$] would be less than 0.10, so pH would be greater than 1.00
27. equal to; molarity
28. stronger; stronger
29. $>$
30. solvolysis; hydrolysis
31. weaker
32. (1) strong; (2) weak; weak; the strengths of the parent acid and base
33. $2H_2O \rightleftharpoons H_3O^+ + OH^-$
34. $NH_4^+ + H_2O \rightleftharpoons NH_3 + H_3O^+$. NH$_4^+$ is the conjugate acid of the weak base NH$_3$.
35. $F^- + H_2O \rightleftharpoons HF + OH^-$. F$^-$ is the conjugate base of the weak acid HF.
36. $X^- + H_2O \rightleftharpoons HX + OH^-$.
37. neutral. The example discussed in the chapter is NaCl. There are many other correct answers, such as NaNO$_3$, KBr, RbClO$_4$, and CaI$_2$.
38. basic. Examples discussed in the chapter include NaCH$_3$COO and NaCN. Many other examples are possible.
39. acidic. Among the examples in this chapter are NH$_4$NO$_3$ and NH$_4$Cl.
40. Several examples are listed in the chapter: NH$_4$CH$_3$COO, NH$_4$CN, and NH$_4$F.
41. K_a for the acid; K_b for the base from which the salt is derived.
42. $C_6H_5COO^- + H_2O \rightleftharpoons C_6H_5COOH + OH^-$; basic; 1.6×10^{-10}. Remember that the hydrolysis constant, K_b, for the conjugate base of a weak acid with ionization constant K_a is given by $K_aK_b = K_w$.
43. $(CH_3)NH_3^+ + H_2O \rightleftharpoons (CH_3)NH_2 + H_3O^+$; acidic; 2.0×10^{-11}. The hydrolysis constant, K_a, for the conjugate acid of a weak base with ionization constant K_b is given by the relationship $K_aK_b = K_w$.
44. $C_6H_5COO^- + H_2O \rightleftharpoons C_6H_5COOH + OH^-$;
    $(CH_3)NH_3^+ + H_2O \rightleftharpoons (CH_3)NH_2 + H_3O^+$; basic; $K_{base} > K_{acid}$. Be sure that you understand how to reason this out (Section 18-10). Compare the K values for the hydrolyses that you calculated in Questions 42 and 43. From these you can see that the hydrolysis of benzoate occurs (slightly) more readily than the hydrolysis of methylammonium ion. This makes the solution slightly basic.
45. product
46. acidic. Section 18-11 discusses the hydrolyses of such ions. Note that this type of reaction can only cause the solution to become acidic, never basic.
47. charge.
48. strong; weak; basic; $CH_3COO^- + H_2O \rightleftharpoons CH_3COOH + OH^-$
49. strong; strong; neutral; none
50. strong; strong; neutral; none
51. weak; strong; acidic; $NH_4^+ + H_2O \rightleftharpoons NH_3 + H_3O^+$
52. weak; strong; acidic; $(CH_3)_2NH_2^+ + H_2O \rightleftharpoons (CH_3)_2NH + H_3O^+$
53. strong; weak; basic; $AsO_4^{3-} + H_2O \rightleftharpoons HAsO_4^{2-} + OH^-$. Probably HAsO$_4^{2-}$ hydrolyzes further, but the reaction shown is the predominant one.
54. strong; weak; basic; $NO_2^- + H_2O \rightleftharpoons HNO_2 + OH^-$
55. strong; strong; neutral; none

56. strong; weak; basic; $F^- + H_2O \rightleftharpoons HF + OH^-$
57. strong; strong; neutral; none
58. strong; strong; neutral; none
59. strong; strong; neutral; none
60. strong; weak; basic; $CO_3^{2-} + H_2O \rightleftharpoons HCO_3^- + OH^-$. This reaction is accompanied by some hydrolysis of HCO_3^-.
61. weak; strong; acidic; $NH_4^+ + H_2O \rightleftharpoons NH_3 + H_3O^+$
62. strong; weak; basic; $CN^- + H_2O \rightleftharpoons HCN + OH^-$
63. weak; strong; acidic; $NH_4^+ + H_2O \rightleftharpoons NH_3 + H_3O^+$
64. insoluble; strong; acidic; $Al^{3+} + 2H_2O \rightleftharpoons Al(OH)^{2+} + H_3O^+$
65. insoluble; strong; acidic; $Fe^{3+} + 2H_2O \rightleftharpoons Fe(OH)^{2+} + H_3O^+$
66. insoluble; strong; acidic; $Bi^{3+} + 2H_2O \rightleftharpoons Bi(OH)^{2+} + H_3O^+$
67. strong; weak; basic; $S^{2-} + H_2O \rightleftharpoons HS^- + OH^-$. The HS^- ion also hydrolyzes to some extent.
68. insoluble; strong; acidic; $Be^{2+} + 2H_2O \rightleftharpoons Be(OH)^+ + H_3O^+$
69. weak; weak; basic; $NH_4^+ + H_2O \rightleftharpoons NH_3 + H_3O^+$ *and*
    $CN^- + H_2O \rightleftharpoons HCN + OH^-$. K_b for NH_3 is greater than K_a for HCN, so the K_a for NH_4^+ is less than K_b for CN^-. Thus the second hydrolysis reaction occurs to a greater extent, making the solution *basic*. See the discussion in Section 18-10.
70. weak; weak; acidic; $NH_4^+ + H_2O \rightleftharpoons NH_3 + H_3O^+$ *and*
    $OCN^- + H_2O \rightleftharpoons HOCN + OH^-$. Now K_b for NH_3 is less than K_a for HOCN, so K_a for NH_4^+ is greater than K_b for CN^-. Thus the first hydrolysis reaction occurs to a greater extent, making the solution *acidic*. See the discussion in Section 18-10.

Multiple Choice

1. (d). Remember the lists of strong acids and strong bases. Also review the solubility rules. HNO_2 is a weak acid, so it is only slightly dissociated into ions in water solution.
2. (c)
3. (a)
4. (a)
5. (c). Because HNO_3 is a strong acid, $[H_3O^+] = 0.04$; $pH = -\log(0.04) = 1.4$
6. (d)
7. (e)
8. (a). This must be solved by the methods in Section 18-4. See Example 18-10.
9. (c)
10. (a)
11. (e). The others all depend on $K_w = 10^{-14}$, and this is only true at 25°C. Remember that K_w, like all equilibrium constants, changes with temperature. See Table 18-2.
12. (d)
13. (a)
14. (b)

15. (d). The hydrolysis reaction $CN^- + H_2O \rightleftharpoons HCN + OH^-$ occurs. This raises $[OH^-]$, lowers $[H_3O^+]$, and thus raises pH. Reasoning this way, it is not necessary to carry out the calculation of pH to find the answer.

16. (c). Remember that this is the only common case of a weak base-weak acid salt in which the acid and base strengths are essentially equal. See Section 18-10.

17. (a). K_a for HNO_2 (4.5×10^{-4}) is greater than K_b for NH_3 (1.8×10^{-5}). Thus NH_4^+ would hydrolyze to a greater extent than NO_2^-, and the resulting solution would be acidic. See Section 18-10.

18. (b)

19. (c)

20. (d). This is the hydrolysis reaction of the base CH_3COO^-. For any conjugate acid-base pair, $K_a K_b = K_w = 1.0 \times 10^{-14}$.

19 Ionic Equilibria II: Buffers and Titration Curves

Chapter Summary

LeChatelier's Principle (Chapter 17) tells us that the addition from an outside source of one of the products to a reaction at equilibrium results in "shifting the equilibrium to the left," i.e., tending to suppress the reaction. When the reaction is the ionization of a weak acid (or base) and the product added is the conjugate base (or acid) from an appropriate soluble salt, this shift is referred to as the **common ion effect** (Section 19-1). The resulting solution, which is able to absorb modest amounts of acid or base and thus to resist changes in pH, is called a **buffer solution.**

The buffering action (Section 19-2) of such a solution may be understood as follows. Consider a solution of an acid, either weak or strong, to which we then add some additional acid from an external source. If the original solution contained only a strong acid, its conjugate base would be far too weak to react with the added acid, so the pH would change markedly. If the original solution contained a weak acid, its conjugate base would be strong, but there would be only a tiny bit of that strong base present to react with the added acid, so the pH still would change markedly. But suppose the original solution had been made by including *both* the weak acid HA *and* a separate source of its conjugate base A⁻ (Section 19-2). There would then be high concentrations of both the undissociated acid and its conjugate base. This solution would be able to absorb added acid by reacting with the base A⁻. On the other hand, if we introduced a little OH⁻ the weak acid HA could ionize additionally to provide more H_3O^+. Thus, such a solution is far more resistant to large pH changes than most other solutions. A solution made by dissolving a weak base and a separate source of its conjugate acid shows similar buffering action. Buffer solutions of desired pH are made either by mixing appropriate solutions or by dissolving the appropriate salt in an acidic or in a basic solution (Section 18-8).

The calculations of Sections 19-1 through 19-3 will help you to understand buffer solutions more clearly. A very useful relationship for use in buffer calculations is the **Henderson-Hasselbalch equation** for acid/salt buffers (and the analogous equation for base/salt buffers later in the section). In application of this equation and related ones, the concentrations to be used are the total concentrations that were added of weak acid, weak base, or salt. But do not be content with plugging values into equations. Be sure you understand how buffer solutions work and the chemical reactions and equilibrium calculations associated with them. Then you will better understand the nature of acid-base equilibria in aqueous solution.

Acid-base indicators are discussed in Section 19-4. As you see, these useful substances are either weak acids or weak bases that undergo a marked color change when they are converted to their conjugate bases or conjugate acids, respectively.

As we slowly add a solution of a strong base to a solution of a strong acid **(titration)** we are gradually neutralizing the acid. Before a stoichiometric amount of base has been added, the solution is still acidic; at the **equivalence point,** the acid and base have just neutralized each other; and beyond this point the solution is basic. Section 19-5 shows how we can represent this information quantitatively in a graph that we call a **titration curve.** We see in this section how an indicator may provide a visual signal when the end point of a titration has been reached. Of course, an analogous discussion would apply to the titration of a strong base by a strong acid solution. The calculations in this section provide useful practice in stoichiometry.

Now that we understand hydrolysis reactions, we can consider the titration curves that describe the **neutralization of weak acids with strong bases** (Section 19-6). We see that before the equivalence point, a buffer is formed so we can calculate pH as we did in Section 19-1. At the equivalence point, the solution is that of a strong base-weak acid salt, so the calculations can be done as in Section 18-8. Similar considerations apply to **titration of a weak base with a strong acid.**

Each of these two types of titration curve can be considered in four parts. Be sure you see how pH calculations in Sections 19-5 and 19-6 fall into the four categories listed in the Problem-Solving Tip on page 742. The pH changes in weak acid/weak base titration curves are far less extreme than when one component is strong (Section 19-).

The various acidic or basic solutions of Chapters 18 and 19 are summarized in Section 19-8. Study Table 19-7 carefully.

Study Goals

1. Describe the common ion effect in terms of LeChatelier's Principle. Use this effect to understand and explain buffering action. (Section 19-1; Exercises 5 through 8, 12, and 18 through 20)
2. Know how buffer solutions are prepared. Be able to calculate $[H_3O^+]$, $[OH^-]$, pH, and pOH in buffer solutions of the acid/salt type and the base/salt type. (Sections 19-2 and 19-3; Exercises 9 through 11, 13 through 17, and 30 through 37)
3. Be able to calculate the changes in the quantities of Study Goal 12 when specified amounts of acid or base are added to the buffer solution. (Section 19-2; Exercises 21 through 29)
4. Describe why acid-base indicators change color over a definite pH range and why this range differs for different indicators. (Section 19-4; Exercises 38 through 45)
5. Calculate concentrations and pH values of solutions resulting from the mixture of any volumes of specified concentrations of a strong acid and a strong base. (Alternatively, the acid or the base may be added as a pure substance.) Use the results of these calculations to plot titration curves for titrations of strong acids and strong bases. (Section 19-5; Exercises 46 through 50)
6. Calculate concentrations and pH values of solutions resulting from the mixing of any volumes of any concentrations of acids and bases when one of these may be weak. Plot titration curves for weak acid/strong base or weak base/strong acid titrations. (Section 19-6; Exercises 51 through 56)
7. Recognize and answer a variety of questions based on the concepts in this chapter. (Mixed Exercises, 57 through 60)
8. Answer conceptual questions based on this chapter. (Conceptual Exercises 61 through 64)
9. Apply concepts and skills from earlier chapters to the ideas of this chapter. (Building Your Knowledge exercises, 65 through 69)

Some Important Terms in This Chapter

Write the meanings *in your own words*. Check the Key Terms list and the chapter reading. Then rewrite your definitions, still in your own words, to improve them. Study other new terms, and review terms from preceding chapters if necessary.

common ion effect

buffer solution

acid-salt buffer

base-salt buffer

Henderson-Hasselbalch equation
   a. for acid-salt buffer

   b. for base-salt buffer

indicator

titration

titration curve

equivalence point

end point

Preliminary Test

This test will check your understanding of basic concepts and types of calculations. Be sure to practice *many* of the additional textbook exercises, including those indicated with the Study Goals.

Short Answer

Answer with a word, a phrase, a formula, or a number with units as necessary.

1. In a solution that is 0.10 M in acetic acid and 0.10 M in sodium acetate, the [H_3O^+] is _____ (greater than, less than, equal to) the [H_3O^+] in a solution of 0.10 M acetic acid.
2. There are two common kinds of buffer solutions; one is made by dissolving _____ and _____, and the other is made by dissolving _____ and _____.
3. The utility of a buffer solution is that it _____ when moderate amounts of acid or base are added to the solution.
4. The Henderson-Hasselbalch equation for pH in a weak acid buffer is _____.
5. An acid-base indicator changes color when the _____ changes sufficiently.
6. For most indicators, one color is distinct if the ratio of [In^-]/[HIn] is greater than about _____ and the other color is seen distinctly if this ratio is less than about _____; thus most indicators change color over a range of about _____ pH units.
7. After the start of titration of a weak acid with a strong base, but before the equivalence point is reached, the solution acts as a _____.
8. A procedure in which a solution containing one reactant is slowly added to a solution of unknown concentration of another reactant until the reaction between them is complete is called _____.
9. The point in a titration at which chemically equivalent amounts of acid and base have reacted is called the _____; the point at which the indicator changes color is known as the _____.
10. We can describe the titration curve for the reaction in which a strong acid solution is titrated by a strong base as follows: first there is a region in which the pH is relatively constant at a _____ value, followed by a region where the pH _____, and then a region where the pH is relatively constant at a _____ value.

Multiple Choice

_____ 1. A buffer solution contains 0.56 M acetic acid and 1.00 M sodium acetate. K_a for acetic acid is 1.8×10^{-5}. The [H_3O^+] in this solution would be closest to which of the following?
   (a) $10^{-3} M$ (b) $10^{-4} M$ (c) $10^{-5} M$ (d) $10^{-6} M$ (e) $10^{-7} M$

_____ 2. The acid form of an indicator is yellow and its anion is blue. The K_a value for this indicator is 10^{-5}. What will be the approximate pH range over which this indicator changes color?
   (a) 3–5   (b) 4–6   (c) 5–7   (d) 8–10   (e) 9–11

_____ 3. What will be the color of the indicator in Question 14 in a solution of pH 3?
   (a) red   (b) orange   (c) yellow   (d) green   (e) blue

____ 4. Which of the following indicators would be most appropriate to use for the detection of the equivalence point of the titration of the weak base ammonia with a strong acid such as HCl?
    (a) methyl red   (b) neutral red   (c) phenolphthalein
    (d) Any of the preceding would be equally useful.
    (e) None of the indicators mentioned would be suitable.

____ 5. At the equivalence point in the titration of 0.10 M acetic acid with 0.20 M sodium hydroxide, which of the following would be true?
    (a) pH = 7.
    (b) For every 1 mL of acid used, 2 mL of base will have been used.
    (c) For every 2 mL of acid used, 1 mL of base will have been used.
    (d) For every 1 mL of acid used, 1 mL of base will have been used.
    (e) For every 1 g of acid used, 2 g of base will have been used.

____ 6. When we titrate a strong acid with a strong base, the pH at the equivalence point is
    (a) 7.   (b) less than 7.   (c) greater than 7.
    (d) Cannot be predicted unless we know which acid and base are used.

____ 7. When we titrate a weak base with a strong acid, the pH at the equivalence point is
    (a) 7.   (b) less than 7.   (c) greater than 7.
    (d) Any of the preceding answers may be correct, depending on the K_b value of the base.

____ 8. When we titrate a weak acid with a strong base, the pH at the equivalence point is
    (a) 7.   (b) less than 7.   (c) greater than 7.
    (d) Any of the preceding answers may be correct, depending on the K_a value of the acid.

____ 9. When we titrate a weak base with a weak acid, the pH at the equivalence point is
    (a) 7.   (b) less than 7.   (c) greater than 7.
    (d) Any of the preceding answers may be correct, depending on the K_a value of the acid and the K_b value of the base.

> Be sure to work many of the textbook exercises involving buffers and titrations.

Answers to Preliminary Test

Do not just look up answers. Think about the reasons for the answers.

Short Answer

1. less than. No calculations are needed; think about the common ion effect. Of course you could do these two calculations if you wish. (See Sections 18-4 and 19-1.)
2. a weak acid; a soluble salt of this weak acid; a weak base; a soluble salt of this weak base.
3. resists changes in pH
4. $pH = pK_a + \log\frac{[salt]}{[acid]}$
5. pH (or $[H_3O^+]$)
6. 10; 0.1; 2.0
7. buffer
8. titration
9. equivalence point; end point
10. low; increases rapidly over a very small volume of base added; high

Multiple Choice

1. (c). The calculation is done as shown in Section 19-3.
2. (b). An indicator changes color when $[In^-]/[HIn] = 1$ (Section 19-4). This condition is satisfied when $[H_3O^+] = K_a$. For this indicator, this is at $[H_3O^+] = 10^{-5}\ M$, or pH = 5. The color change becomes discernible from about 1 pH unit below this to about 1 pH unit above it, or from pH 4 to pH 6.
3. (c). At pH = 3, there is considerable excess H_3O^+. In this excess H_3O^+, the indicator equilibrium is shifted to less dissociation, or predominantly to the form HIn. This form is yellow.
4. (a) From Figure 19-5 we see that the equivalence point occurs at about pH 5. Of the indicators listed, methyl red has a color change at a pH nearest this. See Table 19-4.
5. (c). If you thought the answer was (a), study Section 19-6. If you picked one of the other incorrect answers, think more carefully about the stoichiometry and about the relation between number of moles and volume of solution.
6. (a). See Section 19-5.
7. (b). See Figure 19-5 and the associated discussion in Section 19-6.
8. (c). See Figure 19-4 and the associated discussion in Section 19-6.
9. (d)

20 Ionic Equilibria III: The Solubility Product Principle

Chapter Summary

This chapter continues our study of equilibria in aqueous solutions. Here we consider equilibria involving *slightly soluble ionic compounds*. This is in many ways the simplest of the types of ionic equilibria. It is the basis for many useful laboratory techniques, such as separation by fractional precipitation and the dissolving of precipitates. Recall from your study of acid-base equilibria that some acids are so strong (completely ionized) that equilibrium considerations do not apply. In similar fashion, the equilibrium descriptions of this chapter do not apply to the dissolving of soluble substances. (But you will need to know how to find the concentrations of ions in such solutions.) Before you begin your study of this material, be sure you *know* the solubility guidelines, Section 4-2.5 and Table 4-8. Be aware, however, that the substances classified in those rules as "insoluble" are actually very slightly soluble. It is these types of compounds that we are considering in Chapter 20.

As you shall see, the reactions studied here can be viewed in either of two ways: as the dissolving of slightly soluble substances to form ions in solution (**dissolution**) or as the ions in solution coming together to form ionic solids (**precipitation**). From either point of view, the equation for the reaction may be written as **solid $\rightleftharpoons$ dissolved ions.** We can define an equilibrium constant for this equilibrium process (Section 20-1). As we saw in our discussion in Chapter 18 of heterogeneous equilibria, the solid does not appear in the expression for the equilibrium constant. This equilibrium constant is given the symbol K_{sp} and is referred to as the **solubility product constant,** or sometimes simply the **solubility product.** The remainder of the chapter contains many applications of the **solubility product principle.**

In Section 20-2, we see how the value of K_{sp} for any slightly soluble substance is determined from an experimental measurement of its solubility. In carrying out these calculations, remember the following points: (1) The measured solubility, which is often stated in terms of grams of substance dissolving to make a specified volume of **saturated** solution, must be converted to a molar basis. (2) You must use the stoichiometry of the dissolution process to find out the concentration in moles/liter of each ion present. Solubilities of solids in liquid are usually quite dependent on temperature, so values of K_{sp} can vary considerably as temperature changes.

Several uses of K_{sp} are detailed in Section 20-3. Pay close attention to the reasoning in the examples in this section. All subsequent sections of this chapter will apply the calculations you learn about in Section 20-3. The major uses of the solubility product principle are these: (1) *To calculate the solubility,* in terms of moles or grams of a substance that will dissolve to make a specified amount of saturated solution. This calculation can also let us find the concentration of each ionic species in the saturated solution. (2) *To calculate the concentrations of ions that can exist together in solution.* A very useful extension of this is deciding whether or not a precipitate will form at specified concentrations. The reaction quotient (Section 17-4), here denoted as Q_{sp}, is very useful in arriving at this decision. (3) *To calculate the concentration of an ion necessary to initiate precipitation* of a substance from a solution that contains other ions. (4) *To calculate the concentrations of ions remaining in solution* after precipitation has occurred. This last type of calculation is useful in predicting the efficiency of precipitation methods of recovering valuable metals and removing objectionable ions from solution.

One application of the calculations from Section 20-3 is in designing methods to separate the ions in a mixture in solution. This technique, called **fractional precipitation** (Section 20-4) makes use of K_{sp} differences among salts having an ion in common. Examples 20-8 and 20-9 illustrate this with the silver halides. Notice which types of calculations from Section 20-3 we are using here.

An important aspect of chemical equilibrium is that in a mixture, *all species must be at equilibrium with respect to all relevant reactions.* For example, this principle was used throughout the discussions of Chapters 18 and 19, where the equilibrium involving the dissociation of water and the resulting $[H_3O^+]/[OH^-]$ ratio was affected by other reactions taking place in the same solution. This principle is emphasized again in Section 20-5 for simultaneous equilibria that involve slightly soluble compounds. Such calculations usually involve one species that is involved in more than one reaction. That species therefore affects the equilibria of several processes. In Examples 20-10 through 20-12, OH^- is involved both in the equilibrium for the dissolution or precipitation of $Mg(OH)_2$ and also in that for the ionization of NH_3 as a weak base in aqueous solution. Thus, both equilibrium constant expressions must be satisfied. Of course, OH^- is also involved in a third reaction, that with H_3O^+ to form water (and the reverse of this reaction), so that reaction must also be at equilibrium. This equilibrium condition may also be used, for example, to calculate $[H_3O^+]$ or pH of the solution.

Section 20-6, which discusses methods of **dissolving precipitates,** is merely an application of the ideas presented in Section 20-5. Three methods are presented: (1) converting an ion into a weak electrolyte, (2) converting an ion to another species (usually by oxidation or reduction), and (3) forming complex ions. In each method, the essential strategy is to lower the concentration of one of the products (ions) of the dissolution process. In the first and third methods, other equilibria are used to help remove one of the ions from solution; in the second, a less reversible reaction is used, for instance the oxidation of S^{2-} to elemental sulfur. You should see that each of these methods is an application of LeChatelier's Principle. The latter part of Section 20-6 also includes the application of equilibrium principles to still another type of reaction, the **formation/dissociation of complex ions**. We usually express the equilibrium constant for such reactions as K_d, referring to the reaction written as a dissociation.

This chapter brings us to the end of four chapters that deal with chemical equilibrium and its applications. Many different types of reactions have been studied, each with its own characteristic terminology and notation. However, you should recognize that it is the same set of principles that we have been applying—(1) the general formulation of the concentration or pressure relationships in the equilibrium-constant expression and (2) LeChatelier's Principle. Examine your understanding of these chapters by summarizing the different types of reactions or processes we have studied and by comparing the different terminologies we have used to discuss these two principles for the various reaction types.

<u>Study Goals</u>

1. Know the solubility rules. By application of these rules, be able to determine which ionic solids are soluble and which are slightly soluble ("insoluble"). (Review Section 4-2, part 5 and Table 4-8; review Exercises 82 through 92 of Chapter 4)
2. Write the solubility product expression for any slightly soluble compound that ionizes when it dissolves in water. (Section 20-1; Exercises 1 through 6)

3. Calculate the solubility product constant for a compound, given its molar solubility or the amount of the substance that dissolves to give a specified volume of saturated solution. (Section 20-2; Exercises 8 through 13)
4. Given its solubility product constant, calculate the solubility of a compound. Given solubility product constants for a series of slightly soluble compounds, determine which is most and which is least soluble by comparing molar solubilities. (Section 20-3; Exercises 14 through 18 and 21 through 25)
5. Describe and apply the common ion effect as it pertains to solid-dissolved ion equilibria. (Section 20-3; Exercises 19, 20, and 26)
6. Given appropriate solubility product constants, determine whether precipitation will occur (1) when two solutions of given concentrations of specified solutes are mixed, or (2) when a specified amount of a solute is added to a solution of specified concentration of another solute. (Section 20-3; Exercises 27 through 30)
7. Given appropriate solubility product constants, determine how much or what concentration of a given solute must be added to a solution of known concentration of another solute (1) to cause precipitation of a compound to begin and (2) to remove all but a desired amount, concentration, or percent of a particular dissolved ion. (Section 20-3; Exercises 31 through 33)
8. Describe the process of fractional precipitation. Given appropriate solubility product constants, perform related calculations. (Section 20-4; Exercises 34 through 41)
9. Given appropriate equilibrium constants, perform calculations involving determinations like those listed above for systems that simultaneously involve solid-dissolved ion equilibria *and* one or more kinds of other previously studied equilibria. (Section 20-5; Exercises 42 through 53)
10. Describe and give specific examples of three general methods of dissolving precipitates. Discuss each in terms of LeChatelier's Principle. Explain why the processes work. (Section 20-6; Exercises 54 through 59)
11. Recognize and solve various kinds of problems using the concepts in this chapter. (Mixed Exercises, 60 through 63)
12. Answer conceptual questions based on this chapter. (Conceptual Exercises 64 through 68)
13. Apply concepts and skills from earlier chapters to the ideas of this chapter. (Building Your Knowledge exercises, 69 through 72)

Some Important Terms in This Chapter

Write the meanings *in your own words*. Check the Key Terms list and the chapter reading. Then rewrite your definitions, still in your own words, to improve them. Study other new terms, and review terms from preceding chapters (especially Chapters 17 through 19) if necessary.

solubility product principle

solubility product constant

molar solubility

precipitate

Q_{sp}

fractional precipitation

complex ion

dissociation constant

Preliminary Test

This test will check your understanding of basic concepts and type of calculations. It is very important to get much additional practice by working problems, so be sure to do *many* of the textbook Exercises. This is especially true for fractional precipitation and dissolution of precipitates.

Short Answer

1. The number of moles of a substance that will dissolve to give one liter of a saturated solution is called the _____ of the substance.
2. The symbol for the solubility product constant is _____.
3. The expression for the solubility product of $CuCO_3$ is _____.
4. The expression for the solubility product of $Mn(OH)_2$ is _____.
5. The expression for the solubility product of $Mg_3(PO_4)_2$ is _____.
6. When we mix together, from separate sources, the ions of a slightly soluble ionic substance, the salt will begin to precipitate if Q_{sp} _____ K_{sp}, and will continue to precipitate until Q_{sp} _____ K_{sp}.
7. Insoluble metal hydroxides are _____ (more, less) readily soluble in acidic solution than in basic solution.
8. Lead chloride, $PbCl_2$, is much more soluble in warm water than in cold water. This tells us that the dissolution reaction of lead chloride is _____.
9. The dissociation constant for a complex ion in solution is represented by the symbol _____.

10. Three methods for dissolving a precipitate of a slightly soluble ionic substance are

_____,

_____, and

_____.

11. Given the following values of equilibrium constants:

$$K_{sp}(FeS) = 4.9 \times 10^{-18} \quad \text{and} \quad K_d([Fe(CN)_6]^{4-}) = 1.3 \times 10^{-37},$$

complete the following: The equation for the dissolution of FeS(s) in a solution that contains cyanide ions, CN⁻, is _____. The value of K for this reaction is _____, which tells us that it would be _____ to dissolve FeS by addition of a soluble compound such as NaCN that dissociates completely into Na⁺ and CN⁻.

12. Given the following values of equilibrium constants:

$$K_{sp}(CdS) = 3.6 \times 10^{-29} \quad \text{and} \quad K_d([Cd(CN)_4]^{2-}) = 7.8 \times 10^{-18}$$

complete the following: The equation for the dissolution of CdS(s) in a solution that contains cyanide ions, CN⁻, is _____. The value of K for this reaction is _____, which tells us that it would be _____ to dissolve CdS by addition of a soluble compound such as NaCN that dissociates completely into Na⁺ and CN⁻.

13. In order to dissolve a precipitate we must adjust concentrations of the constituent ions in the solution until Q_{sp} is _____ (less than, greater than, equal to) K_{sp}; this is usually done by _____ from the solution.

14. To prevent precipitation of Cu(OH)₂ from a solution that is 0.10 M in Cu(NO₃)₂, the [OH⁻] must be kept lower than _____.

15. Another way to express the result of Question 14 is to say that the pH must be kept _____.

Multiple Choice

_____ 1. Which of the following ionic chlorides is least soluble in water?
(a) NH₄Cl   (b) AgCl   (c) BaCl₂   (d) KCl   (e) NaCl

_____ 2. In a solution obtained by dissolving silver sulfate, Ag₂SO₄, in water, the concentration of silver ion is observed to be 2 x 10⁻⁵ M. What would be the molar concentration of sulfate ion in this solution?
(a) 1 x 10⁻⁵ M (b) 2 x 10⁻⁵ M (c) 3 x 10⁻⁵ M (d) 4 x 10⁻⁵ M
(e) Cannot be answered without knowing the value of K_{sp} for Ag₂SO₄.

_____ 3. The molar solubility of PbSO₄ is 1.14 x 10⁻⁴ mol/L. What is the value of K_{sp} for PbSO₄?
(a) 1.16 x 10⁻¹⁴   (b) 2.68 x 10⁻⁴   (c) 3.45 x 10⁻²
(d) 3.75 x 10⁻⁷   (e) 1.80 x 10⁻⁸

_____ 4. The solubility of PbF₂ in water is 0.0021 mol/L. K_{sp} for PbF₂ is closest to which of the following values?
(a) 2.1 x 10⁻³   (b) 2.0 x 10⁻⁴   (c) 4.4 x 10⁻⁶
(d) 9.2 x 10⁻⁹   (e) 3.7 x 10⁻⁸

_____ 5. AgCl would be least soluble in
(a) pure water.   (b) 0.1 M CaCl₂.   (c) 0.1 M HCl.   (d) 0.1 M HNO₃.
(e) AgCl would be equally soluble in all of these.

_____ 6. The value of K_{sp} for $SrSO_4$ is 2.8 x 10^{-7}. What is the solubility of $SrSO_4$ in mol/L?
(a) 7.6 x 10^{-7} M (b) 1.4 x 10^{-7} M (c) 5.3 x 10^{-4} M
(d) 5.7 x 10^{-3} M (e) 1.3 x 10^{-8} M

_____ 7. The value of K_{sp} for CuI is 1.0 x 10^{-12}. What is the solubility of CuI expressed in grams per liter?
(a) 9.0 x 10^{-6} g/L   (b) 1.0 x 10^{-12} g/L   (c) 1.9 x 10^{-4} g/L
(d) 1.0 x 10^{-6} g/L   (e) 1.9 x 10^{-6} g/L

_____ 8. Iron(II) hydroxide is a slightly soluble base. In which of the following would $Fe(OH)_2$ be most soluble? It is not necessary to know K_{sp} for $Fe(OH)_2$.
(a) 0.1 M HCl   (b) 0.1 M KOH   (c) 0.1 M $Ca(OH)_2$. (d) 0.1 M $FeCl_2$.
(e) $Fe(OH)_2$ would be equally soluble in all of these.

_____ 9. Values of K_{sp} for some slightly soluble sulfides are as follows:

Sulfide	K_{sp}
CdS	3.6 x 10^{-29}
CuS	8.7 x 10^{-36}
PbS	8.4 x 10^{-28}
MnS	5.1 x 10^{-15}

Which of the following ions must exist at lowest concentration in a solution in which the sulfide ion concentration had been fixed at some fairly large constant value?
(a) Cd^{2+} (b) Cu^{2+} (c) Pb^{2+} (d) Mn^{2+}
(e) Cannot be answered without knowing the sulfide ion concentration.

_____ 10. If NaCl is added to a 0.010 M solution of $AgNO_3$ in water, at what $[Cl^-]$ does precipitation of AgCl begin to occur? K_{sp} for AgCl is 1.8 x 10^{-10}.
(a) 1.0 x 10^{-10} M (b) 1.8 x 10^{-6} M (c) 1.8 x 10^{-8} M
(d) 1.8 x 10^{-12} M (e) 1.8 x 10^{-10} M

- -

Suppose we have a solution that is 0.10 M in Ca^{2+} and also 0.10 M in Pb^{2+}. We wish to separate these ions by fractional precipitation of one of them as the carbonate. We do this by adding a source of carbonate ion, such as solid Na_2CO_3. Questions 11 through 13 refer to this solution. Use the K_{sp} values 4.8 x 10^{-9} for $CaCO_3$ and 1.5 x 10^{-13} for $PbCO_3$. Neglect any change in volume due to addition of the sodium carbonate.

_____ 11. Which will begin to precipitate from solution first, $CaCO_3$ or $PbCO_3$, and what will be the CO_3^{2-} when this begins (whether or not we can see it)?
(a) $CaCO_3$, 4.8 x 10^{-8} M (b) $CaCO_3$, 4.8 x 10^{-9} M (c) $PbCO_3$, 1.5 x 10^{-12} M
(d) $PbCO_3$, 1.5 x 10^{-13} M (e) $PbCO_3$, 7.2 x 10^{-21} M

_____ 12. In the solution of Question 11, suppose we keep adding Na_2CO_3. Eventually the second salt will begin to precipitate. When this second salt does begin to precipitate, what will be the concentration of the first metal ion left in solution?
(a) 0.10 M (b) 1.5 x 10^{-12} M (c) 3.2 x 10^{-5} M
(d) 3.1 x 10^{-6} M (e) 1.3 x 10^{-5} M

_____ 13. When the condition in Question 12 is reached, that is, when the second salt begins to precipitate, what percent of the first metal ion will have been removed from the solution?
(a) 4.8%   (b) 50%   (c) 95.7%   (d) 99.997%   (e) all of it

- -

_____ 14. We have a solution that is 0.0010 M Pb^{2+}, 0.0050 M Ag^+, and 0.0020 M K^+. If the concentration of sulfide ion, S^{2-}, is slowly increased in this solution, which metal sulfide will precipitate first? Refer to a table of K_{sp} values.
    (a) PbS   (b) Ag_2S (c) K_2S
    (d) All would begin to precipitate at the same time.
    (e) None would precipitate at any reasonable sulfide concentration.

_____ 15. K_{sp} for $Fe(IO_3)_3$ is 10^{-14}. We mix two solutions, one containing Fe^{3+} and one containing IO_3^-. At the instant of mixing, $[Fe^{3+}] = 10^{-4}$ M and $[IO_3^-] = 10^{-5}$ M. Which of the following statements is true?
    (a) A precipitate forms because $Q_{sp} > K_{sp}$
    (b) A precipitate forms because $Q_{sp} < K_{sp}$.
    (c) No precipitate forms because $Q_{sp} > K_{sp}$.
    (d) No precipitate forms because $Q_{sp} < K_{sp}$.
    (e) None of the preceding statements is true.

_____ 16. Magnesium hydroxide is a slightly soluble hydroxide with $K_{sp} = 1.5 \times 10^{-11}$. What is the pH of a saturated solution of magnesium hydroxide?
    (a) 3.51   (b) 10.51   (c) 5.39   (d) 8.60   (e) 7.00

Answers to Preliminary Test

Short Answer

1. molar solubility
2. K_{sp}
3. $[Cu^{2+}][CO_3^{2-}]$
4. $[Mn^{2+}][OH^-]^2$
5. $[Mg^{2+}]^3[PO_4^{3-}]^2$
6. >; =
7. more
8. endothermic. Recall the applications of LeChatelier's Principle as you learned them in Sections 17-6 through 17-8. It is very hard to measure the value of ΔH^0 for a reaction that proceeds as slightly as does the dissolution of lead chloride in water. Can you design an experiment based on measuring the solubility of lead chloride at several temperatures to determine ΔH^0 for this reaction? Think about Section 17-13.
9. K_d
10. converting an ion to a weak electrolyte; converting an ion to another species by oxidation or reduction; forming a complex ion involving one of the constituent ions.
11. $FeS(s) + 6CN^-(aq) \rightleftharpoons [Fe(CN)_6]^{4-} + S^{2-}(aq)$; 3.8×10^{19}; very easy.
    First we write out the equations, reversing the one for the dissociation of $[Fe(CN)_6]^{4-}$:

$$FeS(s) \rightleftharpoons Fe^{2+}(aq) + S^{2-}(aq) \qquad K= 4.9 \times 10^{-18}$$
$$Fe^{2+}(aq) + 6CN^-(aq) \rightleftharpoons [Fe(CN)_6]^{4-}(aq) \qquad K= 1/(1.3 \times 10^{-37}) = 7.7 \times 10^{36}$$

(remember to invert K when you reverse the second equation). Then we recall that adding the equations requires that we multiply their K values:

$$FeS(s) + 6CN^-(aq) \rightleftharpoons [Fe(CN)_6]^{4-}(aq) + S^{2-}(aq) \qquad K = 3.8 \times 10^{19}$$

This very large value of K tells us that the dissolution goes essentially to completion.

12. $CdS(s) + 4CN^-(aq) \rightleftharpoons Cd(CN)_4^{4-} + S^{2-}(aq)$; 4.6×10^{-12}; difficult. Reason as outlined in the answer to Question 11. Now the K value for the dissolution reaction is very small.

13. less than; removing one of the constituent ions

14. $1.26 \times 10^{-10}\ M$. The condition for precipitation to begin (barely) is

$$Q_{sp} = (0.10)[OH^-]^2 = 1.6 \times 10^{-19}$$

Solve this for $[OH^-] = 1.26 \times 10^{-10}$. If $[OH^-]$ is kept less than this, then Q_{sp} would be less than K_{sp}, and the precipitate would not form.

15. less than 4.10. In order to keep $[OH^-]$ less than 1.26×10^{-10} we would have to keep $[H_3O^+]$ greater than 7.94×10^{-5}. This corresponds to pH less than 4.10.

Multiple Choice

1. (b). Remember the solubility rules from Section 4-2, part 5 and Table 4-8. AgCl is one of the compounds seen in Chapter 20 as an example of a slightly soluble salt.

2. (a). For every mole of Ag^+ that goes into solution, one-half mole of SO_4^{2-} goes in. Look at the formula of the salt.

3. (e). Each mole of $PbSO_4$ that dissolves gives 1 mole of Pb^{2+} and 1 mole of SO_4^{2-} in solution. So in the saturated solution the concentrations of the ions would be $[Pb^{2+}] = 1.14 \times 10^{-4}\ M$ and $[SO_4^{2-}] = 1.14 \times 10^{-4} M$. Then $K_{sp} = [Pb^{2+}][SO_4^{2-}] = 1.30 \times 10^{-8}$.

4. (e) Did you remember that 2 moles of F^- go into solution for every mole of Pb^{2+}?

5. (b). This would have the highest $[Cl^-]$ to begin with, so less AgCl could dissolve before the value of K_{sp} would be reached by Q_{sp}. Note that actual calculations based on the K_{sp} of AgCl are not necessary.

6. (c). Solve by the methods of Section 20-3, in the way illustrated in Example 20-3.

7. (a). Solve as in Question 6, but then remember to convert from mol/L to g/L.

8. (a). The H_3O^+ present in this solution would help to remove OH^- as it is formed by dissolution of $Fe(OH)_2$. According to LeChatelier's Principle, this would help to shift the equilibrium for the dissolution process to the right.

9. (b). No calculation is needed here, because all K_{sp} expressions would have the same algebraic form. If some of the salts had different formulas, for example some XY and some X_2Y, then you would need to calculate concentrations. See text Exercises 40 and 41.

10. (c)

11. (c). Calculate the concentration of CO_3^{2-} necessary to begin to precipitate each, using the method shown in Section 20-3 and Examples 20-5 and 20-7. This calculation leads to the conclusion that $CaCO_3$ will begin to precipitate when $[CO_3^{2-}] = 4.8 \times 10^{-8}\ M$ and $PbCO_3$ will begin when $[CO_3^{2-}] = 1.5 \times 10^{-12}\ M$. Because the latter is the smaller concentration, it will be reached first. $PbCO_3$ will begin to precipitate first.

12. (d). When $[CO_3^{2-}]$ reaches $4.8 \times 10^{-8}\ M$ and $CaCO_3$ begins to precipitate, $[Pb^{2+}]$ must be down to $(1.5 \times 10^{-13})/(4.8 \times 10^{-8}) = 3.1 \times 10^{-6}\ M$.

13. (d). From the results of Question 12, we know that only $3.1 \times 10^{-6}\ M$ of the original $0.10\ M$ Pb^{2+} remains. Thus the percentage of lead ions removed is $\dfrac{3.1 \times 10^{-6}}{0.10} \times 100 = 0.003\%$. In other terms, $(100 - 0.003)\% = 99.997\%$ has been removed. We might call this "getting the lead out."

14. (b). Use the methods of Section 20-3 to calculate the concentration necessary to begin precipitation of PbS and of Ag_2S. You should remember from the solubility rules that K_2S, like all common salts of the alkali metals, is a soluble salt. To precipitate PbS, $[S^{2-}]$ needed is $8.4 \times 10^{-25}\ M$; for Ag_2S to precipitate, $[S^{2-}]$ needed is $2.0 \times 10^{-47}\ M$. It is clear that Ag_2S would precipitate first.

15. (d). If you got answer (a), you probably forgot to raise $[IO_3^-]$ to the third power.

16. (b). Remember that each mole of $Mg(OH)_2$ that dissolves gives two moles of OH^-. You can use K_{sp} and the methods of Section 20-3 to find the molar solubility of $Mg(OH)_2$ to be $1.6 \times 10^{-4}\ M$. Then $[OH^-] = 3.2 \times 10^{-4}$. In any aqueous solution, $[H_3O^+][OH^-] = 1.0 \times 10^{-14}$, so $[H_3O^+] = 3.1 \times 10^{-11}$. This corresponds to pH $= 10.51$.

21 Electrochemistry

Chapter Summary

As we learned in Chapter 4, oxidation-reduction (redox) reactions always involve transfer (actual or apparent) of electrons. **Electrochemistry** is the study of this type of reaction, with emphasis on the electron transfer. In **electrochemical cells,** we cause the electron transfer to take place through an external circuit rather than directly between reacting species. In Chapter 21, we shall learn about the conceptual aspects of electrochemistry and its relation to the spontaneity of reactions. We shall also learn some of electrochemistry's practical applications.

The central ideas of electrochemistry are summarized in the following questions: How can we tell whether a particular electron transfer process is spontaneous? If it is not, how can we make it occur by supplying electrical energy from an outside source? If it is spontaneous, how can we use this reaction as a source of electrical energy? These questions suggest that our electrochemical studies can conveniently be approached in two parts. The first (Sections 21-3 through 21-7) concerns those reactions in which we use an external source of electrical energy to cause a nonspontaneous reaction to occur. Electroplating and electrolytic refining of substances are applications of this aspect of electrochemistry. The second part of our study of electrochemistry (Sections 21-8 through 21-25) involves those reactions in which we use the spontaneity of the reaction to make it serve as a source of electrical energy, with applications such as batteries and fuel cells.

Sections 21-1 and 21-2 present some of the basic terminology of electrochemistry—**conduction, electrolytic conduction, electrode, anode, cathode,** and so on. Such terms will be used throughout this chapter. A review of Sections 4-4 (**oxidation numbers**) and 4-5 (**oxidation-reduction reactions**) would be a useful prelude to your study of Chapter 21. Concentrate especially on the way terms such as oxidation, reduction, oxidizing agent, and reducing agent are used to describe to electron transfer.

Throughout this chapter, we shall be describing oxidation-reduction reactions in terms of **half-reactions.** In Chapter 4, we saw that the idea of separating the oxidation process from the reduction process was an aid in balancing redox reactions. At that point, we talked about reactions with direct electron transfer between reactant species, so the "separation" was mental, largely a bookkeeping device. In most of the electrochemical studies of this chapter, we arrange the experiment so that these half-reactions take place in physically separate parts of the **cell.** The electrons are transferred through an external circuit, such as a wire.

The first major topic of Chapter 21 is **electrolysis,** in which we cause redox reactions that are spontaneous to occur by our supplying electrical energy from outside. As you study the three sections that discuss the electrolysis of molten sodium chloride (21-3), of aqueous sodium chloride (21-4), and of aqueous sodium sulfate (21-5), be sure that you understand how such terms as oxidizing agent, substance oxidized, anode, cathode, oxidation half-reaction, reduction half-reaction, and so on, are used. You should be able to apply these terms to an electrolytic cell, given the observations associated with the operation of the cell.

As we study the three electrolytic processes just mentioned, two types of questions occur: (1) How much electrical charge must be supplied to cause a specified amount of reaction to occur? (2) Why do different reactions take place in the different electrolytic cells? That is, in the electrolysis of aqueous sodium chloride, why is water reduced in preference to sodium ions (which can, however, be reduced in molten NaCl)? Why, in aqueous NaCl, are chloride ions oxidized in preference to water? This second type of question, "Which of several possible oxidation or reduction process is favored?", will be addressed later in the chapter, beginning in Section 21-15. The first question is the subject of Section 21-6.

The main idea of **Faraday's Law of Electrolysis** is that a specific amount of electrical charge delivered to the cell corresponds to a given number of electrons. Thus, there is a relationship between the amount of electrical charge and the extent of redox reaction that it causes. In particular, one **faraday** of charge, which is 96,485 coulombs, corresponds to one **mole** of electrons. Calculations related to Faraday's Law, including the relation of amount of charge to time and strength of current, are illustrated in Examples 21-1 and 21-2. Applications of the principles of electrolysis are discussed in the next two sections. In Section 21-7, we see how electrolysis may be used in such processes as electrolytic refining and electroplating.

We now turn to the opposite aspect of electrochemistry, that of using spontaneous reactions to produce electrical energy, in **voltaic** or **galvanic cells**. We accomplish this by separating halves of the redox reaction, forcing the electron transfer to take place through the external circuit. Section 21-8 describes the construction of simple voltaic cells for measurement of voltage (electrical potential). The next two sections illustrate the use of such experimental arrangements to study redox reactions. For now we are working only with cells in which all reactants are at standard conditions—solutions are $1\ M$ in each solute, gases are at 1 atm partial pressures. In Section 21-9, we study the zinc-copper cell. Think about the *possible* results of redox processes in such a cell: metallic copper could be oxidized ($Cu \rightarrow Cu^{2+} + 2e^-$) or copper (II) ions could be reduced ($Cu^{2+} + 2e^- \rightarrow Cu$); this could be accompanied by zinc ions being reduced ($Zn^{2+} + 2e^- \rightarrow Zn$) or metallic zinc being oxidized ($Zn \rightarrow Zn^{2+} + 2e^-$), respectively. In our experiments with this cell, we observe that Cu^{2+} is reduced and Zn is oxidized and that the initial cell voltage (i.e., while both solutions are still $1\ M$) is 1.10 V. By comparison, in the copper-silver cell, Section 21-10, Cu is oxidized and Ag^+ is reduced, with an initial cell voltage of 0.46 V. As a result of such experiments as these, we see that the magnitude of the cell potential is a measure of the spontaneity of the redox reaction, a notion that we will extend later in the chapter. For now, we use such results to rank the relative strengths of the oxidizing agents and of the reducing agents in the cells.

First, we must rank oxidizing and reducing strengths on a quantitative basis. We accomplish this with the determination and tabulation of **standard electrode potentials**. These are discussed and applied in the next eight sections of the chapter. Our first requirement is a reference strength, to which we can refer all other tendencies to be oxidized or reduced. For this, we arbitrarily choose the **standard hydrogen electrode (SHE**, Section 21-11) to have an "electrode potential" of exactly 0.0000... V. We can then construct a cell with any other electrode versus the SHE and measure the cell potential. This strategy is applied in Section 21-12 for the Zn/Zn^{2+} electrode and in Section 21-13 for the Cu/Cu^{2+} electrode, each measured against SHE. In these experiments, we are measuring separately, but against the same reference, the quantitative strengths of the oxidizing and reducing agents, by measuring the cell voltages. In each case we have some other electrode vs. the SHE (which has exactly 0 V electrode potential), so the cell voltage is called the **standard electrode potential** (E^0) for that other electrode or half-cell.

In this way, we build up the **electromotive series** or **activity series,** Section 21-14. Remember that by convention we always tabulate these potentials for half-reactions as *reductions*

(Table 21-2). The more positive the E^0 value for a particular half-reaction, the easier that reduction is to accomplish. As a way of helping you remember this convention, just look at the table and find some reaction that you know to be easy (or hard) and notice its sign. For instance, in Table 21-2 the reduction of elemental fluorine to fluoride ion is shown as having $E^0 = +2.87$ V. If you recall that fluorine has a very high affinity for electrons (Section 6-4) or that it is the most electronegative element (Section 6-6), it will help you remember that this reduction is a very easy one. This should enable you to recall that more positive values of E^0 correspond to greater ease of the process. Alternatively, recall that it is easy to oxidize alkali metals to their cations (these metals have low ionization energies, Section 6-3), and therefore *hard* to reduce the cations to the free metals. Observing that these metals have very negative E^0 values will remind you of the convention -- the lower (more negative) the E^0 value, the *harder* it is to carry out that reduction. Here we are only tabulating standard electrode conditions of 1 M for solutions and 1 atm for gases. Corrections for different concentrations or partial pressures will be encountered later in the chapter.

The next four sections of the chapter deal with uses of the knowledge summarized in the electromotive series. One important use, detailed in Section 21-15, is to **determine whether a particular redox reaction would be spontaneous.** To do this, we write E^0 for the possible half-reactions (from the electromotive series). Then we add E^0 for reduction and E^0 for oxidation to determine E^0_{cell} for the overall redox process. The more positive the E^0_{cell} value, the more favorable the process would be at these conditions. Notice that we are not necessarily dealing here with the reaction taking place in an electrochemical cell, because the same conclusions apply to the same reaction in any single solution at standard conditions. As you study Examples 21-4 and 21-5 in this section and in your working of such problems, be very careful of the conventions. If possible, write the half-reaction with the more positive E^0 value as a reduction and reverse the other half-reaction while changing the sign of its E^0 value. The resulting E^0_{cell} will then be positive, indicating that the reaction is spontaneous in the direction written. (If a reaction has a negative E^0_{cell}, then the reaction is spontaneous in the reverse direction.) It is important to remember, though, that we do *not* change the magnitude of E^0 when we reverse the half-reaction or when we multiply it by some factor. This is because E^0 is a measure of the *relative tendency* of the reaction to occur. This is different from the case in thermodynamic calculations where, for example, ΔH^0 is a measure of the *amount* of heat produced or consumed when the reaction proceeds by a certain amount.

In Section 21-16, we see how these techniques can also be applied to oxidation and reduction processes that take place in solution at **inert electrodes.** There are no new concepts here, but remember that we are still talking only about 1 M solutions. You can see in this section, however, that what we have developed is consistent with all of the observations cited earlier (Sections 21-3 through 21-6) about which reactions occurred preferentially in the various electrolytic cells. The very practical topic of corrosion and corrosion protection is actually an application of electrochemical principles, as emphasized in the discussion of Sections 21-17 and 21-18.

Up to this point in the chapter, all electrode potentials and cell potentials have been standard ones (E^0, E^0_{cell}), corresponding to solution concentrations of 1 M and gas partial pressures of 1 atm. In Section 21-19, we see that we can take these standard conditions as references and then make corrections for the effects of other concentrations. This is accomplished by the **Nernst equation**; be sure that you know the exact meaning of the various quantities in this important equation, especially n (the number of electrons transferred in the half-reaction) and Q (the reaction quotient, Section 17-3). As you will see in the samples of this section, the approach used is to first apply the Nernst equation to correct each electrode potential to experimental concentrations. Then we combine these E (no longer E^0) values to get E_{cell} (not E^0_{cell}). Then E_{cell} is an indication of whether the reaction, at the *specified concentrations*, can proceed. Positive values of E_{cell} indicate that the reaction is spontaneous.

252

Another important use of the Nernst equation is to use measured cell voltages to determine concentrations of ions in solution (Section 21-20).

We have now seen three different ways to express the spontaneity of a reaction: (1) the standard Gibbs free energy change, ΔG^0, in Chapter 15, (2) equilibrium constant, K, in Chapter 17, and (3) E^0_{cell} in Chapter 21. You should know the criteria for spontaneity of the forward reaction in terms of all three of these quantities. It should be clear that these are all related quantities; we have already seen (Section 17-12) the relation between ΔG^0 and K. In Section 21-21, we relate E^0_{cell} to the other two indicators of spontaneity. The ideas of this section are important at two levels. First, three important concepts from different areas of chemistry are now connected, which gives an important insight into the fundamental unity of the science and the validity of our descriptions. Second, on a more practical level, it provides us several different ways to derive numerical quantities or results that we may desire. For instance, for some reactions, it may be easier to measure E^0_{cell} than to determine K from experimentally measured concentrations or to determine ΔG^0 from calorimetric measurements or from tables that might not contain all necessary values; for other reactions, one of the latter two determinations may be easier. The relationships in this section then provide us with the means to convert our knowledge of the value of any one of these three quantities into the other two. (Of course, cell potentials apply only to redox reactions.) Examples 21-12 through 21-15 emphasize calculations of this nature.

The final sections of Chapter 21 are concerned with the use of voltaic cells as practical sources of electrical energy. Such cells are referred to as either (1) **primary voltaic cells**, in which the electrochemical reaction cannot be reversed once the reactants have been mixed or (2) **secondary voltaic cells**, also called **reversible cells**, in which the electrochemical reaction can be reversed by the action of a direct current from an external source—"recharging." The applications discussed in the text are the **dry cell** (Section 21-22) as an example of a primary voltaic cell, the automobile **lead storage battery** (Section 21-23), and the rechargeable nickel-cadmium, or **nicad** cell (Section 21-24) exemplifying a secondary voltaic cell. When the reactants are continuously supplied to the voltaic cell, it is referred to as a **fuel cell**. The hydrogen-oxygen fuel cell (Section 21-25) is an example.

Study Goals

1. Distinguish between: (a) metallic conduction and electrolytic (ionic) conduction; (b) oxidation and reduction; (c) electrolytic cells and voltaic cells; (d) anode and cathode. (Section 21-1 through 21-3; Key Terms; Exercises 1, 2, 5, and 6)

2. Understand the general concepts of electrolysis. (Sections 21-3 through 21-7)

3. Given the components of the electrodes and observations of what happens at the electrodes when an electrolytic cell is in operation, be able to do the following: (a) Describe the operation of the cell. (b) Write balanced oxidation and reduction half-reactions. (c) Write balanced chemical equations for the overall reaction. (d) Construct a simplified diagram of the cell (such as Figure 21-4), including designation of anode and cathode, positive and negative electrode, direction of electron flow in the external circuit, and direction of migration of ions within the cell. (Sections 21-2 through 21-5; Exercises 10 through 15)

4. Understand Faraday's Law of Electrolysis; perform calculations to relate the amount of electricity passing through an electrolytic cell to the amount (mass or gas volume) of a specified reactant consumed or product formed in the cell. Relate this information to the atomic weight of an ion consumed or produced. (Section 21-6; Exercises 16 through 34)

5. Describe the processes of electrolytic refining of impure metals and electroplating. (Sections 21-3 and 21-7)

6. Understand the general concepts of voltaic cells. (Sections 21-8 through 21-10; Exercises 35, 36, 41, and 42)

7. Describe what is meant by standard electrochemical conditions. (Section 21-8; Exercise 45)

8. Be able to perform the same operations as in Study Goal 2 for voltaic cells, including the direction of migration of ions in a salt bridge. (Figure 21-7 is a typical such diagram.) Also be able to write and interpret the short notation of voltaic cells. (Sections 21-9 through 21-13; Exercise 37 through 40, 43, and 44)

9. Describe the standard hydrogen electrode. Illustrate how a table of standard electrode potentials (also known as electromotive series or activity series) is constructed and used. (Sections 21-11 and 21-14; Exercise 46 and 47)

10. Assess relative strengths of oxidizing and reducing agents from a table of their standard electrode potentials (the electromotive or activity series). Use these results to assess whether a reaction is spontaneous or nonspontaneous. (Sections 21-15 and 21-16; Exercises 48 through 71)

11. Describe the process of corrosion and methods by which corrosion of a specified metal can be prevented. (Sections 21-17 and 21-18; Exercises 72 and 73)

12. Apply the Nernst equation to determine electrode potentials and cell potentials under nonstandard conditions. (Section 21-19; Exercises 74 through 86)

13. Use the Nernst equation to determine concentrations of ions in solution. (Section 21-20; Exercises 87 through 91)

14. Summarize the restrictions on the values of E^0_{cell}, ΔG^0, and the equilibrium constant, K, for (a) spontaneous reactions, (b) nonspontaneous reactions, and (c) reactions at equilibrium under standard electrochemical conditions. (Section 21-21; Exercises 92 and 93)

15. Perform calculations to relate standard cell potentials, standard Gibbs free energy changes, and equilibrium constants for redox reactions. (Section 21-21; Exercises 94 through 98)

16. Describe and distinguish among primary voltaic cells, secondary voltaic cells, and fuel cells. Give examples, balanced electrode half-reactions, and overall reactions for each. (Sections 21-22 through 21-25; Exercises 99 through 105)

17. Be sure you can recognize and solve various kinds of problems based on the material in this chapter. (Mixed Exercises, 106 through 115)

18. Answer conceptual questions based on this chapter. (Conceptual Exercises 116 through 119)

19. Apply concepts and skills from earlier chapters to the ideas of this chapter. (Building Your Knowledge exercises, 120 through 125)

Some Important Terms in This Chapter

Write the meanings *in your own words*. Check the Key Terms list and the chapter reading. Then rewrite your definitions, still in your own words, to improve them. Study other new terms, and review terms from preceding chapters (especially Chapter 4) if necessary.

oxidation

reduction

electrochemistry

metallic conduction

electrolytic (ionic) conduction

electrode

anode

cathode

electrolysis

electrolytic cell

Faraday's Law of Electrolysis

faraday

voltaic cell

primary voltaic cell

secondary voltaic cell

fuel cell

electrode potential

standard electrode

standard electrode potential

cell potential

standard cell potential

Nernst equation

corrosion

sacrificial anode

galvanizing

Preliminary Test

Practice many additional Exercises from the text.

True-False

_____ 1. All electrochemical reactions are oxidation-reduction reactions.

_____ 2. In both electrolytic cells and voltaic cells, reduction occurs at the cathode.

_____ 3. One coulomb of electricity is the amount of charge carried when a current of 1 ampere flows for 1 second.

_____ 4. One faraday of electricity is the amount of charge of Avogadro's number of electrons.

_____ 5. According to Faraday's Law of Electrolysis, the same amount of electricity will deposit 1 mole of any metal from a solution of one of its ionic salts.

_____ 6. The same amount of electricity that would deposit 107.87 g Ag from an Ag^+ solution would also liberate 11.2 L of dry H_2 gas, measured at STP.

_____ 7. One faraday of electricity is required to reduce 1 mole of Ag^+ to Ag.

_____ 8. One faraday of electricity is required to oxidize 1 mole of Ag to Ag^+.

_____ 9. One faraday of electricity is required to reduce 1 mole of Cu^{2+} to Cu.

_____ 10. One faraday of electricity is required to produce 8.00 g of O_2 gas by electrolysis of an aqueous solution.

_____ 11. During the electrolysis of aqueous sodium chloride, the solution gradually becomes more basic.

_____ 12. During the electrolysis of aqueous sodium sulfate, the solution gradually becomes more basic.

_____ 13. A battery is an electrolytic cell.

_____ 14. In a voltaic cell, the electrochemical reaction proceeds spontaneously.

_____ 15. In a voltaic cell, we can just mix the reactants together and then dip electrodes of the appropriate metals into the solution.

Short Answer

1. Electrochemical cells in which nonspontaneous reactions are forced to occur by the input of electrical energy are called _____ cells.

2. Electrochemical cells in which a spontaneous reaction serves as a source of electrical energy are called _____ cells.

3. An electrode that provides a surface at which the oxidation or reduction half-reaction can occur, but which is not itself a reactant or product of the electrochemical reaction, is called a(n) _____.

4. In an electrolytic cell, the electrode at which oxidation takes place is called the _____; in a voltaic cell, the electrode at which oxidation takes place is called the _____.

5. One coulomb is the amount of charge on _____ (a number) electrons.

6. The cell that is used for the commercial electrolytic production of sodium metal is named the _____ cell. In this cell the processes that occur are:

   reduction half-reaction: _____ ;

   oxidation half-reaction: _____ ;

   overall cell reaction: _____ .

7. In the commercial electrolysis cell for refining of metals, the purified copper is deposited at the _____ .

8. In the commercial electrolytic process for the refining of copper, the net cell reaction is _____ .

9. In an electrolysis, we observe that 0.5 mole of palladium is deposited by the same current, flowing for the same duration of time, that can also deposit 1 mole of silver. The charge on the palladium ions in this solution is _____ .

10. In simple voltaic cells such as those used for measurement, the electrode that gains in weight would be the _____ .

11. In discussions of electrochemistry, the initials SHE stand for _____ .

12. In the standard hydrogen electrode, the purpose of the platinum metal is to provide _____ .

13. When the SHE acts as the anode in an electrochemical cell, the half-reaction that takes place at this electrode is _____ ; when it acts as a cathode the half-reaction is _____ .

14. In the cell $Cr|Cr^{3+}(1.0\ M)\,\|\,H^+(1.0\ M),H_2(1\ atm)|Pt$, the SHE acts as the _____ .

15. A voltaic cell consists of a standard hydrogen electrode connected by a salt bridge and a wire to an electrode consisting of a strip of Cd metal dipping into a 1 M solution of $Cd(NO_3)_2$. When the cell produces current, the electrons flow through the wire from the _____ electrode to the _____ electrode. In this cell, the _____ electrode is acting as the cathode.

16. Liquid bromine _____ (will, will not) oxidize silver metal to Ag^+ in aqueous solution.

17. The process in which metals are oxidized by O_2 in the presence of moisture is called _____ .

18. If, for a given process, E^0_{cell} can be shown to be positive, then ΔG^0 for that process will be _____ , K will be _____ , and the process as written will be _____ .

19. In the Leclanché cell, the carbon rod is the _____ , at which the half-reaction taking place is _____ . The other electrode is provided by the _____ , at which the half-reaction taking place is _____ .

20. In the lead storage battery, sulfuric acid acts as the _____ .

21. During the operation (discharging) of the lead storage battery, the Pb electrode is the _____ and the PbO_2 electrode is the _____ .

Multiple Choice

_____ 1. The substances that we classify as electrolytes will conduct electricity when dissolved in water because
(a) they are polar covalent substances.
(b) they are nonpolar covalent substances.
(c) they dissociate in solution to give ions that are attracted to the electrodes having the same signs as those of the ions.
(d) they dissociate in solution to give ions that are attracted to the electrodes having the opposite signs from those of the ions.
(e) they cause water to dissociate into H_3O^+ and OH^- ions.

_____ 2. In order to obtain 9,000 coulombs in an electrolytic process, how long would a constant current of 18 amperes be required to flow?
    (a) 200 seconds   (b) 500 seconds   (c) 1.6×10^5 seconds
    (d) 50 seconds   (e) 0.002 seconds

_____ 3. The electrode in an electrolysis cell that acts as a source of electrons to the solution is called the _____; the chemical change that takes place at this electrode is called

_____.

    (a) anode, oxidation   (b) anode, reduction
    (c) cathode, oxidation   (d) cathode, reduction
    (e) Cannot tell unless we know what species is being oxidized or reduced.

- -

The next three questions, 4 through 6, refer to the electrolysis of molten sodium bromide, NaBr. (It may help you to diagram the cell.)

_____ 4. During this electrolysis, sodium ions move
    (a) to the anode, which is positively charged.
    (b) to the anode, which is negatively charged.
    (c) to the cathode, which is positively charged.
    (d) to the cathode, which is negatively charged.
    (e) through the wire to the battery.

_____ 5. What is the half-reaction that takes place at the cathode?
    (a) $2Br^- \rightarrow Br_2 + 2e^-$ (b) $Br_2 + 2e^- \rightarrow 2Br^-$ (c) $Na^+ + e^- \rightarrow Na$
    (d) $Na \rightarrow Na^+ + e^-$ (e) $2H_2O + 2e^- \rightarrow 2OH^- + H_2$

_____ 6. What is the half-reaction that takes place at the anode?
    (a) $2Br^- \rightarrow Br_2 + 2e^-$
    (b) $Br_2 + 2e^- \rightarrow 2Br^-$
    (c) $Na^+ + e^- \rightarrow Na$
    (d) $Na \rightarrow Na^+ + e^-$
    (e) $2H_2O + 2e^- \rightarrow 2OH^- + H_2$

- -

_____ 7. Which of the following is *not* produced during the electrolysis of aqueous sodium chloride?
    (a) Na   (b) NaOH   (c) Cl_2 (d) H_2
    (e) All of the preceding are produced in this electrolysis.

_____ 8. During the electrolysis of aqueous sodium chloride, the solution gradually
    (a) disappears.   (b) gets more basic.   (c) gets more acidic.
    (d) dissolves the anode.   (e) dissolves the cathode.

_____ 9. How many coulombs must be supplied in order to plate out 1.07868 g of Ag from an Ag^+ solution?
    (a) 96,485   (b) 96,485/1.07868   (c) 96.485
    (d) 96,485 x 1.07868   (e) 964.85

_____ 10. How many faradays of electricity are required to perform the reduction of 2.0 g of Sn^{4+} to Sn^{2+}?
    (a) 0.017   (b) 0.034   (c) 0.068   (d) 0.008   (e) 2.000

_____ 11. in $Cr_2(SO_4)_3$, chromium is in the +3 oxidation state. What mass of chromium will be deposited by electrolysis of a solution of $Cr_2(SO_4)_3$ for 60 minutes using a steady current of 10 amperes?

    (a) 3.25 g   (b) 6.50 g   (c) 17.3 g   (d) 0.187 g   (e) 0.373 g

_____ 12. An aqueous solution of an unknown salt of ruthenium is electrolyzed by a current of 2.50 amperes passing for 50 minutes. This results in the formation of 2.618 g of ruthenium metal at the cathode. What is the oxidation state of ruthenium in this solution?

    (a) +1   (b) +2   (c) +2   (d) –2   (e) –3

_____ 13. The electrolysis of an aqueous solution of a cobalt salt proceeds until exactly 8.000 g of O_2 gas has been liberated at the anode. At this time, 29.467 g of Co has been deposited at the cathode. What is the charge on the cobalt ion in the solution?

    (a) 1+   (b) 2+   (c) 3+   (d) 1–   (e) 2–

_____ 14. In a voltaic cell, the salt bridge
    (a) is not necessary in order for the cell to work.
    (b) allows mechanical mixing of the solution.
    (c) allows ions to flow from the salt bridge into the two half-cells.
    (d) is tightly plugged with a firm agar gel in order to keep all ions separate.
    (e) drives electrons from one half-cell to the other.

_____ 15. Refer to the table of standard electrode (reduction) potentials in the text, Table 21-2. The most stable ions of the Group IB metals are Cu^{2+}, Ag^+, and Au^{3+}. Which of the following statements regarding the ease of reduction of these ions to the respective metals is true?
    (a) Au^{3+} is easier to reduce than Cu^{2+}, which is easier to reduce than Ag^+.
    (b) Cu^{2+} is easier to reduce than Au^{3+}, which is easier to reduce than Ag^+.
    (c) Au^{3+} is easier to reduce than Ag^+, which is easier to reduce than Cu^{2+}.
    (d) Cu^{2+} is easier to reduce than Ag^+, which is easier to reduce than Au^{3+}.
    (e) Ag^+ is easier to reduce than Au^{3+}, which is easier to reduce than Cu^{2+}.

_____ 16. Refer to the table of standard electrode reduction potentials in the text, Table 21-2. Which of the following statements about the Group IB metals is true?
    (a) Cu is easier to oxidize than Au.
    (b) Au is easier to oxidize than Ag.
    (c) Ag is easier to oxidize than Cu
    (d) Au is easier to oxidize than Cu.
    (e) Nothing can be decided about ease of oxidation from a table of reduction potentials.

_____ 17. According to the table of standard reduction potentials, which of the following metals is most easily oxidized?

    (a) Cd   (b) Cu   (c) Fe   (d) Ni   (e) Zn

_____ 18. Which of the following is the strongest oxidizing agent in 1 M aqueous solutions?

    (a) MnO_4^- (in acid solution)   (b) Ag^+ (c) Cu^{2+}
    (d) Ca^{2+} (e) Li^+

_____ 19. Which of the following is the strongest oxidizing agent? (All solutions are 1 M, gases are 1 atm)

    (a) $Cl_2(g)$ (b) $Br_2(l)$ (c) Li(s)   (d) $Li^+(aq)$ (e) $NO_3^-(aq)$

_____ 20. Would a reaction occur if a piece of pure copper were dipped into a 1 M $FeCl_2$ solution?

    (a) No   (b) Yes, forming Fe metal   (c) Yes, forming Fe^{3+}

The next five questions, 21 through 25, refer to the following electrochemical cell. Each half-cell contains an aqueous solution. Necessary values of standard electrode reduction potentials may be obtained from Table 21-2 in the text.

___ 21. What is the standard cell potential for this cell?
    (a) +0.740 V   (b) –0.740 V   (c) +0.066 V   (d) –0.066 V   (e) 0 V

___ 22. Which of the following statements about the half-cell processes is true?
    (a) Cu^{2+} is reduced at the anode.   (b) Cu^{2+} is reduced at the cathode.
    (c) Cd^{2+} is reduced at the anode.   (d) Cd^{2+} is reduced at the cathode.
    (e) This reaction is not an oxidation-reduction reaction.

___ 23. When this cell is in operation, electrons flow
    (a) through the wire from the Cu electrode to the Cd electrode.
    (b) through the wire from the Cd electrode to the Cu electrode.
    (c) through the salt bridge from the Cu half-cell to the Cd half-cell.
    (d) through the salt bridge from the Cd half-cell to the Cu half-cell.
    (e) No other answer is correct, because electrons do not flow in an electrochemical cell.

___ 24. Which of the following reactions describes the net reaction that takes place in this cell?
    (a) $Cu + Cd^{2+} \rightarrow Cu^{2+} + Cd$ (b) $Cu + Cd \rightarrow Cu^{2+} + Cd^{2+}$
    (c) $Cu^{2+} + Cd^{2+} \rightarrow Cu + Cd$ (d) $Cu^{2+} + Cd \rightarrow Cu + Cd^{2+}$

___ 25. What is ΔG° for the reaction $Cu^{2+} + Cd \rightarrow Cu + Cd^{2+}$?
    (a) –71.34 kJ/mol   (b) –142.7 kJ/mol   (c) +142.7 kJ/mol
    (d) –34.10 kJ/mol   (e) +71.34 kJ/mol

Answers to Preliminary Test

True-False

1. True
2. True
3. True
4. True

5. False. This would depend on the number of electrons necessary to reduce one mole of the metal—that is, on the charge of the ion.
6. True. See Table 21-1. This amount of electricity is one faraday. The amount of H_2 liberated is 1.008 g or $\frac{1}{2}$ mole. At STP, one mole of any gas occupies 22.4 L (Chapter 12).
7. True
8. True
9. False. Each Cu^{2+} ion requires 2 electrons in this process. Two faradays would be required.
10. True. The half-reaction is $2H_2O \rightarrow O_2 + 4H^+ + 4e^-$. Thus, 4 faradays are required to produce 1 mole (32.0 g) of O_2. This means that 1 faraday would produce $\frac{1}{4}$ mol O_2, or 8.0 g.
11. True. Remind yourself what the cathode half-reaction is in this cell (Section 21-4).
12. False. At the cathode, OH^- is produced, but H^+ is produced in equal amount at the anode. These neutralize one another (or in the terminology of Chapter 19, the $[H_3O^+]/[OH^-]$ balance is not disturbed).
13. False
14. True
15. False. The two half-reactions must be forced to take place in two separate places so that the electrons can be transferred externally. If we just mixed the reactants together and dipped the metals into the solution, the same overall redox reaction would occur, but we would not have an electrochemical cell.

Short Answer

1. electrolytic
2. voltaic
3. inert electrode
4. anode; anode
5. 6.241×10^{18}. Avogadro's number of electrons (6.022×10^{23} electrons) corresponds to 96,485 coulombs. So one coulomb would correspond to $(6.022 \times 10^{23})/96,485$ electrons.
6. Downs; $Na^+ + e^- \rightarrow Na(l)$; $2Cl^- \rightarrow Cl_2(g) + 2e^-$; $2Na^+ + 2Cl^- \rightarrow 2Na(l) + Cl_2(g)$
7. cathode
8. none. Copper is oxidized from the impure electrode (anode) and redeposited at the cathode.
9. +2. You should be able to figure this out *without* calculation. This will test whether you really understand the ideas behind those calculations.
10. cathode. This is the electrode at which metal ions are reduced to metal.
11. standard hydrogen electrode
12. a surface at which the electron transfer can take place.
13. $H_2 \rightarrow 2H^+ + 2e^-$; $2H^+ + 2e^- \rightarrow H_2$. Remember that the cathode is the electrode at which reduction occurs.
14. cathode. From the table of standard reduction potentials, we see that H^+ is easier to reduce than is Cr^{3+}, so the SHE is the cathode. At the anode, Cr metal is oxidized to Cr^{3+}.
15. cadmium; hydrogen; hydrogen. From the table of standard reduction potentials, we see that H^+ is more easily reduced than Cd^{2+}. The electrons flow through the wire to the SHE, where they reduce H^+ to H_2. Reduction occurs at the cathode.
16. will. We want to see whether the reaction $Br_2 + 2Ag \rightarrow 2Ag^+ + 2Br^-$ will occur. From Table 21-2 we write the half-reactions so that they add to give the reaction we wish to test; we remember that when we reverse the half-reaction for $Ag^+|Ag$, we must change the sign of its E^0.

Though we multiply the second half-reaction by 2 *we do not* multiply its E^0 value! Then we add the reactions, adding the E^0 values to get E^0_{cell} .

$$
\begin{array}{lll}
Br_2 + 2e^- \rightarrow 2Br^- & E^0 & = +1.08 \quad V \\
\underline{2(Ag \rightarrow Ag^+ + e^-)} & \underline{E^0} & \underline{= -0.799 \quad V} \\
Br_2 + 2Ag \rightarrow 2Ag^+ + Br^- & E^0_{cell} & = +0.28 \quad V
\end{array}
$$

E^0_{cell} for this proposed reaction is positive, so it is spontaneous in the direction written.

17. corrosion
18. negative; greater than 1; spontaneous
19. cathode; $2NH_4^+ + 2e^- \rightarrow 2NH_3 + H_2$; zinc container; $Zn \rightarrow Zn^{2+} + 2e^-$
20. electrolyte
21. anode; cathode

Multiple Choice

1. (d)
2. (b). number of coulombs = (number of amperes) × (number of seconds)
3. (d)
4. (d)
5. (c)
6. (a)
7. (a). See the discussion in Section 21-4. Remember that the alkali metals are much too active (easily oxidized) to exist in contact with water.
8. (b). The cathode half-reaction in this cell is the reduction of water to H_2. This also gives OH^-. The half-reaction is $2H_2O + 2e^- \rightarrow 2OH^- + H_2$.
9. (e). This is 0.01000 mol Ag^+ to be reduced. Each mole Ag^+ requires 1 faraday or 96,485 coulombs.
10. (b). The reduction half-reaction is $Sn^{4+} + 2e^- \rightarrow Sn^{2+}$. Thus, 2 faradays are required to reduce 1 mole of 118.7 g of Sn^{4+} to Sn^{2+}. Applying the unit factor $\dfrac{2 \text{ faradays}}{118.7 \text{ g } Sn^{4+} \text{ reduced}}$, we find that 0.034 faraday is required to reduce the desired 2.0 g of Sn^{4+}.
11. (b)
12. (c). The number of coulombs is $50 \text{ min} \times 60 \dfrac{s}{min} \times 250 \dfrac{C}{s} = 7,500 \text{ C}$. If 7,500 coulombs deposits 2.618 grams, then 96,500 coulombs would deposit 33.7 g of ruthenium. This is 33.7/101.07 = 0.333 mol of Ru deposited by 1 mole of electrons, so each Ru^{n+} ion must require three electrons. Thus the ion is Ru^{3+}.
13. (b). In the anode reaction, $2H_2O \rightarrow O_2 + 4H^+ + 4e^-$, it requires 4 × 96,500 coulombs to liberate 1 mol of 32.0 g O_2. Thus the reaction involves 1 mole of electrons. Since this liberates 29.467/58.933 = 0.500 mol Co, each Co^{n+} ion must require two electrons. Thus the ion is Co^{2+}.
14. (c)
15. (c). The standard electrode potentials for reduction of each of these ions to the free metal are $Au^{3+} = +1.50$ V; $Ag^+ = +0.799$ V; $Cu^{2+} = +0.337$ V. More positive reduction potential means that reduction is easier to accomplish.

16. (a). Reasoning from the reduction potentials, the hardest reduction to carry out for the Group IB ions is Cu^{2+}/Cu. Thus, the reverse of this process, the oxidation, would be the easiest of those listed.

17. (e). Each reaction appears as a reduction. The harder the reduction, the easier the reverse reaction (oxidation). The half-reaction involving $Zn^{2+} + 2e^- \rightarrow Zn$ is hardest to carry out as a reduction. Therefore, its reverse, the oxidation of Zn metal, would be easiest.

18. (a). Referring to the table of reduction potentials, Table 21-2, we see that all species listed are on the left and are being reduced in half-reactions as written. Of the species listed, MnO_4^- in acid solution is most easily reduced. When something is reduced, it is acting as an oxidizing agent. Thus, permanganate ion is the strongest oxidizing agent listed here.

19. (a). Same reasoning as for Question 18. The only wrinkle here is to figure out where Li(s) fits as an oxidizing agent. To act as an oxidizing agent, it would have to be reduced, and we know that alkali metals rarely gain electrons. Thus, Li(s) is a very poor oxidizing agent.

20. (a). Because the Cu metal could only be oxidized ($Cu \rightarrow Cu^{2+} + 2e^-$), Fe^{2+} could not be oxidized to Fe^{3+}, so answer (c) is eliminated. Might Fe^{2+} be reduced ($Fe^{2+} + 2e^- \rightarrow Fe$)? For this reduction, E^0 is -0.44 V. For the Cu oxidation, E^0 is -0.34 V. Adding these electrode potentials, we see that the only possible reaction, $Fe^{2+} + Cu \rightarrow Fe + Cu^{2+}$, has a standard cell potential of $E^0_{cell} = -0.78$ V. The negative value tells us that the reaction will *not* occur.

21. (a)

22. (b)

23. (b)

24. (d)

25. (b). From the answer to Question 21, E^0_{cell} is $+0.740$ V for the reaction as written. Convert using $\Delta G^0 = -nFE^0_{cell}$ to find $\Delta G^0 = -142.7$ kJ. This negative value means that this reaction is spontaneous as written. This is also indicated by the positive value of E^0_{cell}.

22 Metals I: Metallurgy

Chapter Summary

Throughout the first 21 chapters, the emphasis has been on developing a fundamental understanding of chemistry through broad concepts and basic principles. In the rest of the text we study in more detail the specific reactions and properties of various elements and groups of elements—the general area of chemistry called **descriptive chemistry.** As you study each of these chapters, keep in mind the basic principles that you have studied. Go back whenever necessary to review appropriate earlier chapters or sections. In this way, the principles you have learned will become more meaningful and useful as you see how they relate dozens of specific properties, reactions and processes. For instance, in this chapter we use the ideas of oxidation states, redox reactions, anions and cations, acidity and basicity, amphoterism, basic salts, densities, melting points, solubility rules, crystallization, melting point depression, electrolysis, activity series, relative thermodynamic stability, K_{sp} values, and complex ion formation. In your study of each of these descriptive chemistry chapters, it will probably help you to organize the material in your own words, emphasizing the general reactions or processes of the chapter, and especially the similarities and differences of the various elements or compound types being considered.

In Chapter 22, you will study metals from several aspects—the importance of metals and other trace elements in life processes, the distinctions between properties of metals and nonmetals, the occurrence of metals, the methods of obtaining and using free metals, and conservation of metals. You should recall (see Study Goal 1) the general distinguishing properties, both physical and chemical, of **metals** and **nonmetals.** Review (Chapter 6) the locations in the periodic table of general classes of elements and be familiar with the terminology of representative elements, transition elements, and so on. As pointed out in Section 22-1, some metals have sufficiently high positive reduction potentials (see Section 21-16 for a review of what this means) that they can occur in nature in the uncombined **native state.** Most metals are so easily oxidized that they are found in nature only in compounds; when this is true, they are always in positive oxidation states and methods for obtaining the free metals must involve **reduction.** The terms **ore, mineral,** and **gangue,** which will be used many times later in the chapter, are introduced in this section. We can classify ores according to the anion associated with the metal; methods of extraction of the metals depend on which compounds of the metal are present (Table 22-1). This also depends on the solubilities of the compounds; review the solubility guidelines, Section 4-2, part 5 and Table 4-8.

The next portion of the chapter, Sections 22-2 through 22-4, is concerned with the general principles and techniques of **metallurgy**—the commercial purification and use of free metals. The central chemical process in metallurgy is often one of reduction of the metal from a positive oxidation state in a compound. However, the metallurgy also involves separating the components of very complex heterogeneous mixtures. As described in Section 22-2, **pretreatment** of ores to remove the gangue (pronounced "ganj") may take advantage of differences in densities and hydrophilic properties (as in the **flotation** method), or converting the metal compound to a more easily reduced form, often the oxide (as in the **roasting** of sulfides), or both. A second stage of the metallurgical process is the reduction of the metal ores to the free metals (Section 22-3). This can be done in a variety of ways, either by oxidation of the anion away from the metal as in the roasting of cinnabar to produce mercury, by chemical reduction of the ore (which is often an oxide at this stage) with coke or carbon monoxide, by chemical reduction with hydrogen or another reducing agent, or by electrolytic reduction. The general reduction processes are summarized in Table 22-2, with specific applications cited in Table 22-3. The reduction process provides impure metals, which are then **refined** or purified by the methods described in Section 22-4. Metals are often used in the form of **alloys**, in which they are mixed with other elements to give desired changes in their physical and chemical characteristics.

Sections 22-5 through 22-9 discuss in detail the metallurgies of five of the important metals—magnesium, aluminum, iron, copper, and gold. As you study these sections, remember that the metallurgical process described for each metal must involve both (1) processes by which compounds eventually produce the free metals and (2) removal of other unwanted substances, including unwanted components of the ores and the other products of the reduction reactions and other chemical processes. One thing you should recognize is that tremendous amounts of energy must be applied for many of these processes, both for melting huge amounts of material and sometimes for electrolytic reduction.

The Chemistry in Use essay at the end of the chapter discusses some important practical aspects of the utilization, conservation, and recycling of metals.

Study Goals

1. Summarize the differences in physical and chemical properties of metals and nonmetals. (Review Sections 4-1, 6-7, 6-8, and 10-11; introduction to Chapter 22; Exercise 1)
2. Relate the electronic structures of metals to their tendencies to be found combined or uncombined in nature and to the methods that must be used to obtain them from their ores. (Sections 22-1 through 22-4; Exercises 2 and 3)
3. Know the anions that are commonly found combined with metals in ores and give examples of ores containing each. (Section 22-1; Exercise 4)
4. Describe the general sequence of steps used in obtaining metals from their ores. Discuss and illustrate these processes in detail for specific metals. Be able to write balanced equations for important reactions. (Sections 22-2 through 22-4; Exercises 5 through 17)
5. Describe in detail the separation processes and chemical reactions involved in the metallurgies of Mg, Al, Fe, Cu, and Au. (Sections 22-5 through 22-9; Exercises 18 through 33)
6. Carry out calculations relating metallurgy to the concepts of earlier chapters. (Review Chapters 2, 3, 15, and 21; Building Your Knowledge exercises 34 through 46)

Some Important Terms in This Chapter

Write the meanings *in your own words*. Check the Key Terms list and the chapter reading. Then rewrite your definitions, still in your own words, to improve them. Study other new terms, and review terms from preceding chapters if necessary.

transition metals

native state

ore

gangue

metallurgy

flotation

refining

roasting

smelting

slag

Hall-Héroult process

pig iron

cast iron

steel

zone refining

alloying

Preliminary Test

As in other chapters, this test will check your understanding of basic concepts and types of calculations. Be sure to answer *many* of the additional textbook exercises.

True-False

Mark each statement as true (T) or false (F).

_____ 1. There are more metallic elements than nonmetallic elements.
_____ 2. Metallic character decreases with increasing numbers of electrons in the valence shell.
_____ 3. For elements with the same number of electrons in the valence shell, metallic character decreases with increasing numbers of electron shells.
_____ 4. Metals usually have higher electrical conductivity than nonmetals.
_____ 5. Most metals are solids at room temperature.
_____ 6. Metals are usually fairly electronegative.
_____ 7. Free metals are often good oxidizing agents.
_____ 8. Most metals are found free in nature.
_____ 9. Roasting of sulfide ores is one of the most environmentally clean methods presently used in metallurgy.
_____ 10. The metallurgy of all metals involves reduction of a compound to the free metal.
_____ 11. Gold often occurs in the native state, so all that is ever done to recover it is to carry out a mechanical separation from the other components of the ore.

Short Answer

Answer with a word, a phrase, a formula, or a number with units as necessary.

1. Elements with properties intermediate between those of metals and those of nonmetals are called _____.
2. The most common human trace element deficiency is a dietary shortage of _____.
3. The metal that is present in insulin and many enzymes is _____.

4. When the oxides and hydroxides of metals dissolve in water, they give solutions that are _____ or _____.

5. The most common ores of metals, except those in which the element occurs in the native state, have the metal combined with one or more of the following anions: _____, _____, _____, _____, _____, _____.

6. The concentrating of an ore before the reduction stage is referred to as _____ of the ore.

7. In the pretreatment of ores in the cyclone separator, the property that is used to effect the separation is differences in _____.

8. The pretreatment technique in which hydrophobic materials are separated from other components of a mixture by suspending them in water is called the _____ method; this method often involves coating the particles with _____ and blowing a stream of _____ through the suspension.

9. In the pretreatment process known as roasting, metal sulfides are converted to _____ by _____.

10. An example of a metal that can be obtained from the sulfide ore by the simple process of roasting is _____; the sulfide ore from which it is obtained is called _____ and has the formula _____.

11. The reduction of metal compounds to the free metal, especially by chemical means, is called _____.

12. The formula of a carbonate that is converted to the oxide by heating to drive off carbon dioxide is _____.

13. The very active metals are usually reduced by _____ methods.

14. The coke that is used in reduction processes in metallurgy is _____.

15. An example of a reaction in which hydrogen is used as the reducing agent in the production of a metal from an oxide is _____.

16. The name of the common ore of aluminum is _____. The commercial process most widely used for the production of aluminum from this ore is called the _____ process.

17. The relatively large amount of sand, soil, clay, rock, and other material that accompanies the minerals in ores is called _____.

18. In the blast furnace process for the metallurgy of iron, limestone is added with the crushed ore. The purpose of this limestone (the flux) is to react with the silica in the gangue to form calcium silicate, which melts at these high temperatures. This molten material is referred to as _____.

19. Two metals that are commercially produced by the electrolytic reduction of the molten chloride salts are _____ and _____.

- -

Each of the following equations, Questions 20 through 33, represents a reaction that was discussed in Chapter 22. Each reaction is used in the production of some metal. For each equation, tell what metal or metals M could represent. The possible metals are Al, Au, Ca, Cu, Fe, Hg, Mg, Na, Ni, Pb, and Zn.

20. $MCO_3(s) \xrightarrow{heat} MO(s) + CO_2(g)$ _____

21. $M(OH)_2(s) \xrightarrow{heat} MO(s) + H_2O(g)$ _____

22. $2MS(s) + 3O_2(g) \xrightarrow{\text{heat}} 2MO(s) + 2SO_2(g)$ _____

23. $M_2S(s) + O_2(g) \xrightarrow{\text{heat}} 2M(s) + SO_2(g)$ _____

24. $MS(s) + O_2(g) \xrightarrow{\text{heat}} M(l) + SO_2(g)$ _____

25. $MO_2(s) + C(s) \xrightarrow{\text{heat}} M(l) + 2CO(g)$ _____

26. $MO(s) + C(s) \xrightarrow{\text{heat}} M(s) + CO(g)$ _____

27. $M_2O_3(s) + 3CO(g) \xrightarrow{\text{heat}} 2M(l) + 3CO_2(g)$ _____

28. $M_2O_3(s) + 3C(s) \xrightarrow{\text{heat}} 2M(l) + 3CO(g)$ _____

29. $2M_2O_3(l) \xrightarrow[\text{electrolysis}]{\text{heat}} 2M(l) + 3O_2(g)$ _____

30. $2MCl(l) \xrightarrow[\text{electrolysis}]{\text{heat}} 2M(l) + Cl_2(g)$ _____

31. $2MCl_3(l) \xrightarrow[\text{electrolysis}]{\text{heat}} 2M(l) + 3Cl_2(g)$ _____

32. $4M(s) + 8CN^-(aq) + O_2(g) + 2H_2O(l) \longrightarrow 4[M(CN)_2]^-(aq) + 4OH^-(aq)$ _____

33. $Ca(OH)_2(s) + M^{2+}(aq) \longrightarrow Ca^{2+}(aq) + M(OH)_2(s)$ _____

Multiple Choice

_____ 1. Which of the following metals is most likely to occur in the native state?
(a) platinum   (b) sodium   (c) magnesium (d) iron   (e) mercury

_____ 2. Which of the following metals is produced in large quantities by treatment of its salts as found dissolved in sea water?
(a) sodium   (b) magnesium (c) mercury (d) gold   (e) iron

_____ 3. Which of the following is not a method for reduction of ores?
(a) electrolysis   (b) reaction with coke   (c) reaction with hydrogen
(d) roasting   (e) zone refining

_____ 4. Which of the following metals is produced primarily by electrolytic reduction of its molten oxide?
(a) gold   (b) iron   (c) copper   (d) platinum   (e) magnesium

_____ 5. The purification process known as zone refining depends on the fact that
(a) many impurities will not fit into the lattice of a pure crystalline substance.
(b) different substances have different densities.
(c) some metals are more easily reduced than others.
(d) some salts are soluble in water.
(e) some particles are wet by oil but not by water.

_____ 6. Which of the following metals finds use in lightweight structural alloys because of its low density?
(a) gold   (b) iron   (c) copper   (d) platinum   (e) magnesium

_____ 7. Purification of which of the following metals involves precipitation of a slightly soluble hydroxide?
(a) gold   (b) iron   (c) copper   (d) platinum   (e) magnesium

_____ 8. In the metallurgy of iron, the metal obtained from the separation of the slag and the molten metal is called

    (a) gold   (b) iron   (c) copper   (d) platinum   (e) magnesium

_____ 9. The purpose of adding sand to the roasted copper ore in the reverberatory furnace is to

    (a) oxidize the copper oxide to copper metal.

    (b) provide structural strength to the copper metal.

    (c) aid in the melting for the subsequent electrolytic refining of copper.

    (d) react with the limestone impurity to form a molten silicate glass.

    (e) cause the metal to conduct so it can be electrolyzed.

Answers to Preliminary Test

Do not just look up answers. Think about the reasons for the answers.

True-False

1. True. See Chapter 6 and Section 22-1.
2. True. Remember that metallic character decreases going left to right across a period in the periodic table.
3. False. Remember that metallic character increases going down a group in the periodic table. If you need help to remember this trend, look at Group IVA, which goes from carbon (a nonmetal) through silicon and germanium (metalloids) to tin and lead (metals). Review Section 4-1.
4. True
5. True. Mercury is a liquid at room temperature, and cesium and gallium melt slightly above room temperature. All other metals have higher melting points.
6. False. To be quite electronegative means to have a tendency to gain electrons; metals tend to lose electrons, having low ionization energies.
7. False. Because metals tend to lose electrons, they are easily oxidized; hence they act as reducing agents.
8. False. Only a few are sufficiently inactive that they are found in the native state. Can you name some of these metals?
9. False. Read Section 22-2 regarding the SO_2 formed in the process.
10. False. Some metals occur in the free or native state, and require no reduction.
11. False. As described in Section 22-9, additional gold can be recovered by amalgamation with mercury followed by distillation of the mercury, or by oxidation in the presence of cyanide ion to form the soluble complex ion $[Au(CN)_2]^-$, followed by electrolytic reduction.

Short Answer

1. metalloids
2. iron
3. zinc
4. basic; amphoteric
5. oxide; sulfide; chloride; carbonate; sulfate; silicate (sometimes phosphate). These answers could appear in any order.

6. pretreatment
7. density
8. flotation; oil; air
9. oxides; heating in the presence of oxygen
10. mercury; cinnabar; HgS
11. smelting. In such processes, the metal is often obtained in molten (liquid) form.
12. $CaCO_3$. This conversion is used both in the production of calcium (Section 22-3) and to produce CaO for use in the production of magnesium (Section 22-5).
13. electrochemical (or electrolytic). You may wish to reread Section 21-3 regarding the production of sodium metal by electrolysis of molten sodium chloride in the Downs cell. This method is used to produce the metals of Groups IA and IIA, as well as some others.
14. impure carbon
15. $WO_3 + 3H_2 \rightarrow W + 3H_2O$. See Table 22-3. This method is also used for molybdenum.
16. bauxite; Hall
17. gangue. Can you pronounce this word?
18. flux
19. magnesium; sodium. These two have been discussed—magnesium in Section 22-9 and sodium in Section 21-3. As pointed out in Table 22-3, this method is also used for calcium and the other active Group IA and IIA metals.

As you think about the reactions in Questions 20 through 33, remember to take into account the stoichiometry in identifying the metals. For example, even though the Group IIA metals are produced by electrolysis of their molten chloride, neither the equation in Question 30 nor that in Question 31 could represent this process, because these metal chlorides have the formula MCl_2.

20. Ca
21. Mg
22. Zn, Ni, Pb
23. Cu
24. Hg
25. Sn
26. Zn
27. Fe
28. Fe. Both this reaction and that in Question 27 take place in the blast furnace.
29. Al
30. Na (as well as the other Group IA metals)
31. Al
32. Au
33. Mg

Multiple Choice

1. (a). This is discussed in Section 22-1. The way to tell is by comparing reduction potentials for the cations of these metals. Metals with higher reduction potentials are less active and are more likely to occur in the native state. The list of metals inactive enough to occur in the native state is sufficiently short that you should remember which metals are included.

2. (b)
3. (e). This is a method for purifying metals once they have been reduced.
4. (c). The metallurgy of aluminum is discussed in detail in Section 22-6.
5. (a)
6. (e)
7. (e)
8. (b)
9. (d). The calcium silicate that is formed also dissolves the iron(II) oxide impurity.

23 Metals II: Properties and Reactions

Chapter Summary

In this descriptive chemistry chapter, we discuss some properties and reactions of the **representative metals** and of selected **d-transition metals**. All members of Group IA (the alkali metals) and Group IIA (the alkaline earth metals), as well as some of the lower (heavier) members of Groups IIIA, IVA, and VA, have predominantly metallic properties. Each group of metals is discussed in this chapter from three aspects—(1) **properties and occurrence**, (2) **reactions**, and (3) **uses of the elements and their compounds**.

Descriptive chemistry contains much detail, but the textbook presents it so that you can learn it with a systematic approach. The following suggestions may help you to organize and remember the material in each of the remaining chapters. First, do not first read the entire chapter in detail, front to back—go over it first lightly, just to see the main points of the presentation. Second, outline the chapter in detail for yourself, perhaps one group of elements at a time, noting particular examples and more details of the main ideas that were pointed out. Third, try to remember something about the occurrence of the elements in each group—at least which are the most common and which are quite rare—and the major sources of the elements of the group. Fourth, as you study each group, try to see the trends in the general chemical and physical properties, such as those summarized in Table 23-1 for the alkali metals. Such trends and similarities are more important to remember and to understand than the particular numerical values. For example, it is not as important to remember that sodium metal melts at 97.5°C as it is to recall that all alkali metals melt at relatively low temperatures and that the melting points decrease with increasing atomic number—and to relate these observations to atomic properties. Also, be sure that you see and understand exceptions to these trends. Often, even the significant exceptions are systematic enough to help you remember them. For example, the somewhat anomalous properties of the first members of many groups and the resulting **diagonal similarities** discussed in the latter part of Section 23-2. Try to organize your study of the specific chemical reactions of the elements of each group along these same general lines. In the discussion of each group of elements, a table summarizes the reactions of the metals of that group (e.g., Table 23-2 for the alkali metals), but you also need to remember when one or two members of a group display a different reactivity than other elements in the group. The interesting Chemistry in Use essay "Trace Elements and Life" early in this chapter discusses the biological importance of a few elements, most of them metals.

The **alkali metals, Group IA**, are discussed in Section 23-1 through 23-3. You should be able to name these metals without reference to the text. All of these melt at relatively low temperatures and are relatively low-density metals whose properties are determined by their ns^1 outer shell configurations. All are very electropositive, with low first ionization energies and very negative reduction potentials. These metallic properties all increase going down the group. Thus, all alkali metals are very reactive elements, often reacting so vigorously and with evolution of so much heat that they can be dangerous, as in their reactions with water. They are never found free in nature, but occur in many stable salts, all with the metals in the +1 oxidation state. In their reactions (Section 23-2), the alkali metals are all strong reducing agents. The oxides and hydroxides of all of these metals are strongly basic. In your study of Group IA, be sure to notice that the physical and chemical properties of lithium are anomalous in nearly all respects, presumably as a consequence of its very small size and resulting high charge density. Some of the many uses of alkali metal compounds, especially those of sodium, are listed in Section 23-3.

The metals of **Group IIA**, the **alkaline earth metals** (can you name them?) are discussed in Sections 23-4 through 23-6. All of these metals have the outer configuration ns^2. They are not as reactive as the alkali metals, but are still reactive enough that they are never found free in nature. The lighter elements, especially Be, have more tendency to form covalent substances than any of the alkali metals. Calcium and magnesium are the especially abundant members of the group. As you see in Section 23-5, the types of reactions of these elements are similar to those of the alkali metals, except for stoichiometry. Notice especially the trends in basic and amphoteric nature of the oxides of Groups IA to IIIA, summarized in this section. Several important uses of compounds of the Group IIA metals, especially of the very common calcium and magnesium, are mentioned in Section 23-6.

Each of the other groups of elements in this chapter is a post-transition metal group—i.e., it is to the right of the transition metals in the periodic table. Each of these groups ranges from predominantly nonmetallic elements at the top of the group to the metallic ones at the bottom. Only elements with significantly metallic properties are discussed in this chapter. The corresponding discussion of the nonmetallic elements of these groups appears in Chapter 24.

Periodic table Groups IIIA and IVA each consist of elements ranging from nonmetallic to metallic properties. Section 23-7 deals with periodic trends in these two groups. You will see that gradations in properties in the post-transition metal groups are more extreme than they are in Group IA or IIA. All Group IIIA metals can exhibit +3 oxidation state, but with predominantly covalent bonding. The elements of this group that are to the right of a transition metal series (Ga, In, Tl) also exhibit increasingly stable +1 oxidation states going down the group, often with marked ionic character. This is due to the tendency, which increases going down the group, for the elements to lose or share only their outer p electrons, with the s electrons remaining nonionized or unshared. This **inert s-pair effect** is common to all post-transition metals. This section also discusses the properties and chemistry of **aluminum**, the most important of the post-transition metals. As you will see, many of its important properties depend on its intermediate, or metalloid, character. Hydrolysis reactions (Chapter 18) play an important role in the aqueous chemistry of aluminum.

In the latter portion of Chapter 23, we undertake a survey of the major aspects of the properties and chemistry of the **d-transition metals.** There are many of these elements and each has a quite varied chemistry, so we will not study all of the elements in detail; that would (and does!) require several quite large volumes. Rather, we are concerned here with summarizing the major trends in properties and behavior of these elements. We will look more closely at a few compounds of only one of them, chromium.

The d-transition elements (usually called just the transition metals) are the ten columns of elements located between Groups IIA and IIIA in the periodic table. Their general properties are summarized in Section 23-8.

When a group of elements has such a widely varied chemistry as do the transition elements, we must organize the reactions or compounds of the elements in some way; as we have seen before, a very convenient organization is in terms of the **oxidation state** of the element. You should pay attention to the discussion in Section 23-9 with regard to the existence of several oxidation states for each transition metal and the relative reducing and oxidizing tendencies of the element in its various oxidation states. It will help to remember that in its simple ions, each of the transition metals loses its outer s electrons (or sometimes electron, as in Cu^+), and may also lose one or more d electrons. It may also be helpful to observe that, for most odd-numbered groups, at least some of the most important oxidation states are odd, whereas for even groups, several of the principal oxidation states are even (though this is by no means an infallible rule but only a guide to your memory). As is very clear from Table 23-7, the tendency to form several different oxidation states is much higher near the middle of each series of transition metals than near the ends. Moving down within a group, higher oxidation states become more stable and common, which is an opposite trend from that observed among the representative elements. Other trends regarding correlations of acidity, basicity, and covalent character with oxidation state and with position in the group are parallel to those already seen for the representative elements.

As mentioned earlier, the chemistry of the transition metals is far too vast and too varied to cover in detail in any one chapter. However, it is illustrative to look a little more closely at one of these metals. The next section does this, going into some detail regarding the reactions and properties of the oxides and hydroxides of chromium (Section 23-10). The hydroxides of transition metals in high oxidation states are acidic, with acidity increasing with oxidation state, just as was seen earlier for the representative nonmetals. Compounds with chromium in its highest oxidation state (+6) are very strong oxidizing agents.

Study Goals

1. Review the general physical and chemical characteristics of metals. Know where the representative metals are located in the periodic table. Be able to write electronic configurations for metals and their ions. (Review Sections 4-1, 5-17, and 6-8; introduction to Chapter 23; Exercises 1 through 5 and 11 through 13)

2. Describe, with examples, what is meant by diagonal similarities (diagonal relationships). (Section 23-2; Exercises 23, 24, and 29)

3. Account for the trends in periodic properties of the alkali metals, Group IA. Give several examples of typical reactions of the Group IA metals. Know some of the uses of the alkali metals and their compounds. (Sections 23-1 through 23-2; Exercises 7 through 10, 14 through 20, 27, 29, and 30)

4. Account for the trends in periodic properties of the alkaline earth metals, Group IIA. Give several examples of typical reactions of the Group IIA metals. Know some of the uses of the alkaline earth metals and their compounds. (Sections 23-4 through 23-6; Exercises 14 through 18, 21, 22, 26, 28, and 30)

5. Summarize and illustrate, with properties and reactions, the trends in metallic and nonmetallic characteristics of the post-transition metals. Summarize and illustrate the trends in basic and acidic characteristics of their oxides, hydroxides, and salts. Know some of the properties and uses of aluminum. (Section 23-7)

6. Summarize the general properties of the d-transition metals. Explain why the properties of successive transition metals in one period vary less dramatically than those of representative metals. (Section 23-8; Exercises 31 through 36)

7. Write out the electron configurations of the first row (Period 4) transition metals and their common ions. Explain why the Period 5 and Period 6 transition metals contain more exceptions to the expected, or Aufbau, electron configurations than do the Period 4 metals. (Review Sections 5-17 and 5-18; Section 23-9; Exercises 34 and 35)

8. Know trends in oxidation states, occurrence, and stabilities for the d-transition metals. (Section 23-9; Exercises 37 through 39)

9. Show how the properties and reactions of the oxides and hydroxides of chromium demonstrate the ideas of this chapter. Be familiar with the structures, properties, and reactions of compounds of chromium. (Section 23-10; Exercise 40)
10. Answer conceptual questions based on this chapter. (Conceptual Exercises 41 through 45)
11. Apply principles learned earlier to perform calculations concerning the representative metals, the *d*-transition metals, and their compounds. (Review earlier chapters as needed—especially Chapters 2, 3, 12, 15, 19, 20, and 21; Exercise 31; Building Your Knowledge exercises 46 through 49)

Some Important Terms in This Chapter

alkali metals

alkaline earth metals

post-transition metals

diagonal similarities

acidic anhydride

basic anhydride

amphoterism

inert *s*-pair effect

transition metal

oxidation state

Preliminary Test

True-False

Mark each statement as true (T) or false (F).

_____ 1. All transition elements are metals and all representative elements are nonmetals.
_____ 2. Of the alkali metals, only the heavier ones are sufficiently inactive to be found free in nature.
_____ 3. Alkali metals can easily form either 1+ or 2+ ions because they have such low ionization energies.
_____ 4. Francium is so rare an element that its properties have not even been characterized well.
_____ 5. Because they are easily oxidized by atmospheric oxygen, the alkali metals are usually stored under water.
_____ 6. In many ways, the compounds of lithium resemble those of magnesium more than they do those of the other alkali metals.
_____ 7. Because they have only ns^1 electrons in the outer shell, alkali metals generally show the +1 oxidation state and never the +2.
_____ 8. Because they have ns^2 configurations in their outer shell, alkaline earth metals frequently show either the +1 or the +2 oxidation state in their compounds.
_____ 9. All alkali metals react readily with water at room temperature to produce hydroxides and hydrogen.
_____ 10. All alkaline earth metals react readily with water at room temperature to produce oxides or hydroxides and hydrogen.
_____ 11. Because of its ease of oxidation, aluminum is rarely used in the pure state.
_____ 12. The Group IIIA metals are sufficiently less reactive than those of Groups IA and IIA that they (Group IIIA) frequently occur free in nature.
_____ 13. All of the representative elements are solids at room temperature.
_____ 14. Oxides and hydroxides of an element in a higher oxidation state are always more acidic and covalent than oxides and hydroxides of the same element in a lower oxidation state.
_____ 15. Nonmetallic character decreases as the oxidation state of a particular metal increases.
_____ 16. The transition metals are generally more reactive than the Group IA and IIA representative elements.
_____ 17. The d-transition metals discussed are in B groups of the periodic table.

Short Answer

Answer with a word, a phrase, a formula, or a number with units as necessary.

1. The names and symbols of the metals of Group IA are _____, _____, _____, _____, _____, and _____.
2. As a common group name, the elements of Group IA are referred to as the _____.

3. The names and symbols of the metals of Group IIA are _____, _____, _____, _____, _____, and _____.

4. As a common group name, the elements of Group IIA are referred to as the _____.

5. The names and symbols of the metals of Group IIIA are _____, _____, _____, and _____.

6. The names and symbols of the metals of Group IVA are _____ (a metalloid), _____, and _____.

7. The name and symbol of the metal of Group VA _____.

8. The two most commonly occurring metals of Group IA are _____ and _____.

9. The melting points of the alkali metals _____ going down the group.

10. All of the metals of Group IA have quite _____ reduction potentials, which tells us that they are very _____ to reduce to the metallic state.

11. Denoting the metal by M, the formula of the normal oxides of the Group IA metals is _____, the formula for the peroxides of these metals is _____, and the formula for their superoxides is _____.

12. The only Group IA metal that combines with nitrogen to form a nitride is _____.

13. When burned in limited O_2, the heavier alkali metals form compounds called _____, having formula _____; in the presence of excess O_2, the metal compounds formed are called _____ and have formula _____.

14. The formulas and uses of two compounds of sodium are _____ _____ (formula, use) and _____ _____(formula, use).

15. Aqueous solutions of the oxides and hydroxides of all Group IA elements are strongly _____.

16. In general, oxides of the representative metals become decreasingly acidic going _____ across the periodic table and going _____ within a group of the periodic table.

17. One use of beryllium is _____.

18. The substance that is used to coat the gastrointestinal tract for obtaining x-ray photographs is _____.

19. Until recently, the metal used in electrical wiring was primarily _____, but it is probable that _____ will eventually replace it.

- -

Each of the equations in Questions 20 through 27 represents a reaction of one or more of the representative elements (M), from Groups IA or IIA, as discussed in Chapter 23. For each equation, tell whether the reaction is typical of the metals of Group IA, IIA. Some reactions may be limited to only one or a few of the metals of the group; if so, you should indicate that. Some other reactions may be applicable to metals from more than one group. Be careful of the stoichiometry.

20. $4M + O_2 \rightarrow 2M_2O$ (limited O_2) _____

21. $6M + N_2 \rightarrow 2M_3N$ _____

22. $M + O_2 \rightarrow MO_2$ _____

23. $2M + 2H_2O \rightarrow 2MOH + H_2$ _____

24. $M + 2H_2O \rightarrow M(OH)_2 + H_2$ _____

25. $M + X_2 \rightarrow MX_2$ (X = halogen) _____

26. $2M + X_2 \rightarrow 2MX$ (X = halogen) _____

27. $M_2O + H_2O \rightarrow 2[M^+ + OH^-]$ _____

> Questions 20 through 27 emphasize only a few of the fairly general reaction types of the Groups IA and IIA elements. Tables 23-2 and 23-4 in the text include many more reactions of some generality. In addition, the reactions that may be specific to one or a few metals should also be studied—e.g., the many reactions demonstrating acidic, or basic behavior of oxides, hydroxides, and salts.

- -

28. The electron configuration of Ti is _____.
29. The electron configuration of V is _____.
30. The electron configuration of Cr is _____.
31. Because many transition metals and metal ions have one or more unpaired electrons, they are _____.
32. The three common oxidation states of chromium are ____, ____, and ____.
33. An example of a compound in which chromium has an oxidation number of +2 is _____.
34. An example of a compound in which chromium has an oxidation number of +3 is _____.
35. Two ions in which chromium has an oxidation state of +7 are _____ and _____ (name and formula for each).
36. Chromate and dichromate ions in solution exist in a pH-dependent equilibrium, for which the equation is _____.
37. A common (but potentially hazardous) laboratory glassware cleaning solution is made by adding concentrated _____ to a concentrated aqueous solution of _____.

Multiple Choice

____ 1. Which of the following oxides is most basic?
(a) B_2O_3 (b) Ga_2O_3 (c) BeO   (d) CaO   (e) K_2O

____ 2. Which of the following oxides is most acidic?
(a) B_2O_3 (b) Ga_2O_3 (c) BeO   (d) CaO   (e) K_2O

____ 3. Which of the following is not a substance that includes calcium?
(a) quicklime   (b) gypsum   (c) slaked lime   (d) mortar   (e) baking soda

____ 4. Which of the following is the most abundant metal in the earth's crust?
(a) aluminum   (b) sodium   (c) silicon   (d) oxygen   (e) iron

____ 5. What metal has the largest liquid-state temperature range of any element?
(a) gallium   (b) mercury   (c) aluminum   (d) potassium   (e) bromine

____ 6. The tendency for the post-transition metals to retain their *s* electrons nonionized or unshared is referred to as
(a) the diagonal similarity.   (b) the heat of hydration.   (c) the inert *s*-pair effect.
(d) amphoterism.   (e) stoichiometry.

_____ 7. Which of the following is not a general property of the transition elements?
   (a) They form many complex ions.
   (b) They often exhibit multiple oxidation states.
   (c) They are metals.
   (d) They form their most stable ions by attaining a noble gas configuration.
   (e) Their ions and compounds are often colored.

_____ 8. The colors exhibited by many transition metal ions and compounds are caused by
   (a) light absorption of wavelengths in the visible region of the spectrum.
   (b) light emission of wavelengths in the visible region of the spectrum.
   (c) inability to form positive ions.
   (d) the different color of electrons in a d-orbital from those in a p-orbital.
   (e) the large number of oxidation states available to these elements.

_____ 9. Which $3d$-transition element exhibits the largest number of oxidation states?
   (a) Mn   (b) Cr   (c) Fe   (d) An   (e) Ni

_____ 10. Which of the following oxides of chromium is the strongest oxidizing agent?
   (a) CrO   (b) Cr_2O_3 (c) CrO_3

_____ 11. Which of the following oxides of chromium is most acidic?
   (a) CrO   (b) Cr_2O_3 (c) CrO_3

Answers to Preliminary Test

Do not just look up answers. Think about the reasons for the answers.

True-False

1. False. It is true that all transition elements are metals, but not all representative elements are nonmetals. All of the elements of Groups IA and IIA, as well as several of those of Groups IIIA, IVA, and VA have pronounced metallic character.
2. False. All alkali metals are far too reactive to occur in the native state. The heavier ones are even more active than the lighter ones.
3. False. These metals have low first but very high second ionization energies. Thus, they do not form 2+ ions by ordinary chemical means. Think about what electron configuration would need to be disturbed for one of these metals to go from a 1+ to a 2+ ion.
4. True
5. False. They react with water to form hydrogen in a manner ranging from "readily" for Li, through "vigorously" for Na, to "explosively" for K, Rb, and Cs. They are stored under anhydrous nonpolar liquids such as mineral oil.
6. True. Do you know why this is so? (Do not answer "diagonal similarity"—this is just another way of saying that it is so, but not why.)
7. True
8. False. Alkaline earth metals (Group IIA) do not exhibit the 1+ oxidation state, because their second ionization energies are also quite low.
9. True

10. False. The Group IIA metals are sufficiently less reactive than those of Group IA that only the heavier alkaline earth metals react with water at room temperature. Magnesium reacts with water only as steam. Even when red hot, beryllium will not react with pure water. See Section 23-5.

11. False. In fact, the easily formed, hard, unreactive, transparent oxide coating protects the underlying metal from further oxidation.

12. False. These are still so reactive that they never occur in the native state. Aluminum occurs mainly in aluminosilicate minerals (from which it cannot easily be extracted) and as the oxide. Gallium, indium, and thallium occur mainly as the sulfide, but are much rarer than aluminum.

13. True. Mercury is the only metal that is liquid at room temperature. It is in Group IIB, so it is not a representative metal.

14. True. To help you remember and understand this, read Section 19-5 again. There you learned that the hydrolysis of small, highly charged metal ions (that is, the action of their hydrates as acids) is due to the strong pull of the highly charged metal for the electrons in the water molecule bonded to the metal.

15. False

16. False

17. True

Short Answer

1. lithium (Li); sodium (Na); potassium (K); rubidium (Rb); cesium (Cs); francium (Fr)
2. alkali metals
3. beryllium (Be); magnesium (Mg); calcium (Ca); strontium (Sr); barium (Ba); radium (Ra)
4. alkaline earth metals
5. aluminum (Al); gallium (Ga); indium (In); thallium (Tl). Boron is not considered to have significant metallic properties.
6. germanium (Ge); tin(Sn); lead (Pb). Carbon and silicon are not metals.
7. antimony (Sb); bismuth (Bi). The lighter elements of this group are not metals.
8. sodium (Na); potassium (K)
9. decrease
10. negative; difficult. Review the meaning of reduction potentials, Section 21-14.
11. M_2O; M_2O_2; MO_2. Do you remember which metals form which of these oxides and under what conditions? See Table 23-2.
12. lithium (Li)
13. normal oxides, M_2O; superoxides, MO_2. Remember that only K, Rb, and Cs form superoxides.
14. Many possible answers to this question appear in the first paragraph of Section 23-2.
15. basic
16. right to left; down
17. as windows for X-ray tubes. See the photo on page 930.
18. $BaSO_4$
19. copper; aluminum
20. Group IA (all metals)
21. Group IA (Li only). Magnesium also reacts with nitrogen to form nitrides, but the stoichiometry is different. Can you write that equation and the formulas of the Group IIA nitrides?

22. Group IA (K, Rb, Cs only, in the presence of excess O_2—these are the Group IA superoxides); Group IIA (Ba only—this is barium peroxide); Group IVA (all metals—these are the normal oxides of Group IVA). Be sure you see that the formulas for the various oxides (normal, per-, and super-) differ for different groups.
23. Group IA. (The heavier alkali metals undergo this reaction in a dangerously explosive manner)
24. Group IIA (Ca, Sr, Ba at room temperature, Mg only at high temperature, Be not at all)
25. Group IIA; Group IVA (Pb only)
26. Group IA
27. Group IA
28. $[Ar]3d^24s^2$
29. $[Ar]3d^34s^2$
30. $[Ar]3d^54s^1$. This irregularity in electron configuration is related to to the special stability of the d^5 configuration. Copper is an analogous exception to the usual order, as a result of the stable d^{10} configuration.
31. paramagnetic. The relationship between unpaired electrons and paramagnetism was first discussed in Chapter 5. This property is a useful aid in confirming electron configurations.
32. +2; +3; +6
33. Some examples are seen in Table 23-8, including CrO, $Cr(OH)_2$, and various salts such as $CrCl_2$.
34. See Table 23-8. Examples are Cr_2O_3, $Cr(OH)_3$, and various salts such as $CrCl_3$.
35. chromate (CrO_4^{2-}); dichromate ($Cr_2O_7^{2-}$)
36. $2CrO_4^{2-} + 2H^+ \rightarrow Cr_2O_7^{2-} + H_2O$
37. H_2SO_4; $K_2Cr_2O_7$.

Multiple Choice

1. (e). Trends in acidic and basic character of oxides were discussed in Section 6-8, part 1.
2. (a)
3. (e). Can you write the formula for each of these substances?
4. (a). Both silicon and oxygen are more abundant than aluminum in the earth's crust, but neither is classified as a metal.
5. (a)
6. (e)
7. (d). Very few transition metal ions have noble gas configurations. Can you name some that do?
8. (a)
9. (a)
10. (c). Remember that strengths of oxidizing agents increase with increasing oxidation number for the same element.
11. See Table 23-8.

24 Some Nonmetals and Metalloids

Chapter Summary

Chapter 24 briefly presents the descriptive chemistry of some representative **nonmetallic** elements and some **metalloids.** First, we shall learn about the group of quite unreactive elements at the extreme right of the periodic table, the **noble gases** (Sections 24-1 and 24-2). Group VIIA, the **halogen group**, which consists of very reactive nonmetals, is the subject of Sections 24-3 through 24-7. Sulfur, a very important element from Group VIA, is discussed in Sections 24-8 through 24-12. The two elements at the top of Group VA, nitrogen and phosphorus, are covered in Sections 24-13 through 24-17. The chapter closes (Section 24-18) with a brief section on silicon, from Group IVA.

Because of their rarity on earth (they were formerly called the rare gases) and their lack of occurrence in any natural compounds due to their very low reactivity, the noble gases were quite late to be discovered. The discovery, isolation, and occurrence of these gases are described in Section 24-1. It is interesting to observe that argon, a noble gas, is actually the third most abundant element in the earth's atmosphere (about one percent), many times more abundant there than hydrogen. (Of course, hydrogen is very abundant on earth, but most of it appears in compounds, of which the vast bulk is water.) Section 24-2 also gives the physical properties of the noble gases. Because of their very low interatomic attractions, all have low melting and boiling points, increasing with increasing atomic number (i.e., going down the group). Several uses of the noble gases are given in Table 24-2. As you might expect, most of these uses depend on the relative chemical inertness of these elements. Some applications involve the emission of light under the influence of an electric field. Some of the uses of helium are based on its unusually low density. Hydrogen, the only gas that is less dense than helium at the same conditions, has many disadvantages due to its high reactivity, especially with oxygen.

In the 1960s it was discovered that some of these elements, previously thought to be totally inert to chemical reaction, could form compounds. It is to be noted, however, that the only known reactions of the noble gases are with extremely powerful oxidizing agents. Their compounds all involve the noble gas elements in positive oxidation states. The first compounds of noble gases to be made, and those best characterized to date, are the **xenon-fluorine compounds** (Section 24-2). In the known compounds of this class, XeF_2, XeF_4, and XeF_6, the hybrid orbitals involve Xe d orbitals, which accounts for the deviation from the octet rule. No compounds are known of argon, neon, or helium, the lighter members of the group.

The next few sections concern the properties, reactions, and compounds of the elements of Group VIIA, the **halogens.** As is described in Sections 24-3 and 24-4, the halogens are all nonmetals, displaying a greater similarity in properties than any other group except the noble gases and possibly the Group IA metals. Table 24-4 emphasizes the quite regular gradation in properties of the halogens. In most of their compounds, these elements exhibit the −1 oxidation state, but, except for F, they can also show +1, +3, +5, and +7 states.

The occurrence, production, and uses of the individual halogens are detailed in Section 24-4. As we expect from their quite high reactivities, these elements do not occur free in nature; they appear most commonly as halide salts containing X^- (X is the halogen). (1) **Fluorine** is produced mostly by the electrolysis of molten potassium hydrogen fluoride. The free element is used as a powerful oxidizing agent. Many fluorocarbons (compounds involving C–F bonds) are extremely stable, leading to a variety of uses as refrigerants, lubricants, plastics, aerosol propellants, and coating agents. The chemistry of the familiar use of fluoride in preventing tooth decay is explained. (2) **Chlorine,** produced primarily by electrolysis, is used extensively in extractive metallurgy, as a disinfectant, in making chlorinated hydrocarbons, and in bleaching agents. (3) **Bromine,** one of the two liquid elements at room temperature, is a very volatile, corrosive substance. It is produced in the free state primarily by extraction from sea water or brine wells and by displacement of Br^- by Cl^- (already discussed in Section 4-8). Compounds of bromine find use as light-sensitive substances (such as silver bromide) in photographic emulsions, as a sedative, and as a soil fumigant. (4) **Iodine,** a volatile black crystalline solid, is obtained from dried seaweed or from $NaIO_3$ impurities in nitrate deposits. Its biological importance in the function of the thyroid gland was discussed in the "Trace Elements and Life" essay in Chapter 23. Iodine is used as an antiseptic, and its compounds find use as antiseptics and germicides.

Section 24-5 describes some reactions of the free halogens. These are sufficiently strong oxidizing agents to react with most other elements. Oxidizing strengths range from F_2, a very vigorous agent, to I_2, a quite mild one. The reactions with iron and copper detailed early in this section exemplify this, but you can see the same trend by studying the remarks in this section. Notice that the displacement reaction with other halogens, $X'_2 + 2X^- \rightarrow 2X'^- + X_2$, is limited to X' being lighter than X, consistent with the order of oxidizing power of the diatomic free elements.

The most important and generally useful compounds of the halogens are the **hydrogen halides, HX**. The preparation and properties of these compounds are summarized in Section 24-6. These compounds can be prepared by direct reaction of the elements H_2 and X_2, but at greatly varying rates, ranging from the fastest for F_2 to the slowest for I_2. Hydrogen halides, which are covalent substances, react with water to give the **hydrohalic acids,** of which HF(aq) is a weak acid and HCl(aq), HBr(aq), and HI(aq) are strong. The strengths of the last three, which cannot be discriminated in water, are seen to increase going down the group. Some uses of the hydrohalic acids are also described in this section.

Section 24-7 is concerned with the other important class of inorganic halogen compounds, the **oxoacids (ternary acids)** and their salts. You may wish to review the naming of ternary acids and their salts, Section 4-12. These compounds are conveniently organized and studied according to the oxidation state of the halogen. Note that fluorine rarely exists in an oxidation state other than –1 in a compound; this is because the most electronegative element in a compound is always assigned the negative oxidation number (Section 4-4) and fluorine is the most electronegative element. Stability of higher oxidation states increases with increasing halogen size and atomic number.

The next several sections are concerned with the three most common heavier Group VIA elements: **sulfur, selenium,** and **tellurium**. (Polonium, a very rare radioactive element, is not well characterized by chemists.) There is considerable variation in the properties of this group of elements, much more so than for the halogens. This is the case when we move further from the extreme ends of the periodic table. By looking at the group properties presented in Table 24-6, you can see that although most properties undergo a trend within the group, the big jump in properties is between oxygen (which has no valence shell d orbitals) and sulfur (which does). For instance, oxygen is the only gaseous element in Group VIA—the other members of the group are solids. The other elements of the group only differ by about 0.5 electronegativity units altogether, whereas oxygen is one full unit higher than sulfur; the boiling point changes by more than 600 degrees from oxygen to sulfur and much less after that; oxygen exhibits almost exclusively the -2 oxidation state, whereas for the members of the group, $+2$, $+4$, and $+6$ are also common. You can see other such irregularities in this table and as you study the chapter in more detail. Recall that the chemistry of the most nonmetallic Group VIA element, oxygen, was discussed in Section 6-8.

The occurrence, properties, and uses of these elements are discussed in Section 24-1. Because of its ability to bond to itself, sulfur can exist in several forms in each of the three physical states. As you learn here, the properties of the two solid and the various liquid forms of sulfur depend on the extent of S—S bonding present. Selenium, a rare element, can also exhibit several different solid forms and exists in various molecular forms in the vapor phase. Several uses of elemental selenium are mentioned, including some that depend on its sensitivity to light. Not surprisingly, selenium, like the even more rare tellurium, occurs mainly in sulfide ores.

The general reactions of the Group VIA nonmetals are summarized in Table 24-2, Section 24-9. As you study this table, notice that not all elements of the group undergo all of these reactions. All of the Group VIA elements form covalent hydrides H_2E (E represents a Group VIA element), with E in the -2 oxidation state. Sections 24-9 and 24-10 are concerned with the preparation, properties, and reactions of the heavier Group VIA hydrides, both as compounds and in aqueous solution. Just as for the hydrogen halides (Section 24-6), acidity of the Group VIA hydrides increases going down the group.

All of the heavier Group VIA elements form oxides, Section 24-11. The most important of these are the dioxides, EO_2, and the trioxides, EO_3. The tendency toward increasing metallic character going down a group is shown dramatically by the stable dioxides of the VIA elements. As has been mentioned before, SO_2 is a serious atmospheric pollutant that is formed by the combustion of sulfur-containing fossil fuels and by the roasting of sulfide ores. Some of the chemistry of this noxious, dangerous gas and some methods for removing it from flue gases are discussed in this section.

The oxides of the Group VIA elements are acid anhydrides: (1) the dioxides, EO_2, which dissolve in water to give the "-ous" acids, and (2) the trioxides, EO_3, which are the anhydrides of the "-ic" acids. Section 24-12 is concerned with the chemistry of the oxoacids of sulfur, selenium, and tellurium. The most important of these are sulfurous and sulfuric acids and their salts, the sulfites and sulfates, respectively.

In Group VA, sometimes referred to as the **nitrogen family**, the properties of the elements range from the quite nonmetallic **nitrogen** and **phosphorus**, through predominantly nonmetallic arsenic and more metallic **antimony**, to metallic **bismuth**. The latter was discussed with the other representative metals in Chapter 23. In this chapter we learn more about the two lightest nonmetallic members of this group: nitrogen and phosphorus.

As you can see in Table 24-7, the elements of this group show the expected trends in properties. All of the elements of Group VA, including nitrogen, show an unusually large range of oxidation states in their compounds. Thus, their chemistry is quite varied and you will find it convenient to organize your study of these elements by oxidation number of the Group VA elements. These elements also show chemical trends with which you are probably familiar by now—for instance, the regular variation from acidic to basic properties of the oxides formed from Group VA elements in the same oxidation state. This trend is illustrated by the oxides N_2O_3, P_4O_6, As_4O_6, Sb_4O_6, and Bi_2O_3.

Although nitrogen is the most abundant element in the atmosphere, it occurs in the form of its compounds as only a minor fraction of the earth's crust (Section 24-13). All living matter contains nitrogen, especially in proteins and in nucleic acids. In the nitrogen cycle, atmospheric nitrogen is converted by bacteria into a form that is chemically available to higher living systems. Nitrogen in the form of N_2 is one of the most stable elements except for the noble gases and the noble metals. It does, however, have a very rich and interesting chemistry, forming a wide variety of compounds. The versatility of the chemistry of nitrogen is seen in the observation that it can exhibit all oxidation states from −3 to +5.

Ammonia, NH_3, is commercially and chemically the most important of the binary compounds of nitrogen with hydrogen (Section 24-14). The industrial production of ammonia by the Haber process was discussed in detail in Section 17-7 in connection with the concepts of chemical equilibrium and kinetics; this would be a good time for you to review that discussion. As we have seen (Chapters 10 and 18), aqueous solutions of ammonia are weakly basic and form **ammonium salts** when neutralized by acids. Ammonia can also act as a Lewis base, both with metal ions (as we shall discuss in more detail in Chapter 25) and with other electron-pair acceptors. Organic compounds called **amines**, which are structurally related to ammonia but have one or more hydrogens replaced by organic groups, are mentioned in this section. These have already been encountered in Section 18-4 as weak bases and will be discussed more generally in Sections 27-11 and 28-7, in the chapters on organic chemistry.

The variety of the chemistry of nitrogen is further shown by the number of oxides it forms, with nitrogen having all positive oxidation states from 1 to 5 in these binary compounds. The preparations, properties, reactivities, and uses of some of these, as well as their interconversions, are described in Section 24-15. These compounds and their rather involved chemistry are of great environmental significance in atmospheric processes, especially those related to the production of photochemical smog (see the Chemistry In Use essay, page 962).

In its **oxoacids** and their salts (Section 24-14), nitrogen displays oxidation states +1 (**hyponitrous acid** and **hyponitrites**), +3 (**nitrous acid** and **nitrites**), and +5 (**nitric acid** and **nitrates**). Of these, the latter two are the most important. The commercial production of nitric acid by the **Ostwald process** is described in this section. Much of the useful chemistry of nitric acid depends on its oxidizing power. You should observe that the reduction products of nitric acid in such reactions depend on the concentration of this acid, with increasing acid concentration generally leading to products with nitrogen in higher oxidation states. Oxidation of metals produces solutions of the metal ions, whereas oxidation of nonmetals yields solutions of oxoacids of the nonmetal. The use and the possible danger of nitrates and nitrites as food additives are also discussed in Section 24-16.

The occurrence, production, and a few uses of phosphorus are the topic of Section 24-17. Phosphorus is an essential nutrient to all living organisms. The natural phosphorus cycle that produces soluble phosphorus compounds is extremely slow, so the largest commercial use of phosphorus is in **fertilizers.** The production of superphosphate of lime also represents the biggest single use of sulfuric acid.

To close our brief look at the nonmetals, we study silicon and a few of its compounds (Section 24-18). Silicon is the second most abundant element in the earth's crust, occurring in vast quantities in **silica** and the **silicate minerals**. It does not occur free in nature. Pure silicon has a structure like that of diamond, but it is less dense and less hard than diamond. Uses of elemental silicon include semiconductors, transistors, solar cells, additives to steel and aluminum alloys, and as the source of silicon in silicone polymers. Even though silicon is in the same group as carbon, it has quite different chemical properties because of (1) its inability to form double bonds, (2) its tendency not to form Si-Si bonds under most circumstances, and (3) the availability of $3d$ orbitals for bonding.

The only important oxide of silicon is silicon dioxide (silica). However, unlike carbon dioxide, which is a gaseous molecular substance, silicon dioxide is a polymeric substance that is a solid at room temperature. The two familiar natural forms of silicon dioxide, quartz and flint, are described in this section. Section 24-18 discusses the widely occurring **silicates**. These include a large variety of compounds, all having SiO_4 tetrahedra linked into chains, sheets, or three-dimensional networks, but with metal ions occupying spaces between the tetrahedra. These silicate minerals can adopt a wide variety of structural arrangements. This section also includes a brief discussion of the formation of **glass** and the dependence of its properties on other elements that may be present with the silicates.

Study Goals

1. Give a few practical uses of the noble gases and the physical and chemical properties on which they depend. (Section 24-1; Exercises 1 through 4)

2. Describe some compounds of xenon. Be able to describe the formation of these compounds and the bonding of the noble gas element in each. (Section 24-2; Exercises 6 through 8)

3. Describe the bonding in the free halogens. Describe the trends in such properties of the halogens as electronegativities, standard reduction potentials, atomic and ionic radii, polarizabilities, and melting and boiling points. Give reasons for the trends. (Section 24-3; Exercises 9 through 11)

4. Describe the occurrence, production, and uses of the halogens. (Section 24-4; Exercise 12)

5. Give several typical reactions of the free halogens. Know how the tendency of the halogens to undergo these reactions varies with position in the group. (Section 24-5; Exercises 13 through 19)

6. Give several methods of preparation of the hydrogen halides and some of their properties. (Section 24-6; Exercises 20 and 21)

7. Compare and explain the trends in acid strengths of the hydrogen halides, both in anhydrous form and in aqueous solution. Describe some uses of these solutions. (Section 24-6; Exercise 22)

8. Know the names and formulas of the halogen oxoacids and their salts. Draw representative structures of perhalate, halate, halite, and hypohalite ions. (Review Section 4-10; Section 24-7; Exercises 23 through 26)

9. Give representative preparations and reactions of the perhalic, halic, halous, and hypohalous acids. Explain the trend in acid strengths of the chlorine oxoacids. (Section 24-7; Exercises 27 and 28)

10. Remember and account for the trends in properties of the Group VIA elements given in Table 24-6. Summarize the occurrence, properties, uses, and general reactions of sulfur, selenium, and tellurium. (Sections 24-8 and 24-9; Exercises 20 through 33)

11. Draw Lewis formulas for compounds of S, Se, and Te. Describe the structures and bonding in such compounds. (Sections 24-10 and 24-11; Exercises 34 and 35)

12. Write equations for the preparations of the Group VIA hydrides. Compare the properties of the VIA hydrides, including the acidic nature of their aqueous solutions. (Sections 24-9 through 24-10; Exercise 38)

13. Compare the structures, properties, preparations, and reactions of the Group VIA dioxides and trioxides. (Section 24-11; Exercises 34, 36, 41, and 46)

14. Write equations for the preparations of, and draw representative structures of sulfur oxoacids and their salts: (a) sulfurous acid and sulfite salts; (b) sulfuric acid and sulfate salts. Be familiar with the properties and reactions of these acids and salts. (Section 24-11; Exercises 34, 35, and 42)

15. Compare the Group VA elements. Account for the unusual stability of elemental nitrogen and the differences in its chemical properties from those of other VA elements. (Chapter introduction, Section 24-13; Exercises 45 through 47 and 49)

16. Summarize the occurrence, properties, preparation, and importance of nitrogen. (Section 24-13; Exercise 45, 49, and 53 through 56)
17. Describe the nitrogen cycle in nature. (Section 24-13; Exercise 8)
18. Give Lewis formulas and structures and assign oxidation numbers for some compounds of nitrogen. Know the preparations and some reactions of these compounds. (Sections 24-14 through 24-16; Exercises 53 through 56, 60, and 62 through 64)
19. Describe the structure and bonding of ammonia and the Haber process for its preparation. Describe (with equations) ammonia as a Bronsted-Lowry base and as a Lewis base. Know the important reactions of ammonia and its salts. (Review Section 17-6; Section 24-14; Exercises 50 through 58)
20. Compare the properties of liquid ammonia and liquid water. (Section 24-14)
21. Be able to draw the general structural formulas for typical amines. Know their most important chemical property. (Section 24-14)
22. Describe the role of oxides of nitrogen in the problem of photochemical smog. (Section 24-15; Chemistry In Use essay, page 884; Exercises 65)
23. Describe, with equations, the preparations of the following oxoacids of nitrogen and their salts: (a) nitrous acid and nitrite salts, (b) nitric acid and nitrate salts. Give Lewis formulas, structures and uses of these compounds. (Section 24-16; Exercises 61 through 66)
24. Summarize the occurrence, production, and uses of phosphorus and some of its compounds. (Sections 24-17; Exercises 46, 47, 49, 55, 57, 66, and 67)
25. Describe the allotropes of phosphorus in words and structural representations. (Section 24-17; Exercise 17)
26. Describe the physical properties, production (with equations), and uses of silicon. Know some reactions of the free element. (Sections 24-18)
27. Compare the structure of silicon dioxide (silica) with carbon dioxide. Understand reasons for differences in structures and properties. Describe the occurrence and some uses of silicon and the silicates. (Section 24-18)
28. Answer conceptual questions based on this chapter. (Conceptual Exercises 69 through 75)
29. Apply concepts and skills from earlier chapters to the ideas of this chapter. (Building Your Knowledge exercises, 76 through 83)

Some Important Terms in This Chapter

Here are some important terms that are used in this chapter. Several of them were first used in earlier chapters. Write the meanings of these terms in your own words. Then look them up in the Key Terms list or in previous chapters.

nonmetal

noble gases

halogens

ionization energy

bond energy

reduction potential

displacement reaction

oxoacid

ternary acid

Frasch process

acid anhydride

contact process

nitrogen cycle

Haber process

photochemical smog

photochemical oxidants

PANs

Ostwald process

allotropes

silica

silicates

Preliminary Test

Be sure to answer additional textbook Exercises. Many of these are indicated in the Study Goals.

True-False

Mark each statement as true (T) or false (F).

_____ 1. All of the noble gases are chemically inert.
_____ 2. Most of the nonmetals are found at or near the right-hand end of the periodic table.
_____ 3. All of the noble gases are gaseous at all conditions of temperature and pressure.
_____ 4. In their compounds, the noble gases always obey the octet rule.
_____ 5. The noble gases react only with very strong reducing agents.
_____ 6. Reactivity of the noble gases increases going down the group.
_____ 7. All of the halogens exist as stable diatomic gases at room temperature and atmospheric pressure.
_____ 8. Even though they are in the same group, the halogens display widely varying properties.
_____ 9. Generally, small ions are harder to polarize than large ones.

_____ 10. The halogens are generally nonmetallic in their properties.

_____ 11. Because of their strengths as oxidizing agents, halogens always show negative oxidation states in their compounds.

_____ 12. Because of its strength as an oxidizing agent, fluorine cannot be produced by electrolysis.

_____ 13. The most abundant natural sources of the halogens are halide salts.

_____ 14. Halogens are so reactive that their binary compounds with other elements are all ionic.

_____ 15. The elements of Group VIA are much more alike than are the elements of Group VIIA.

_____ 16. All of the elements of Group VIA form covalent compounds in which they exhibit the oxidation number –2.

_____ 17. All of the elements of Group VIA form covalent compounds in which they exhibit the oxidation number +6.

_____ 18. The hydrides of all of the Group VIA elements are often used as perfumes in soaps and cosmetics.

_____ 19. Aqueous solutions of hydrogen sulfide, hydrogen selenide, and hydrogen telluride are acidic.

_____ 20. The metallic character of the elements studied in this chapter increases in the order sulfur < selenium < tellurium.

_____ 21. Sulfur is too reactive to occur as the free element, so it is found exclusively in compounds.

_____ 22. The major source of sulfur in naturally occurring compounds is as metal sulfides.

_____ 23. Pure sulfur is always bright yellow, whether solid, liquid, or gas.

_____ 24. Selenium is so rare that it has no significant practical use.

_____ 25. All Group VIA hydrides are liquid at room temperature and atmospheric pressure.

_____ 26. Sulfur dioxide is a gas at room temperature and atmospheric pressure.

_____ 27. Sulfur trioxide is a gas at room temperature and atmospheric pressure.

_____ 28. Sulfurous acid is a diprotic acid.

_____ 29. Sulfuric acid has never been isolated in pure form.

_____ 30. Nitrogen, N_2, is a very reactive molecule because it contains a highly reactive unsaturated triple bond.

_____ 31. Like O_2, N_2 is paramagnetic.

_____ 32. No other element exhibits more oxidation states than does nitrogen.

_____ 33. Amines can be considered as derivatives of ammonia.

_____ 34. Gaseous nitrogen oxide, NO, is paramagnetic.

_____ 35. The oxidizing power of the nitrogen oxides decreases as the oxidation state of the nitrogen increases.

_____ 36. The atmospheric reaction in which nitrogen dioxide reacts with water vapor to produce nitric acid and nitrogen oxide is a disproportionation reaction.

_____ 37. The reason nitrogen can form so many oxides is the availability of d orbitals on nitrogen for bonding.

_____ 38. Phosphorus can exist in more than one solid form.

_____ 39. Phosphorus is so reactive that it must be stored under oil to keep it from reacting with water vapor in the air.

_____ 40. Because of its relatively low reactivity, most of the silicon in the earth's crust occurs in the native, or uncombined, form.

_____ 41. Like carbon dioxide, SiO_2 is a gas.

_____ 42. Because of the availability of d orbitals, silicon can form some stable species in which its coordination number exceeds four.

Short Answer

Answer with a word, a phrase, a formula, or a number with units as necessary.

1. The noble gas whose uses often depend on its low density is _____.
2. There are more compounds of _____ than of any other noble gas.
3. The order of decreasing electronegativity of the halogens is _____.
4. The halogens have _____ standard reduction potentials.
5. The order of increasing strength of the halogens as oxidizing agents is _____.
6. All halogens have the outer-shell electron configuration _____.
7. The order of increasing atomic size of the halogen atoms is _____.
8. The least metallic halogen is _____; the most metallic halogen is _____.
9. As a result of the _____ ease of polarizability of a larger electron cloud, compounds containing iodide ions display _____ covalent character than those containing fluoride ions.
10. The binary compounds of hydrogen and a halogen are called _____.
11. The oxidation numbers exhibited by the halogens in the halogen oxoacids and their salts are ____, ____, ____, and ____.
12. The hydrogen halides, arranged in order of increasing acidity in aqueous solution are _____.
13. The general formula of the alkali metal hypohalites is _____; one of their main uses is as _____.
14. The oxoacids of chlorine, arranged according to increasing acidity, are _____.

As you have seen, we often discuss descriptive chemistry according to the oxidation states that the elements of a group display. In Questions 15 through 43, tell the oxidation state of any noble gas or halogen in the element, compound, or ion.

15. Xe _____
16. Cl_2 _____
17. PtF_6 _____
18. XeF_2 _____
19. XeF_4 _____
20. XeF_6 _____
21. $RbXeF_7$ _____
22. AgI _____
23. $MgBr_2$ _____
24. KF _____

25. $MnCl_2$ _____
26. HClO _____
27. NaClO _____
28. Br^- _____
29. BrO_3^- _____
30. Br_2 _____
31. IO_3^- _____
32. SOF_2 _____
33. HCl _____
34. $CuCl_2$ _____

35. $HClO_4$ _____
36. $NaBrO_3$ _____
37. HOF _____
38. H_5IO_6 _____
39. OCl^- _____
40. ClO_2^- _____
41. $Zn(ClO_4)_2$ _____
42. $NaBrO_4$ _____
43. $NaClO_3$ _____

44. The names and symbols for the elements of Group VIA are _____, _____, _____, _____, and the quite rare element _____.
45. Because it does not have available d orbitals in its outermost shell, oxygen can bond covalently to a maximum of ____ other atoms.
46. Because of the availability of d orbitals in their outer shells, the heavier VIA elements sulfur, selenium, and tellurium, can bond covalently to as many as ____ other atoms.

293

47. The outermost electronic configuration of all Group VIA elements is _____.
48. Of the Group VIA elements heavier than oxygen, _____ is by far the most abundant.
49. All of the Group VIA elements form covalent hydrides having the formula _____, in which the VIA element is represented as E.
50. The major natural source of selenium and tellurium is in _____ ores.
51. The oxidation number of sulfur in the sulfides is ____ .
52. The dioxides of sulfur, selenium, and tellurium, arranged according to increasing covalent (decreasing ionic) character, are _____ (write the formulas).
53. Sulfur dioxide is the anhydride of the acid _____ (name, formula), whereas sulfur trioxide is the anhydride of the acid _____ (name, formula).
54. The name of the ion S^{2-} is _____.
55. The name of the compound SO_2 is _____.
56. The formula for sodium sulfite is _____.
57. The formula for sodium hydrogen sulfate is _____.
58. The formula for sodium sulfate is _____.
59. The formula for sulfurous acid is _____.
60. The formula for sulfuric acid is _____.
61. The formula for the sulfate ion is _____.
62. The three nonmetallic (or predominantly nonmetallic) elements of Group IA are _____, _____, and _____. The two more metallic elements of this group are _____ and _____.
63. Each of the Group VA elements can exhibit at least some of the oxidation states ranging from ____ to ____.
64. The equation for the autoionization reaction of liquid ammonia is _____.
65. The organic compounds in which one or more of the hydrogen atoms of ammonia have been replaced with organic groups are called _____; like ammonia, these compounds are all _____ bases.
66. The name of the major commercial process for producing nitric acid is the _____ process.
67. The two oxosalts of nitrogen that are sometimes used as meat additives to retard oxidation are _____ (name, formula) and _____ (name, formula).
68. Nitric acid is prepared by dissolving _____ (name, formula) in water; nitric acid is a _____ acid and a _____ oxidizing agent (strong or weak?).
69. The existence of elements in several different forms in the same physical state, as exemplified by phosphorus and arsenic, is called _____.

- -

In each of the following Questions, 70 through 84, supply either the name or the formula of the compound or ion, and tell the oxidation state of nitrogen in that compound or ion.

Formula	Name	Oxidation State of Nitrogen
70. NH_3	_____	_____
71. HNO_3	_____	_____

	Formula	Name	Oxidation State of Nitrogen
72.	_____	nitrogen oxide	_____
73.	_____	nitrogen	_____
74.	_____	nitrous acid	_____
75.	N_2O	_____	_____
76.	_____	nitrogen dioxide	_____
77.	_____	dinitrogen tetroxide	_____
78.	NO_3^-	_____	_____
79.	_____	ammonium nitrate	_____
80.	NH_4^+	_____	_____
81.	$Ca(NO_3)_2$	_____	_____
82.	_____	potassium nitrite	_____
83.	_____	potassium nitrate	_____
84.	_____	lead(II) nitrate	_____

In each of the following Questions, 85 through 92, supply either the name or the formula of the compound or ion, and tell the oxidation state of phosphorus or arsenic in that compound or ion.

	Formula	Name	Oxidation State of Phosphorus
85.	$Ca_3(PO_4)_2$	_____	_____
86.	P_4	_____	_____
87.	_____	calcium dihydrogen phosphate	_____
88.	_____	tetraphosphorus trisulfide	_____
89.	H_3PO_3	_____	_____
90.	H_3PO_4	_____	_____
91.	H_3AsO_4	_____	_____
92.	_____	phosphate ion	_____

93. The element that is second to oxygen in abundance in the earth's crust is _____.
94. Some gems and semiprecious stones are crystals of the compound _____ in the form of _____, but containing colored impurities.
95. The class of minerals in which aluminum atoms replace some silicon atoms in the silica structure is called _____.
96. The glass used in bottles and window panes is a fused mixture of _____ (name, formula) and _____ (name, formula).

Multiple Choice

_____ 1. Which of the halogens is conveniently purified by sublimation?
    (a) F_2 (b) Cl_2 (c) Br_2 (d) I_2

_____ 2. Which of the halogens has the lowest melting point?
    (a) F_2 (b) Cl_2 (c) Br_2 (d) I_2

_____ 3. Which of the halogens has the strongest X–X bond in the diatomic element?
    (a) F_2 (b) Cl_2 (c) Br_2 (d) I_2

_____ 4. Which of the halogens has the shortest X–X bond?
    (a) F_2 (b) Cl_2 (c) Br_2 (d) I_2

_____ 5. Which of the following is not a common use of fluorine or one of its compounds?
    (a) additive to toothpaste   (b) refrigerant
    (c) manufacture of plastics   (d) oxidizing agent
    (e) dietary supplement to aid in thyroid function

_____ 6. Which of the following is a likely method for preparing a free halogen?
    (a) $Cl_2 + 2Br^- \rightarrow 2Cl^- + Br_2$
    (b) $Br_2 + 2Cl^- \rightarrow 2Br^- + Cl_2$
    (c) $I_2 + 2F^- \rightarrow 2I^- + F_2$
    (d) $2HF \rightarrow H_2 + F_2$
    (e) $2KBr + I_2 \rightarrow 2KI + Br_2$

_____ 7. Which of the following is not described in the chapter as a common use of some compound of bromine?
    (a) photographic film
    (b) sedative
    (c) soil fumigant
    (d) manufacture of plastics

_____ 8. Compounds of which of the halogens are commonly used as bleaching agents?
    (a) fluorine   (b) chlorine   (c) bromine   (d) iodine

_____ 9. Which of the following is a principal use of hydrofluoric acid and not of any other hydrohalic acid?
    (a) disinfectant   (b) bleaching agent   (c) dietary supplement   (d) agent for etching glass   (e) oxidizing agent

_____ 10. What is the name of $NaClO_2$?
    (a) sodium hypochlorite   (b) sodium chlorite
    (c) sodium chlorate   (d) sodium perchlorate
    (e) sodium paraperchlorate

_____ 11. What is the formula for sodium iodate?
    (a) NaI   (b) NaIO   (c) NaIO$_2$ (d) NaIO$_3$ (e) NaIO$_4$

_____ 12. What is the formula of the only oxoacid that contains fluorine?
    (a) HF   (b) HOF   (c) HFO$_2$ (d) HFO$_3$ (e) HFO$_4$

_____ 13. Which of the elements studied in this chapter is necessary to aid thyroid function?
    (a) fluorine   (b) chlorine   (c) bromine   (d) iodine   (e) xenon

_____ 14. Which of the following salts are used as bleaching agents?
    (a) halides   (b) halites   (c) hypohalites   (d) halates   (e) xenates

_____ 15. Which of the Group VIA elements is not a solid at room temperature and atmospheric pressure?
    (a) oxygen   (b) sulfur   (c) selenium   (d) tellurium   (e) polonium

_____ 16. Which of the following Group VIA elements has the highest first ionization energy?
    (a) O   (b) S   (c) Se   (d) Te

_____ 17. Which of the following is a major method for obtaining sulfur from deposits of the free element?
    (a) strip mining
    (b) pumping out the sulfur after melting with very hot water
    (c) deep shaft mining
    (d) treatment in a blast furnace
    (e) electrolysis

_____ 18. Which of the following is not mentioned in the chapter as a use of selenium?
    (a) as a coloring agent in glass   (b) in photocopying machines
    (c) in solar cells   (d) as a component of gunpowder

_____ 19. Which of the following Group VIA hydrides has the lowest melting point?
    (a) H$_2$O   (b) H$_2$S   (c) H$_2$Se   (d) H$_2$Te

_____ 20. Sulfur trioxide is environmentally significant because
    (a) it is depleting the ozone layer.
    (b) it helps protect streams from excessive contamination by phosphates.
    (c) it dissolves in atmospheric water to produce "acid rain."
    (d) it helps to absorb ultraviolet light that could attack the ozone layer.
    (e) it is formed by the rotting of eggs.

_____ 21. The process in which sulfur dioxide is catalytically oxidized to sulfur trioxide is called the _____ process.
    (a) contact   (b) Frasch   (c) Hall   (d) Downs   (e) Ostwald

_____ 22. The Group VA element with the highest melting point is
    (a) nitrogen.   (b) phosphorus.   (c) arsenic.   (d) antimony.   (e) bismuth.

_____ 23. Nitrogen is made available to living organisms primarily by
    (a) bacterial action in the nitrogen cycle.   (b) the Haber process.
    (c) the Ostwald process.   (d) the use of detergents.
    (e) the oxidizing action of nitric acid.

_____ 24. The multistep procedure by which nitric acid is produced commercially from ammonia is called
    (a) the Haber process.   (b) the Ostwald process.   (c) respiration.
    (d) eutrophication.   (e) the Frasch process.

_____ 25. The most important oxoacid of nitrogen in commercial uses is
    (a) HNO$_3$.   (b) HNO$_2$.   (c) H$_2$N$_2$O$_2$.   (d) H$_3$NO$_4$.   (e) HN$_3$.

_____ 26. The largest single use of phosphorus compounds is in

(a) the fertilizer industry. (b) the detergent industry.
(c) rocket propellants.   (d) insecticides and other poisons.
(e) smog preventive agents.

_____ 27. Clay minerals generally have _____ structures with _____ surface areas.

(a) sheet-like, large   (b) sheet-like, small   (c) globular, large
(d) globular, small   (e) irregular, small

Answers to Preliminary Test

True-False

1. False. The heavier ones have been shown to form compounds.
2. True
3. False. All have quite low boiling and melting points, but all can be condensed and solidified.
4. False. Almost all of the compounds of the noble gases involve *d* orbitals of that element, and thus deviate from the octet rule.
5. False. They react only with very strong oxidizing agents.
6. True
7. False. All are diatomic, but bromine is a liquid, and iodine and astatine are solids.
8. False. The properties show more similarity within this group than in most others of the periodic table.
9. True. See the discussion of the relative polarizability of fluoride and iodide in Section 24-3.
10. True
11. False. All form many stable halides (−1), but all except fluorine commonly exhibit oxidation numbers of +1, +3, +5, and +7. Fluorine shows +1 in a few compounds. These higher oxidation states are discussed, for instance, in Section 24-13 for the oxoacids and oxyanions of the halogens.
12. False. It is because of its strength as an oxidizing agent that no other chemical agent can directly oxidize fluoride ions to fluorine gas, so it is usually prepared electrochemically. Section 24-4 describes a chemical synthesis that was recently discovered.
13. True
14. False. They range from quite ionic, with active metals, to quite covalent, with nonmetals.
15. False. The differences in the properties within a group are more pronounced near the middle of the periodic table than they are near the ends.
16. True
17. False. Oxygen never exhibits an oxidation number as high as +6. Can you tell why? The higher oxidation states +2, +4, and +6 are quite common for the heavier members of Group VIA.
18. False. H_2O is odorless. H_2S, H_2Se, and H_2Te are poisonous, bad-smelling gases. For instance, H_2S is the smell of rotten eggs.
19. True. As discussed in Section 24-10, the strength of these substances as acids increases as we descend within the group.

298

20. True. Remember that the trend throughout the periodic table is that metallic properties increase going down within a group.
21. False. There are large deposits of free sulfur, predominantly as S_8 molecules, for instance, along the U.S. Gulf Coast. See also the photo on page 952.
22. True. The major sulfide ores are galena (PbS), iron pyrites (FeS_2), and cinnabar (HgS).
23. False. At some temperatures, liquid sulfur is dark brown. Read in the first part of Section 24-8 how the properties of sulfur in the solid and liquid phases change with temperature.
24. False. Read the second part of Section 24-8.
25. False. All of these compounds except for water are gases. Remember that water has very unusual properties because of its ability to form hydrogen bonds.
26. True
27. False. It is a liquid.
28. True
29. False. Its properties are described in Section 24-12.
30. False. Nitrogen is very stable and unreactive.
31. False. It is diamagnetic, all electrons being paired. You may wish to review the discussion of bonding in O_2 and N_2 in terms of molecular orbitals, Chapter 9.
32. True. It exhibits all oxidation states from -3 to $+5$.
33. True
34. True. Any molecule that contains unpaired electrons is paramagnetic. NO, which contains an odd number of electrons, must have an unpaired electron.
35. False. If you think about it, this statement just makes no sense, whether or not you know anything about oxides of nitrogen. In order to act as an oxidizing agent, the substance must already be in a higher oxidation state and be able to be reduced.
36. True. The reaction is $3NO_2 + H_2O \rightarrow 2HNO_3 + NO$. See the Chemistry In Use essay, page 962. Determine the oxidation state of nitrogen in these compounds.
37. False. It is reactive with regard to oxidation by atmospheric oxygen, but it does not react with water, so it is stored under water to protect it from air.
38. False. There are no d orbitals in the valence shell ($n = 2$) of nitrogen.
39. True. This is discussed in Section 24-17.
40. False. This element does not occur free in nature. About 87% of the silicon in the earth's crust is in the form of silica (SiO_2) or the silicate minerals.
41. False. It has a polymeric structure, occurring in several solid forms. SiO_2 does not exist as discrete molecules.
42. True. Not many such compounds actually form. One of the few stable such species is the hexafluorosilicate ion, SiF_6^{2-}.

Short Answer

1. helium
2. xenon
3. $F > Cl > Br > I$
4. high. This means that the elements are easily reduced and that they are good oxidizing agents.
5. $I_2 < Br_2 < Cl_2 < F_2$
6. ns^2np^5

7. $F < Cl < Br < I$

8. fluorine; astatine. Metallic character increases going down any group in the periodic table.

9. greater; greater. Study Section 24-3.

10. hydrogen halides

11. +1; +3; +5; +7. Be aware of which halogens commonly exhibit which oxidation states. See Table 24-9.

12. $HF << HCl = HBr = HI$. Water is such a strong base that it does not distinguish among the acid strengths of HCl, HBr, and HI. Read about the leveling effect, Section 10-5.

13. NaXO or NaOX (X = Cl, Br, I); bleaching agents (especially with Cl)

14. $HClO < HClO_2 < HClO_3 < HClO_4$. You learned in Chapter 4 (and saw again in Chapters 10 and 18) that perchloric acid, $HClO_4$, is one of the few common strong acids. For any element, acidity of the oxoacids increases with increasing oxidation number of the central atom.

In answering Questions 15 through 43, it may help for you to remember the following: In free elements, the oxidation number is always zero. You may need to review the determination of oxidation numbers, Section 4-4.

15. Xe 0	25. Cl −1	35. Cl +7
16. Cl 0	26. Cl +1	36. Br +5
17. F −1	27. Cl +1	37. F −1
18. Xe +2; F −1	28. Br −1	38. I +7
19. Xe +4; F −1	29. Br +5	39. Cl +1
20. Xe +6; F −1	30. Br 0	40. Cl +3
21. Xe +6; F −1	31. I +5	41. Cl +7
22. I −1	32. F −1	42. Br +7
23. Br −1	33. Cl −1	43. Cl +5
24. F −1	34. Cl −1	

44. oxygen (O); sulfur (S); selenium (Se); tellurium (Te); polonium (Po)

45. 4

46. 6

47. ns^2np^6

48. sulfur

49. H_2E

50. sulfide. It should not surprise you that sulfur and selenium are found together in the same chemical state.

51. −2

52. sulfur dioxide (SO_2) < selenium dioxide (SeO_2) < tellurium dioxide (TeO_2)

53. sulfurous acid (H_2SO_3); sulfuric acid (H_2SO_4)

54. sulfide. Review naming of compounds in Sections 4-11 and 4-12.

55. sulfur dioxide

56. Na_2SO_3

57. $Na_2S_2O_3$

58. Na_2SO_4

59. H_2SO_3

60. H_2SO_4

61. SO_4^{2-}

62. nitrogen; phosphorus; arsenic; antimony; bismuth

63. −3; +5

64. $2NH_3(l) \rightarrow NH_4^+ + NH_2^-$. This reaction takes place to a much smaller extent than the corresponding one for water ($K_{H_2O} = 10^{-14}$, $K_{NH_3} = 10^{-30}$), so ammonia is a more basic solvent than is water. Liquid ammonia is a very useful solvent for some reaction systems.

65. amines; weak

66. Ostwald

67. sodium nitrite ($NaNO_2$); sodium nitrate ($NaNO_3$). There is, however, considerable concern about the formation of carcinogenic (cancer-causing) compounds from these additives.

68. nitrogen dioxide, NO_2; strong; strong. Dissolving NO_2 in water is part of the Ostwald process. You should be aware that this is an oxidation-reduction process, so NO_2 is not to be considered the "anhydride" of nitric acid in the usual sense. Although nitric acid can be prepared by dissolving N_2O_5 (the oxide that already has nitrogen in the proper oxidation state) in water, this is not the major method of preparation.

69. allotropism. The various forms are called allotropes or allotropic forms.

70. ammonia, −3. Review oxidation numbers (Section 4-4).

71. nitric acid, +5

72. NO, +2

73. N_2, 0. Sometimes this is called "dinitrogen," though not in this text.

74. HNO_2, +3

75. dinitrogen oxide, +1

76. NO_2, +4

77. N_2O_4, +4

78. nitrate ion, +5

79. NH_4NO_3, −3 in the ammonium ion, +5 in the nitrate ion

80. ammonium ion, −3

81. calcium nitrate, +5

82. KNO_2, +3

83. KNO_3, +5

84. $Pb(NO_3)_2$, +5

85. calcium phosphate, +5

86. white phosphorus, 0

87. $Ca(H_2PO_4)_2$, +5

88. P_4S_3, +1.5

89. phosphorous acid, +3

90. phosphoric acid, +5

91. arsenic acid, +5

92. PO_4^{3-}, +5

93. silicon

94. SiO_2; quartz

95. aluminosilicates. These are mentioned in Section 24-18, with a few examples.

96. sodium silicate (Na_2SiO_3); calcium silicate ($CaSiO_3$). The chemistry of glass is quite complicated, but the major features are indicated in Section 24-18.

Multiple Choice

1. (d)
2. (a)
3. (b). This is one of the exceptions to the very regular trends in properties of the elements of Group VIIA. See the bond energies tabulated in Table 24-4.
4. (a). Review atomic radii, Section 6-2.
5. (e)
6. (a). Remember that in the displacement reactions, the heavier halogens replace the lighter ones from halides. This is consistent with our observation that the lighter the halogen, the greater its oxidizing power. Review displacement reactions, Section 4-8.
7. (d)
8. (b)
9. (e)
10. (b)
11. (d). This is one of the principal naturally occurring sources of iodine.
12. (b)
13. (d)
14. (c)
15. (a)
16. (a). This trend of decreasing first ionization energy going down the group prevails in all representative groups of the periodic table. Review Section 6-3.
17. (b). This is called the Frasch process, described in Section 24-8.
18. (d)
19. (b). The anomalously high melting point of water is due to its extensive hydrogen bonding.
20. (c). We have discussed acid rain several times in this text.
21. (a). Can you identify the processes mentioned in the other answers?
22. (c). The melting points of the elements of Group VA reach a maximum in the middle of the group.
23. (a). This is sometimes accomplished with the aid of an enzyme called nitrogenase.
24. (b)
25. (a). This is nitric acid. HN_3 is not an oxoacid and there is no such acid as H_3NO_4.
26. (a)
27. (a). These minerals, which represent an important class of the silicates and aluminosilicates, are discussed in Section 24-18.

25 Coordination Compounds

Chapter Summary

Many compounds are formed by **coordinate covalent bond** formation. This is a process in which one of the atoms provides the pair of electrons to be shared, the other atom providing a vacant orbital. As we saw in Section 10-10, this process can be described in acid-base terms, using the Lewis acid-base theory. Transition metal ions have vacant d orbitals, as well as vacant s and p orbitals, so they often act as Lewis acids, reacting by coordinate covalent bond formation with molecules or ions that have unshared electron pairs. The compounds or polyatomic ions thus formed are referred to as **coordination compounds, coordination complexes**, or **complex ions**. Some representative metal ions can also bond in this way—e.g., Al^{3+}, Sn^{2+}, Sn^{4+}—but even these often use d orbitals in their formation of coordination compounds. In Chapter 25, we shall study some aspects of the formation, nomenclature, structures, and properties of coordination compounds.

Section 25-1 introduces this topic, with many examples to illustrate the importance and some of the history of coordination chemistry. Recall that the stability of these complexes with respect to dissociation is described by the value of K_d, the **equilibrium constant** for the dissociation reaction.

Several transition metals whose hydroxides are amphoteric can dissolve in an excess of strong soluble base. Ammonia, however, is such a weak base that it cannot dissolve most amphoteric hydroxides by the usual method of formation of hydroxo complexes. Yet the hydroxides of some metals do dissolve in an excess of aqueous ammonia, to form soluble **ammine complexes** (Section 25-2).

Several important terms in coordination chemistry are defined and illustrated in Section 25-3. A molecule or ion that acts as the Lewis base in complex formation is called a **ligand**; within a ligand, the atom that actually provides the electron pair is the **donor atom**. A ligand is termed **unidentate**, **bidentate**, and so on, to describe the number of different atoms that ligand can use as donor atoms simultaneously. The metal atom or ion and its ligands (but not any uncoordinated ions) constitute the **coordination sphere**. The **coordination number** is the number of donor atoms (not ligands) to which the metal is bonded. Study the examples in Tables 25-4 and 25-5, which illustrate the use of these terms.

Likewise, a knowledge of and careful adherence to the rules for naming coordination compounds (Section 25-4) along with much practice with examples and exercises, will enable you to write formulas from names and vice versa. Notice that the naming of the compound sometimes must specify the oxidation state of the metal. You should review the rules for assignment of oxidation number in polyatomic molecules and ions (Section 4-4). It will help you to remember that the sum of the oxidation numbers of atoms in a neutral molecule (and also a neutral ligand) is equal to *zero* and that the sum of the oxidation numbers of the atoms in a polyatomic ion (or in an ion acting as a ligand) is equal to the *charge* on that ion.

Once we know the coordination number of the metal in a coordination compound, we can use the **VSEPR theory** (do you recall what these letters mean?—see Section 8-2) to predict the arrangement of donor atoms in the coordination sphere—the structure of the complex. Review how the VSEPR theory is applied (Sections 8-5 through 8-12) before you study Section 25-5, which deals with the structures of coordination compounds. The heart of this discussion is Table 25-7, which is similar to Table 8-4 in the earlier applications of VSEPR theory. Please notice, though, that in coordination chemistry there may be different structures possible for some coordination numbers, especially 4 (tetrahedral or square planar) and 5 (trigonal bipyramidal or square pyramidal). Each of the different geometries around the metal ion corresponds, of course, to a different set of hybrid orbitals.

One point of confusion for some students involves the way structural formulas of coordination compounds are represented. Look, for instance, at the representation of the hexachlorostannate(IV) ion, at the beginning of Section 25-1. The lines that are drawn from one Cl^- ligand to another are not bonds, but are drawn merely to help visualize the shape of the coordination sphere and the arrangement of the ligands; the lines drawn from each Cl^- to the tin(IV) ion are bonds (coordinate covalent bonds). Similar comments apply to the common representations of square planar, trigonal bipyramidal, and square pyramidal structures.

The next portion of the chapter, Sections 25-6 and 25-7, deals with the existence of **isomers**, which are substances having the same number and kinds of atoms, but arranged differently. There are several ways in which the atoms can differ in their arrangements, so several different classes of isomers can exist. These are broadly classified into two major categories. (1) **Structural (constitutional) isomers** involve differences in the bonding arrangement of different donor atoms on the same ligand or differences in bonding beyond a single coordination sphere. Section 25-6 breaks the topic of structural isomers into four subclasses—**ionization isomers**, **hydrate isomers**, **coordination isomers**, and **linkage isomers**—with several examples of each. (2) The differences among **stereoisomers** involve only one coordination sphere and the same ligands and donor atoms. The two subclasses of stereoisomers, known as **geometrical isomers** and **optical isomers** (or **enantiomers**) are described in Section 25-6. The following distinctions between structural isomers and stereoisomers may help you to recognize the broad categories. In structural isomers, the pattern or order of bonds is different, so you should be able to find at least one atom that is bonded to different atoms in the isomers. The rules already given for naming compounds serve to distinguish among various structural isomers. In stereoisomers, on the other hand, the order of bond linkages is the same, but the spatial arrangements of the atoms are different. If this is due to the atoms being linked in different relative positions on the central metal atom, the isomers are geometrical isomers; if it is due only to the "handedness" of the entire arrangement, they are optical isomers. As you find out in your study of stereoisomerism, additional terms are used to describe such isomers; be sure you understand and can use properly such terms, as **cis**, **trans**, **dextro**, **levo**, and **racemic**.

The rest of the chapter is concerned with the bonding in coordination compounds. Section 25-8 presents the very successful **crystal field theory**, illustrated with applications to complexes with coordination number 6. The crystal field theory takes the opposite view that the bonding is entirely (or at least predominantly) ionic and that the d orbitals of the metal, which in an isolated ion would be at the same energy, have their energies disturbed (**degeneracy** is removed) by the presence of the ligands. The amount by which these energies change, called the **crystal field splitting**, depends on the nature and arrangement of the ligands. These ideas are illustrated for the octahedral $[CoF_6]^{3-}$ and $[Co(NH_3)_6]^{3+}$ ions. The magnetic nature of a complex, that is, whether it is diamagnetic or paramagnetic (and the degree of paramagnetism), is often used to propose or verify the bonding description for that complex. As you study these sections, you should become familiar with the meanings and uses of such terms as e_g, t_{2g}, **crystal field splitting**, **low spin complex**, **high spin complex**, and Δ_{oct}.

The great variety of colors exhibited by transition metal coordination compounds can be interpreted in terms of the crystal field theory, as in Section 25-9. According to this interpretation, the different crystal field strengths of various ligands cause different e_g vs. t_{2g} splitting. Thus, in order to promote an electron from one d orbital to another, the various complexes must absorb light of different energy, that is, a different color. The common ligands are ranked in order of increasing crystal field strength, based on a study of the colors of light that they absorb. This ordered list of ligands is called the **spectrochemical series**.

Study Goals

It may help you to write out a brief summary of the ideas of each Study Goal, summarizing the applicable portions of your class notes and text reading. Be sure to answer many of the suggested Exercises related to the Study Goals.

1. Account for the tendency of the d-transition metals to form coordination compounds. (Section 25-1; Exercises 1 and 3)
2. Describe the experiments and reasoning of Werner that allowed him to determine which species of a complex were in the metal's coordination sphere, and therefore, what were the true formulas for coordination compounds. (Section 25-1; Exercises 2, 9, and 10)
3. Write equations that account for the solubility of water-insoluble hydroxides in aqueous ammonia by formation of ammine complexes. (Section 25-2; Exercises 12 through 15)
4. Distinguish among the terms ligand, donor atom, polydentate, unidentate, chelate complex, coordination number, and coordination sphere. (Section 25-3; Exercises 4 through 8 and 11)
5. Know the names and formulas of common ligands as well as their donor atoms. (Section 25-4)
6. Know the general IUPAC rules for naming coordination compounds. Name coordination compounds if given their formulas, and supply formulas if given names. (Section 25-4; Exercises 16 through 21)
7. Sketch structures of coordination compounds of the types ML_2, ML_4 (square planar and tetrahedral), ML_5 (trigonal bipyramidal and square pyramidal), and ML_6. (Sections 25-5 and 25-8; Exercises 22 and 23)
8. Distinguish between structural isomers and stereoisomers. (Sections 25-6 and 25-7; Exercise 28)
9. Describe and give examples of the following kinds of isomers: (a) ionization, (b) hydrate, (c) coordination, (d) linkage, (e) geometrical, and (f) optical. Name the isomers, using appropriate prefixes where necessary. (Sections 25-6 and 25-7; Exercises 24 through 27 and 29 through 31)
10. Describe and relate the following: (a) dextro and levo isomers, (b) racemic mixture, (c) plane polarized light, (d) polarimeter, and (e) optical activity. (Section 25-7; Exercise 29)
11. Draw and distinguish between the t_{2g} and e_g sets of orbitals. (Section 25-8)
12. Describe with words and diagrams the crystal field theory as it applies to octahedral coordination compounds. Explain the reversal in relative energies of t_{2g} and e_g orbitals for the two cases. (Section 25-98 Exercise 34)
13. Distinguish between and give examples of low spin and high spin complexes. Explain the relationship between electron pairing energy and Δ_{oct} with respect to these kinds of complexes. (Section 25-8; Exercises 35 through 39)

14. Describe what is meant by the terms "strong field ligand" and "weak field ligand." Describe the spectrochemical series and its relationship to the colors of coordination compounds. (Sections 25-8 and 25-9; Exercises 37 and 38)
15. Answer conceptual questions based on this chapter. (Conceptual Exercises 40 through 43)
16. Apply concepts and skills from earlier chapters to the ideas of this chapter. (Review appropriate sections of Chapters 2, 3, 12, 18, 20, and 21; Building Your Knowledge exercises, 44 through 49)

Some Important Terms in This Chapter

Write the meanings *in your own words*. Check the Key Terms list and the chapter reading. Then rewrite your definitions, still in your own words, to improve them. Study other new terms and review terms from preceding chapters if necessary.

coordinate covalent bond

coordination compound or complex

ligand

donor atom

coordination number

coordination sphere

structural (constitutional) isomers

stereoisomers

ionization isomers

hydrate isomers

coordination isomers

linkage isomers

geometrical isomers

optical isomers

t_{2g} orbitals

e_g orbitals

crystal field theory

weak field ligand

strong field ligand

spectrochemical series

low spin complex

high spin complex

Preliminary Test

True-False

_____ 1. The higher the value of K_d for a complex, the more stable the complex is.

_____ 2. The properties of transition metal ions that enable them to form coordination complexes are their magnetism and their colors.

_____ 3. Whenever a metal is bonded to four donor atoms in a coordination complex, the coordination geometry must be tetrahedral.

_____ 4. Even though the atoms are arranged differently in geometrical isomers, the properties of these compounds are alike.

_____ 5. When we say that a substance is exhibiting optical activity, we mean that it is colored.

_____ 6. Ligands must be neutral molecules.

_____ 7. Coordination complexes in which the metal is bonded to six donor atoms have octahedral coordination geometry.

_____ 8. The crystal field theory can be applied to complexes only when they occur in crystals.

_____ 9. In an isolated metal atom or ion, the five d orbitals with the same principal quantum number all have the same energy.

_____ 10. When a metal atom or ion is bonded to several ligands, the five metal d orbitals with the same principal quantum number all have the same energy.

_____ 11. The ion Fe^{3+} is referred to as a d^5 ion.

_____ 12. In the crystal field theory description of coordination compounds, ligands that are more electronegative lead to smaller crystal field splitting.

_____ 13. All octahedral complexes of metal ions with d^5 configuration are low spin complexes.

Short Answer

1. A molecule or ion that provides an electron pair for coordinate covalent bond formation is called a Lewis _____; a molecule or atom that accepts the electron pair is called a Lewis _____.

2. The name of the chemist who was responsible for much of the early development of the ideas of coordination chemistry was _____.

3. The molecules or ions that act as Lewis bases in the formation of coordination compounds are called _____.

4. Ligands that can bond through only one donor atom at a time are called _____.

5. Ligands that can bond through two or more donor atoms simultaneously are called _____.

6. A molecule or ion that can act as a ligand in a coordination complex usually has at least one _____.

7. The portion of a compound consisting of the metal ion and its coordinated ligands, but no uncoordinated ions, is called the _____.

For each of the next fifteen questions, 8 through 22, provide either the formula or the name of the ligand, as required, and classify each ligand as unidentate, bidentate, and so on. When a ligand name is required, be sure that you name it as a ligand, and not as a free molecule or ion. Dot formulas are not required.

Formula	Name	Classification
8. NH_3	_____	_____
9. _____	cyano	_____
10. _____	chloro	_____
11. $H_2NCH_2CH_2NH2$	_____	_____
12. _____	diethylenetriamine (dien)	_____
13. PH_3	_____	_____
14. _____	carbonyl	_____
15. Br^-	_____	_____
16. SO_4^{2-}	_____	_____
17. _____	oxo	_____
18. _____	hydroxo	_____
19. $O_2C-CO_2^{2-}$	_____	_____
20. OH_2	_____	_____
21. NO_2^- (bonded through the N)	_____	_____
22. _____	nitrato	_____

23. A ligand that can bond through three donor atoms simultaneously is called _____.
24. When a metal has a coordination number of 2 in a complex, the coordination geometry is _____.

25. When a metal has a coordination number of 6 in a complex, the coordination geometry is _____.

26. The two mirror image forms of a molecule are called _____ of one another; one is identified by the term _____, while the other is called _____.
27. The number of geometrical isomers possible for the square planar metal complex with formula MX_2Y_2 is _____. (Note: M is the metal and X and Y are each unidentate ligands.)
28. The number of geometrical isomers possible for the tetrahedral metal complex with formula MX_2Y_2 is _____. (Note: The symbols have the same meaning as in Question 27.)

For each of the next twelve questions, 29 through 40, provide either the formula or the name of the complex. For each complex, give the oxidation state (Ox. State) of the metal and the coordination number (Coord. No.) of the metal. In naming, do not worry about stereoisomerism, unless sufficient information is explicitly shown.

	Formula	Name	Ox. State	Coord. No.
29.	$[Cu(en)Br_2]$	_____	____	____
30.	_____	hexamminecobalt (III) ion	____	____
31.	_____	hexacyanocobaltate(III) ion	____	____
32.	$[Fe(CO)_5]$	_____	____	____
33.	$[CuCl_5]^{3-}$	_____	____	____
34.	_____	cis-diamminedichloroplatinum(II)	____	____
35.	_____	trans-diamminedichloroplatinum(II)	____	____
36.	$[Cr(OH_2)_2Br_2(NH_3)_2]^+$	_____	____	____
37.	_____	cis-tetraamminediaquacobalt(III) ion	____	____
38.	_____	trans-tetraamminediaquacobalt(III) ion	____	____
39.	_____	tetraamminedichloroplatinum(IV) chloride	____	____
40.	_____	triamminetrichloroplatinum(IV) chloride	____	____

41. The name of the complex ion $[AlCl_6]^{3-}$ is _____; in this complex, the oxidation number of aluminum is ____.

42. The name of the theory of bonding in coordination compounds that is discussed in this chapter is the _____ theory.

43. An experiment that is used to distinguish between inner and outer orbital complexes is to determine how _____ the complex is.

44. According to the crystal field theory, the five originally equivalent d orbitals on a metal become nonequivalent under the influence of the array of ligands. If there are six ligands in an octahedral arrangement, these d orbitals then occur in two groups; the two groups are called, collectively, the _____ orbitals (a set of three orbitals) and the _____ orbitals (a set of two orbitals). For this octahedral arrangement of ligands, the _____ set is at lower energy, with the two sets being separated by an amount of energy designated _____.

45. According to the crystal field theory, the crystal field splitting is larger for ligands that are more able to _____ into vacant metal orbitals; for octahedral complexes, this is expressed quantitatively as a _____ value of Δ_{oct}.

46. Low spin octahedral complexes exist only for metal ions with configurations ____, ____, ____, and ____.

47. The elements whose hydroxides dissolve in excess aqueous ammonia by the formation of soluble ammine complexes are the metals of the _____, _____, _____, and _____ families.

48. According to the crystal field description of bonding, the color that an octahedral complex of a particular metal exhibits depends on the value of _____, which in turn depends on the _____ of the ligands.

49. The ranking of a set of ligands according to increasing crystal field strength, as determined by the study of the colors of several of their complexes, is called the _____.

Multiple Choice

____ 1. What is the coordination number of cobalt in $[Co(NH_3)_6]^{3+}$?
  (a) 2   (b) 3   (c) 4   (d) 6   (e) 9

____ 2. What is the coordination number of cobalt in $[Co(NH_3)_6]Cl_3$?
  (a) 2   (b) 3   (c) 4   (d) 6   (e) 9

____ 3. What is the oxidation number of cobalt in $[Co(NH_3)_6]^{3+}$?
  (a) 0   (b) +2   (c) +3   (d) +4   (e) +6

____ 4. What is the coordination number of cobalt in $[Co(en)_2Cl_2]^+$?
  (a) 2   (b) 3   (c) 4   (d) 6   (e) 9

____ 5. Which of the following types of isomers is not an example of structural isomerism?
  (a) ionization isomers   (b) optical isomers   (c) hydrate isomers
  (d) hydrate isomers   (e) coordination isomers

____ 6. What is the number of geometrical isomers possible for the complex $[Pt(NH_3)_2Cl_4]$?
  (a) 1   (b) 2   (c) 3   (d) 4   (e) 6

____ 7. What is the number of geometrical isomers possible for the complex $[Pt(NH_3)_3Cl_3]^+$?
  (a) 1   (b) 2   (c) 3   (d) 4   (e) 6

____ 8. Which of the following is correctly referred to as a d^8 ion?
  (a) Ti^{3+} (b) Ru^{3+} (c) Ni^{2+} (d) Ag^+ (e) Cr^{3+}

Answers to Preliminary Test

Do not just look up answers. Think about the reasons for the answers.

True-False

1. False. Remember that K_d is the equilibrium constant for the *dissociation* of the complex. A higher value for K_d means that the numerator (product concentration) is higher, which would correspond to *less* stable complexes. Section 20-6 discusses these dissociation constants in quantitative terms.

2. False. They form complexes so readily because of the availability of vacant d orbitals. Properties such as magnetism and color are a result of the bonding in the complexes, not a cause of that bonding.

3. False. Many tetracoordinate metal atoms form square planar complexes.
4. False. A different arrangement of atoms gives different properties. Many examples of this are given in the text chapter, especially regarding the color of the complex.
5. False. Read the second part of Section 25-7.
6. False. Many common ligands are anions, and a few are cations.
7. True
8. False. The term "crystal field" has to do with the symmetry of the arrangements of the ligands and the resulting effects on symmetry of the metal orbitals. Similar symmetry arrangements frequently occur in crystals.
9. True
10. False. In the terminology of the crystal field theory, the degeneracy is removed by the arrangement of the ligands around the metal. In the terminology of the valence bond theory, some of the d orbitals become involved, along with s and p orbitals, in the formation of hybrid orbitals.
11. True. Fe^0 has the outer electronic configuration $4s^23d^6$; on formation of the Fe^{3+} ion, iron has the configuration $4s^03d^5$.
12. True. See the discussion of $[Co(NH_3)_6]^{3+}$ versus $[CoF_6]^{3-}$ in the first part of Section 25-8.
13. False. They can be either high spin or low spin, depending on the nature of the ligands. See Section 25-8.

Short Answer

1. base; acid. Review Sections 10-10 and 25-1.
2. Alfred Werner. In fact, for many years coordination complexes were commonly called Werner complexes. The amminechloroplatinum complexes described in Section 25-1 are now often referred to as Werner complexes.
3. ligands
4. unidentate
5. polydentate
6. unshared pair of electrons
7. coordination sphere
8. ammine; unidentate
9. CN^- (with the carbon acting as the donor atom); unidentate
10. Cl^-; unidentate
11. ethylenediamine (or en); bidentate
12. $NH_2–CH_2–CH_2–\underset{\underset{H}{|}}{N}–CH_2–CH_2–NH_2$; tridentate
13. phosphine; unidentate
14. CO (with carbon as the donor atom); unidentate
15. bromo; unidentate
16. sulfato; unidentate
17. O^{2-}; unidentate or bidentate
18. OH^-; unidentate
19. oxalato; bidentate
20. aqua; unidentate
21. nitro; unidentate
22. NO_3^- (bonded through an oxygen); unidentate

23. tridentate
24. linear
25. octahedral
26. optical isomers (or enantiomers); dextro; levo
27. 2
28. 1. No matter how you arrange the ligands, the two X ligands will be adjacent, as will be the two Y ligands.
29. dibromoethylenediaminecopper(II); +2; 4
30. $[Co(NH_3)_6]^{3+}$; +3; 6
31. $[Co(CN)_6]^{3-}$; +3; 6
32. pentacarbonyliron(0); 0; 5
33. pentachlorocuprate(II) ion; +2; 5. Note the use of the "-ate" ending when the total complex ion has a negative charge.
34.

H_3N, Cl (Pt); +2; 4

H_3N Cl

35.

H_3N, Cl (Pt); +2; 4

Cl NH_3

36. Diamminediaquadibromochromium(III) ion; +3; 6. Notice the application of the rule about alphabetizing ligand names. There is no particular convention about the order in which the ligands are shown in the formula.

37.

$$\left[\begin{array}{c} OH_2 \\ H_3N \quad Co \quad OH_2 \\ H_3N \quad NH_3 \\ NH_3 \end{array} \right]^{3+} \quad ; +3; 6$$

38.

$$\left[\begin{array}{c} OH_2 \\ H_3N \quad Co \quad NH_3 \\ H_3N \quad NH_3 \\ OH_2 \end{array} \right]^{3+} \quad ; +3; 6$$

39. $[Pt(NH_3)_4Cl_2]Cl_2$; +4; 6

313

40. [Pt(NH$_3$)$_3$Cl$_3$]Cl; +4; 6
41. hexachloroaluminate ion; +3. The "-ate" ending is used because the complex ion is negatively charged. We do not specify the oxidation number of aluminum by including (III) in the name, because this is the only oxidation state that aluminum has in compounds.
42. crystal field
43. paramagnetic
44. t_{2g} ; e_g ; t_{2g} ; Δ_{oct}
45. donate electrons; larger
46. d^4; d^5; d^6; d^7
47. cobalt; nickel; copper; zinc
48. Δ_{oct}; crystal field strength. This is discussed in some detail in Section 25-10.
49. spectrochemical series

Multiple Choice

1. (d)
2. (d). The coordination sphere is enclosed in the square brackets so we know that the chloride ions are not bonded to cobalt. Only the nitrogen atoms of the six ammonia molecules (ammine ligands) are bonded to cobalt. This contains the complex ion in Question 1.
3. (c). Each ammine ligand (neutral) has total oxidation number equal to zero. The sum of oxidation numbers in a polyatomic ion must be equal to the charge on the ion, so cobalt must be in an oxidation state of +3.
4. (d). Each (en) ligand is bidentate, containing two donor atoms.
5. (b)
6. (b). These are the *cis* and the *trans* isomers.
7. (b)
8. (c). Remember that when transition metal atoms form ions the two s electrons are lost first, after which some d electrons may be lost also.

26 Nuclear Chemistry

Chapter Summary

Up to now, we have studies chemical reactions, in which elements maintain their identity and interact only through electronic changes. Now we turn our attention to the study of nuclear reactions, in which elements are often converted from one to another by alterations in their nuclei. This area of chemistry, called **nuclear chemistry,** is the subject of Chapter 26. As you read the chapter introduction, pay close attention to the differences between nuclear reactions and ordinary chemical reactions. A preliminary mention is also made of two important types of nuclear reactions, **nuclear fission** and **nuclear fusion**. Both of these will be discussed more fully later in the chapter.

Section 26-1 summarizes the main features of atomic nuclei. Nuclear reactions depend on nuclear stability, which in turn is related to the ratio of number of neutrons to number of protons in the nucleus. This relationship is explained in Section 26-2. The most stable **nuclides** are those with an even number of protons and an even number of neutrons; the least common stable nuclides are those with odd number of each kind of nuclear particle. (These nuclear particles, the neutrons and protons, are collectively called **nucleons**.) Nuclides having either a number of protons or a number of neutrons (or a sum of the two) equal to one of the so-called **magic numbers**—2, 8, 20, 28, 50, 82, or 126—are exceptionally stable. The plot of the number of neutrons versus number of protons for the stable nuclei is called the **band of stability** (Figure 26-1).

The question of the energy relationships in the nucleus of the atom, the **binding energy** of the nucleus, is addressed in Section 26-3. This topic is related to the release of energy accompanying nuclear reactions such as those that occur in atomic bombs and in nuclear reactors. These will be discussed later in this chapter.

The decay or decomposition of unstable nuclei is accompanied by the emission of radiation (**radioactive decay**). The practical reason for carrying out nuclear reactions is often either to make some use of the radiation emitted by nuclear decay or to harness and use the tremendous amounts of energy that are released with many nuclear reactions. The most common types of radiation emitted in nuclear decay processes are discussed in Section 26-4. These various types of radiation may be particles—positive, negative, or neutral—or high-energy electromagnetic radiation, each with a characteristic mass, charge, velocity, and penetrating power as summarized in Table 26-3. Remember that all of these particles are emitted from the **nucleus** of the atom, including the beta particles, which are identical to the extranuclear electrons. The high-energy gamma rays that are often emitted in nuclear decay processes serve to remove excess energy from a nucleus that is in an excited state. This is similar to the emission of light from an atom whose electrons are in an excited configuration.

Each type of decay process causes a change in the atomic number and mass number of the nucleus. These changes are summarized in the following table. Of course, it is not necessary to memorize a table such as this if you just remember the mass number and atomic number for each type of particle that can be emitted or captured (for example, the first column of Table 26-3). Just as for atomic nuclei, the symbol that represents each of these types of radiation can include its atomic number as a left subscript and its mass number as a left superscript.

315

ss	Change in Atomic Number	Change in Mass Number
mission	increase by 1	no change
ron emission	decrease by 1	no change
a emission	decrease by 2	decrease by 4
ton emission	decrease by 1	decrease by 1
neutron emission	no change	decrease by 1
electron capture (*K* capture)	decrease by 1	no change
gamma emission	no change	no change

Section 26-5 gives the conventions for writing equations for nuclear reactions. As you study this chapter, you should check that each equation meets the requirements presented in this section.

Neutron-rich nuclei (those that have a ratio of neutrons to protons too high to be stable) are said to be above the band of stability; these decay so as to decrease the ratio, usually by **beta emission** and less commonly by **neutron emission.** Examples of such nuclei and their characteristic decay processes are discussed in Section 26-6. This section also introduces the writing of equations for nuclear reactions, in a way entirely analogous to the writing of equations for ordinary chemical reactions. In such equations, it is most convenient to write the full **nuclide symbol** for all species involved in the reaction. Neutron-poor nuclei (with a too-low ratio of neutrons to protons) are described as below the **band of stability.** These nuclei decay by processes that increase this ratio— **positron emission, electron capture** (in which an electron from the first, or *K,* shell is captured by the nucleus), or **alpha emission** (especially for heavier nuclei), as discussed in Section 26-7. All nuclei with *atomic number greater than 83* are radioactive and are below and to the right of the band of stability (Section 26-8) so most such nuclides decay by alpha emission. This includes the heavier representative elements, the most recently produced transition metals, and all of the actinides. Some of these heavy nuclides, especially those beyond uranium ($Z = 92$) also disintegrate by spontaneous nuclear fission into nuclides of intermediate masses, giving neutrons as well.

Many studies and applications of nuclear reactions depend on the detection and measurement of the radiation that is emitted. Several common methods of **detection of radiation** are described in Section 26-8—photographic methods, fluorescence methods, cloud chambers, and gas ionization counters. Each of these methods depends on the properties and energies of the various types of radiation; you should know the general basis of each method of detection. Photographic detection and use of a cloud chamber are convenient methods for detecting the presence of radiation, requiring little in the way of sophisticated equipment, but they are tedious and inaccurate to use for quantitative measurements of radiation intensity and energy. Fluorescence counters such as scintillation counters and gas ionization counters both require electronic circuitry for effective use, but both techniques can permit quite sensitive studies of energies and intensities of radiation.

All radioactive decay processes obey temperature-independent **first-order** kinetics (Sections 16-3 and 16-4). Thus, the rates of all such decay processes can be discussed conveniently in terms of the **half-lives** (Section 26-10) of the decaying nuclei. Review the discussion of the order of reactions, the resulting changes in amount of substance present, and the description of reaction rates in terms of the half-life of a reactant (Section 16-4). In the period of one half-life, half of any initial amount of a particular radioactive substance will have decayed. Remember that the half-life of a substance in a first-order process is constant. Examples of the rate calculations for radioactive decay processes and the interpretations and applications of these calculations are also found in this section.

While some unstable nuclides can attain stability by a single decay process, many must undergo a series of such decays, known as **disintegration series.** Three well-known natural series, beginning with $^{238}_{92}U$, $^{235}_{92}U$, and $^{232}_{90}Th$, respectively, are discussed in Section 26-11. A careful study of these series will give you further practice in understanding the results of emission of various types of radioactive particles.

The steady, predictable decay rates of unstable nuclei (**radionuclides**) and the resulting steady production of radioactivity make them very useful for many applications. Some of these uses are discussed in Section 26-12. The technique of **radioactive dating** depends on the known rates of decay of various long-lived nuclei. Analysis of the amount of carbon-14 remaining in very old materials of plant or animal origin allows dating of such objects when they are less than about 50,000 years old. Similar dating techniques are based on the ratio of potassium-40 to argon-40 (articles up to 1 million years old) or on the ratio of uranium-238 to lead-206 (articles several million years old). The calculations associated with such analyses are, of course, just first-order kinetics calculations. Other uses of radionuclides that are outlined in this section include applications in medicine, chemical research, agriculture, and industry.

So far in this chapter we have been discussing processes associated with nuclei that are naturally unstable. It is also possible to bombard nuclei with particles so that they become unstable by collision with the bombarding particle, and then decay. This technique, known as **nuclear bombardment,** can be used to artificially transmute one element to another, as described in Section 26-13. Such processes have resulted in the production of many nuclides that do not occur naturally, including many new elements—technetium, astatine, francium, and the transuranium elements, to name a few. Some of these artificially produced radioactive nuclides decay in a series of steps.

Most of the nuclear processes studied so far have been ones in which an unstable nucleus emits a particle and is transformed into a nuclide of slightly different mass number of atomic number. Two other kinds of nuclear reactions are discussed in the remainder of this chapter. **Nuclear fission** (Section 26-14) is a process in which the unstable nucleus disintegrates into two nuclei of intermediate mass and one or more neutrons. Nuclear fission is common for isotopes of some elements with atomic numbers above 80. This process is accompanied by the release of a vast amount of energy corresponding to the loss of mass that accompanies the reaction.

The two principal applications of nuclear fission reactions, the **atomic bomb** and the **nuclear reactor.** Both depend on establishing of a **chain reaction** in which the neutrons from one nuclear disintegration are absorbed by other nuclei; these then disintegrate in their turn, each such disintegration releasing energy and more neutrons to propagate the process. In the atomic bomb, this chain reaction is initiated by bringing together more than the minimum amount of material necessary to sustain the chain reaction, called the **critical mass,** and then allowing the reaction to proceed at an uncontrolled rate. In the nuclear fission reactor (Section 26-15), the rate of the fission reaction is controlled by other substances that are present to absorb some of the neutrons. The general design features of **light water reactors** are described in this section. Some of the problems associated with the application of nuclear energy to power generation are also discussed.

Nuclear fusion (Section 25-16) is a very energetic process in which two nuclei are combined to make one of higher mass. This process is favorable only for the very light nuclei. Nuclear fusion reactions produce even greater amounts of energy than nuclear fission (the hydrogen bomb, for instance). But these reactions are so difficult to contain and control that development of this nearly limitless energy source is in its infancy, for reasons described in this section.

rize the characteristics that distinguish between nuclear reactions and ordinary chemical
ns. (Chapter introduction and Section 26-1; Exercise 3)

le to compare the calculated mass of an isotope to its observed mass. Relate this mass
ency to the binding energy of the nucleus. (Section 26-3; Exercises 4, 5, 7, 8, 14 through
, and 66)

Give symbols, identities, and properties of the common types of radiation: (a) beta particles, (b) alpha particles, (c) positrons, (d) neutrons, (e) protons, and (f) gamma rays. (Section 26-4; Exercises 20, 21, and 24)

4. Be able to balance equations for nuclear reactions (Section 26-5)

5. Given a plot of number of neutrons versus number of protons, identify stable nuclei, suggest and give specific examples of the kind of radioactive decay most likely for nuclides that are (a) above, (b) below, and (c) beyond and to the right of the band of stability. (Section 26-2 and 26-4 through 26-8; Exercises 9 through 13 and 26)

6. Describe the following methods for the detection of radiation: (a) photography, (b) fluorescence, (c) cloud chamber, and (d) gas ionization. (Section 26-9; Exercise 24)

7. Relate rate constants for radioactive decays to half-lives. (Section 26-10; Exercises 50 and 51)

8. Perform calculations relating the following four variables to calculate the fourth when three are known: (a) an initial amount of radionuclide present, (b) a specified time, (c) the amount of radionuclide remaining after that time, and (d) the rate constant or half-life for the decay. (Section 26-10; Exercises 52 through 62)

9. In nuclear disintegration processes, be able to relate parent nuclide, daughter nuclide, and type of emission. (Section 26-11; Exercises 19, 25, 27 through 32, and 28 through 46)

10. Give several examples of practical uses of radionuclides. (Section 26-12; Exercises 22 and 23)

11. Describe and illustrate the operation of a cyclotron and a linear accelerator. (Section 26-13; Exercise 49)

12. Give specific examples, with balanced equations, of various methods of inducing artificial transmutations of elements. (Section 26-13; Exercises 33 through 43 and 66)

13. Describe and distinguish between nuclear fission and nuclear fusion. Relate the position of a nuclide on a plot of binding energy per nucleon versus mass number to its tendency to undergo fission or fusion. (Sections 26-14 through 26-16; Exercises 1, 25, and 63 through 65)

14. Describe the principles behind the atomic bomb. Tell how ignition of the detonation process is accomplished and how the reaction is sustained. (Section 26-14; Exercise 67)

15. Describe the operation of a light water nuclear reactor. Understand why nuclear fission explosions are impossible in such a setting. (Section 26-15; Exercises 68, 69, and 71)

16. Describe the vapor diffusion method currently used for separating fissionable isotopes for use in nuclear reactors. Relate this method to relative rates of diffusion of gas molecules (Section 12-14). (Section 26-15; Exercise 70)

17. Give the advantages and disadvantages of nuclear fusion reactors as potential future energy sources. (Section 26-16; Exercise 72)

18. Be able to recognize and solve a variety of types of problems based on this chapter. (Mixed Exercises, 73 through 77)

19. Answer conceptual questions based on this chapter. (Conceptual Exercises 78 and 79)

20. Apply concepts and skills from earlier chapters to the material of this chapter. (Building Your Knowledge exercises, 80 and 81)

Some Important Terms in This Chapter

Write the meanings *in your own words*. Check the Key Terms list and the chapter reading. Then rewrite your definitions, still in your own words, to improve them. Study other new terms, and review terms from preceding chapters if necessary.

radioactivity

alpha particle (α)

beta particle (β)

gamma ray (γ)

band of stability

mass deficiency

nuclear binding energy

nuclear fusion

nuclear fission

nuclides

positron

half-life

radioactive dating

parent nuclide

daughter nuclide

artificial transmutation

chain reaction

critical mass

nuclear reactor

Preliminary Test

True-False

_____ 1. In a nuclear reaction, one element can be changed into another.
_____ 2. The electrons surrounding the nucleus are usually involved in nuclear reactions.
_____ 3. The actual mass of an atom is equal to the sum of the masses of its constituent particles.
_____ 4. The particles emitted from the nucleus in nuclear reactions are always positively charged.

_____ 5. Only very heavy nuclei are naturally radioactive.
_____ 6. All nuclei except hydrogen are somewhat unstable because of the very close proximity of the positively charged protons in the nucleus.
_____ 7. Nuclei above the band of stability usually decay by nuclear fission.
_____ 8. No naturally occurring nuclides with atomic number less than 8 are radioactive.
_____ 9. All radioactive decay processes follow first-order kinetics.
_____ 10. The eventual product of each of the three common natural radioactive series is a stable isotope of lead.
_____ 11. Because lead is the product of the three common natural radioactive series, we can tell that all isotopes of lead are stable.
_____ 12. Any amount of a fissionable material such as U-235 can explode if heated to a high enough temperature.
_____ 13. Light water nuclear reactors use a large amount of water, both for cooling and as a moderator to slow the neutrons produced in the reaction.

Short Answer

Answer with a word, a phrase, a formula, or a number with units as necessary.

1. Nuclides having either 2, 8, 20, 28, 50, 82, or 126 neutrons are said to have a _____ of neutrons and are found to be exceptionally _____.

2. Above atomic number 20, most stable nuclei have _____ protons than neutrons.

3. A nucleus having a proton to neutron ratio either higher or lower than stable nuclei is said to lie outside the _____.

4. Of the common types of radiation, the ones having positive charge are the _____, the _____, and the _____. (For each one, give the name of the particle and an acceptable symbol.)

5. The types of radioactive emissions that result in the daughter nucleus having a lower atomic number than the parent nucleus are _____, _____, and _____.

6. The type of radioactive emission that results in a lower mass number of the nucleus but no change in atomic number is _____.

7. The type of nuclear emission that produces no change in either mass number or atomic number of the nucleus is _____.

8. The radioactive particles with the lowest penetrating ability are _____ _____.

9. Nuclei above the band of stability have too _____ a ratio of neutrons to protons; such nuclei usually decay by _____ emission or by _____ emission.

10. The first several steps of the thorium-232 radioactive decay series involve, in order, the loss of an alpha particle, then a beta particle, then another beta particle; the nuclide symbol for the product after these three steps of decay is _____.

11. All nuclides with atomic number greater than _____ are radioactive.

12. When a nuclear reaction occurs, the particles emitted can have different kinetic energies. The kinetic energy of the emitted particle is equal to the energy equivalent of the _____ of products relative to reactants, minus the energy associated with any _____ that are later emitted.

For Questions 13 through 19, tell whether the effect of the change would be to increase (I), decrease (D), or leave unchanged (U) the atomic number and the mass number of the nuclide. Tell the magnitude of any increase or decrease.

Change	Effect on Atomic Number	Effect on Mass Number
13. alpha emission	_____	_____
14. beta emission	_____	_____
15. gamma emission	_____	_____
16. positron emission	_____	_____
17. electron capture	_____	_____
18. proton emission	_____	_____
19. neutron emission	_____	_____

20. The half-life of lead-210, one of the nuclides in the U-238 series, is 20.4 years. Suppose we start with 100 µg of lead-210 and wait for it to decay. After 20.4 years, the mass of lead-210 present would be _____ µg , after 40.8 years it would be _____ µg , and after 61.2 years it would be _____ µg .

21. The effectiveness of cobalt-60 treatment in arresting certain types of cancer depends on the ability of _____ to destroy tissue.

- -

For each of the next fifteen questions, 22 through 36, fill in the nuclide symbol that correctly completes the equation. Describe each reaction with a phrase, such as beta emission, alpha emission, positron emission, electron capture, neutron emission, gamma emission, nuclear fission, or nuclear fusion. For all reactions *except* those that are nuclear fission or nuclear fusion, tell whether the neutron/proton ratio increases (I), decreases (D), or remains unchanged (U) by the process.

Reaction	Description	Change in n/p ratio
22. $_{0}^{1}n \rightarrow {}_{1}^{1}p + $ ____	_____	_____
23. $_{6}^{14}C \rightarrow$ ____ $+ {}_{-1}^{0}\beta$	_____	_____
24. $_{7}^{17}N \rightarrow$ ____ $+ {}_{0}^{1}n$	_____	_____
25. $_{94}^{240}Pu \rightarrow {}_{92}^{236}U + $ ____	_____	_____
26. $_{73}^{186}Ta \rightarrow {}_{74}^{186}W + $ ____	_____	_____

	Reaction	Description	Change in n/p ratio
27.	$^{197}_{80}Hg + ^{0}_{-1}e \rightarrow$ ____	_____	_____
28.	____ $\rightarrow ^{39}_{20}Ca + ^{0}_{-1}e$	_____	_____
29.	____ $+ ^{0}_{-1}e \rightarrow ^{22}_{10}Ne$	_____	_____
30.	$^{236}_{92}U^* \rightarrow ^{236}_{92}U +$ ____	_____	_____
31.	$^{223}_{88}Ra \rightarrow$ ____ $+ ^{4}_{2}\alpha$	_____	_____
32.	$^{1}_{1}p \rightarrow ^{1}_{0}n +$ ____	_____	_____
33.	$^{1}_{1}H + ^{1}_{1}H \rightarrow$ ____ $+ ^{0}_{-1}e$	_____	_____
34.	$^{3}_{2}He + ^{3}_{2}He \rightarrow ^{4}_{2}He + 2$____	_____	_____
35.	$^{235}_{92}U + ^{1}_{0}n \rightarrow ^{93}_{36}Kr +$ ____ $+ 3^{1}_{0}n$	_____	_____
36.	____ $+ ^{1}_{0}n \rightarrow ^{99}_{40}Zr + ^{139}_{54}Xe + 2^{1}_{0}n$	_____	_____

37. For any radioactive decay, the decaying nuclide is called the _____ nuclide and the product is called the _____ nuclide.
38. The sequence of radioactive decay processes by which some radionuclides attain stability in several steps is called a _____.
39. The method of radioactive dating based on the decay of carbon-14 is applicable only to objects up to about _____ years old.
40. The process in which a nuclide is made unstable by hitting it with a high-energy particle is called _____.
41. All nuclides with mass number greater than _____ undergo nuclear fission spontaneously, whereas those between mass numbers _____ and _____ can be induced to do so by bombarding them with particles of relatively low energy.
42. The most stable nuclide is _____.
43. A material that is capable of undergoing nuclear fission is called _____.
44. A nuclear reaction in which two smaller nuclei combine to form a larger nucleus, with the release of energy, is called a _____ reaction.
45. Most stable naturally occurring nuclides have _____ numbers of protons and _____ numbers of neutrons.

Multiple Choice

____ 1. Which of the following processes produces two nuclei of moderate mass from one heavier nucleus?
   (a) alpha emission   (b) nuclear fusion   (c) nuclear fission
   (d) fluorescence   (e) ionization

_____ 2. The observed mass of deuterium, $_1^2H$, is 2.01400 amu. What is the mass deficiency of deuterium?
   (a) 0.00255 amu/atom   (b) 4.03066 amu/atom   (c) 0.126 amu/atom
   (d) 0.127 amu/atom   (e) Deuterium has no mass defect.

_____ 3. What is the nuclear binding energy of one mole of $_{11}^{23}Na$ atoms? Sodium exists in nature as a single isotope, so you may use the observed atomic weight as the observed mass of the Na atom.
   (a) 1.80×10^{10} kJ/mol   (b) 2.70×10^{10} kJ/mol
   (c) 3.92×10^{10} kJ/mol   (d) 9.01×10^9 kJ/mol

_____ 4. Which of the following is not a commonly used method of radiation detection?
   (a) photographic detection
   (b) detection by fluorescence
   (c) use of a cloud chamber
   (d) use of a gas ionization chamber
   (e) bombardment with positive ions

_____ 5. The radiocarbon method of dating would not be directly applicable to determining the age of which of the following objects?
   (a) a piece of charcoal from a fire   (b) a piece of wood
   (c) a piece of animal skin   (d) a woven straw mat
   (e) a flint arrowhead

_____ 6. Measurement based on which of the following radioactive decay processes would be applicable to age determination for objects of greatest age?
   (a) decay of carbon-14 (half life 5,730 years)
   (b) decay of potassium-40 (half-life 1.3 billion years)
   (c) decay of radium-223 (half-life 11.4 days)
   (d) decay of uranium-238 (half-life 4.5 billion years)
   (e) decay of uranium-235 (half-life 710 million years)

_____ 7. The source of high-energy positive ions or high-energy negative ions for nuclear bombardment processes is usually which of the following?
   (a) a particle accelerator   (b) a nuclear fission reactor
   (c) a cloud chamber   (d) an atomic bomb
   (e) a nuclear fusion reactor

_____ 8. Which of the following experimental methods has been used for the production of elements that do not occur naturally?
   (a) nuclear fusion   (b) nuclear bombardment
   (c) gas ionization   (d) natural radioactive decay
   (e) labeling experiments

_____ 9. Which of the following is not a problem associated with the operation of a nuclear fission reactor for the production of electricity?
   (a) possibility of a meltdown
   (b) storage of long-lived radioactive wastes
   (c) a nuclear fission explosion
   (d) thermal pollution
   (e) escape of short-lived radioactive wastes

Answers to Preliminary Test

Do not just look up or memorize answers—be sure you could figure them out yourself.

True-False

1. True. The only nuclear reactions for which this is not true are those in which only a neutron is emitted, and these are not common.
2. False. Most nuclear reactions involve only the nucleus. One kind of nuclear reaction, called electron capture (K capture), involves the capture of an electron from the inner electronic shell (the K shell) by the nucleus, but this type of nuclear reaction is relatively uncommon.
3. False. Some of the mass, called the mass deficiency, is converted into energy. This is the binding energy that holds the nuclear particles together. See the calculations and discussion in Section 26-3.
4. False. Some are positive, some are negative, and some are neutral. See Section 26-4, and be sure that you are familiar with the general properties of the particles that take part in nuclear decay processes.
5. False. Most very heavy nuclei are radioactive, but so are many light nuclei.
6. False. Many nuclei are nonradioactive, meaning that they are stable and do not decay. It is not entirely understood how the strong coulombic proton-proton repulsive forces in the nucleus are overcome, but it is thought that some of the very short-lived subatomic particles that have recently been discovered may play a role. See Section 26-1.
7. False. They usually decay by beta emission and less frequently by neutron emission. The only nuclides that spontaneously undergo fission are isotopes of elements with atomic numbers greater than 80. These are found to the right of the band of stability.
8. False. For instance, the chapter mentions radioactive isotopes of carbon and nitrogen.
9. True
10. True
11. False. Several radioactive isotopes of lead occur in these series. See Table 26-4.
12. False. If too little material is present, the reaction cannot become self-sustaining, because too few neutrons are produced. It is necessary to have a critical mass for an explosion. Study the section on the atomic bomb (Section 26-14) and the comments regarding the impossibility of explosion of a nuclear fission reactor (Section 26-15).
13. True. The design and general features of such a light water reactor are described in Section 26-15.

Short Answer

1. magic number; stable. The same principle is found to apply to the numbers of protons in the nucleus and to the sum of protons plus neutrons.
2. fewer
3. band of stability
4. positron ($_{+1}^{0}\beta$ or $_{+1}^{0}e$); alpha particle (α, $_{2}^{4}\alpha$, or $_{2}^{4}$He); proton ($_{1}^{1}p$ or $_{1}^{1}$H)
5. positron emission; alpha emission; proton emission
6. neutron emission
7. gamma emission

8. alpha particles
9. high; beta; neutron. The first of these, beta emission, is much more common.
10. $^{228}_{90}$Th. You should be able to figure this out without looking at the disintegration series in Table 26-4.
11. 82
12. mass loss; gamma rays

You should be able to answer questions 13 through 19 without looking up the answer, just from knowing the characteristics of the various kinds of radiation.

13. D (by 2); D (by 4)
14. I (by 1); U
15. U; U
16. D (by 1); U
17. D (by 1); U
18. D (by 1); D (by 1)
19. U; D (by 1)

20. 50; 25; 12.5. Radioactive decay is a first-order kinetic process. During any time period equal to one half-life, the amount of material present decreases to one-half of what it was at the beginning of *that* half-life.
21. gamma rays. This is discussed in Example 26-3, which is also an illustration of calculations associated with radioactive decay.
22. $^{0}_{-1}\beta$; beta emission; D
23. $^{14}_{7}$N; beta emission; D
24. $^{14}_{6}$N; neutron emission; D
25. $^{4}_{2}\alpha$; alpha emission; I
26. $^{0}_{-1}\beta$; beta emission; D
27. $^{197}_{79}$Au; electron capture; I
28. $^{39}_{19}$K; beta emission; D
29. $^{22}_{11}$Na; electron capture; I
30. $^{0}_{0}\gamma$; gamma emission; I
31. $^{219}_{86}$Rn; alpha emission; I
32. $^{0}_{-1}\beta$; beta emission; D
33. $^{2}_{1}$H; nuclear fusion. (This also gives off a beta particle, but it is principally nuclear fusion.)
34. $^{1}_{1}$H; nuclear fusion. (This also gives off protons, but it is principally nuclear fusion.)
35. $^{140}_{56}$Ba; nuclear fusion
36. $^{239}_{94}$Pu; nuclear fission
37. parent; daughter
38. disintegration series or radioactive series
39. 50,000
40. nuclear bombardment
41. 250; 225; 250
42. $^{56}_{26}$Fe
43. fissionable

44. nuclear fusion
45. even; even

Multiple Choice

1. (c)
2. (a). Solve this problem as in Example 26-1. This mass deficiency could also be expressed as 0.00255 g/mol.
3. (a). First find the mass deficiency as in Example 26-1. Then convert this to binding energy as shown in Example 26-2.
4. (e)
5. (e). This method applies only to material that originated as a living organism, either plant or animal.
6. (d). It has the longest half-life.
7. (a). This is discussed in the first part of Section 26-13.
8. (b). See Section 26-13 for a discussion of this topic.
9. (c). Read Section 26-15. Reactors use fuels that have neither the correct composition nor the correct physical arrangement ever to achieve the critical mass necessary for an explosion. The other dangers are real.

27 Organic Chemistry I: Formulas, Names and Properties

Chapter Summary

Because of its position in Group IVA and its intermediate electronegativity, carbon can form very stable covalent bonds with as many as four other atoms. Thus, carbon has a striking ability to exhibit extensive **catenation** ("chainmaking") by bonding to other carbon atoms. As a result, it can form a large number and variety of compounds with carbon chains and rings. The study of such compounds involving C–C or C–H bonds comprises a major branch of chemistry known as **organic chemistry.** Our study of descriptive chemistry continues in Chapters 27 and 28 with some of the major features of this extensive and varied area of chemistry. A more detailed, but still introductory, study of organic chemistry is usually a course of a year or more at the college level, so the present study should be viewed as a short preliminary survey of this field.

Chapter 27 deals with the major classes or "families" of compounds of organic chemistry, with emphasis on their structures, sources, general properties, and uses. Because of the wide variety of compounds in organic chemistry, it is important to be able to name compounds systematically, so an introduction to organic nomenclature is an important part of this early study of organic chemistry. Chapter 28 also discusses the main classes of reactions that organic compounds undergo. That chapter also contains a systematic look at the geometries of organic molecules.

The first two sections of Chapter 27 deal with the **saturated hydrocarbons**, or **alkanes**, which contain only single covalent bonds. This means that each carbon is bonded to four other atoms, either carbon or hydrogen, with tetrahedral geometry and sp^3 hybridization of the carbon atoms (Section 8-7). Section 27-1 introduces the structures and terminology of **homologous series** of compounds, including both the **normal** or unbranched hydrocarbons and the **branched** hydrocarbons. Do not be misled by the term "straight chain" when applied to organic carbon chains; this is clarified in a marginal note on page 962. In Chapter 25 you learned about the existence of **isomers,** compounds having the same number of atoms of all types, but with differing arrangements. Isomerism, which is very common in organic chemistry, will be discussed in more detail in Chapter 28. The first type to be encountered, called **constitutional** (or **structural**) **isomerism**, is described in Section 27-1.

As you probably expect, we have developed a systematic way of naming organic compounds. The rules for this nomenclature of saturated hydrocarbons are discussed in Section 27-2. Notice the relation between the name of the parent hydrocarbon and the **alkyl group** obtained by removing one H atom. These alkyl groups can be thought of as building blocks that will appear in many classes of organic compounds. Study carefully the examples for naming organic compounds, and you will find that it is not as complex as it seems at first.

Petroleum consists primarily of hydrocarbons, but of a mixture of many different chain lengths. The refining of petroleum involves the separation of this complex mixture into **fractions** that are suitable for various uses. See the Chemistry in Use essay on this topic.

Hydrocarbons that contain some carbon-carbon bonds other than single bonds are known as **unsaturated hydrocarbons**. There are three classes of such compounds, which are discussed in the next several sections.

Section 27-3 describes the structures and naming of the **alkenes**, which contain carbon-carbon double bonds. Because of the lack of free rotation around a double bond, compounds containing such a bond can exhibit **geometrical isomerism**. Notice both the differences in properties of such isomeric pairs and the additional terminology needed in the nomenclature—terms *cis* and *trans*, which have the same meanings you learned in Chapter 25. The second class of unsaturated hydrocarbons is the **alkynes**, which contain carbon-carbon triple bonds. These are introduced and named in Section 27-4. Sections 27-5 through 27-6 describe the third class of unsaturated hydrocarbons, the **aromatic hydrocarbons**. The origin of this term, which you can see is quite a misnomer, is described in the introductory paragraphs to this segment of the chapter. The bonding in all aromatic compounds is characterized by **delocalization** of π-bonded electrons around a ring (most often a six-membered ring) and a consequent lowering of the energy. This effect is known as **resonance** (review Sections 7-6 and 9-6). This bonding is described in detail for **benzene**, the simplest aromatic hydrocarbon, in Section 27-5. Other aromatic hydrocarbons, presented in Section 27-6, consist of benzene with one or more hydrogen atoms replaced by alkyl groups (e.g., toluene and the three isomeric xylenes) and the fused-ring aromatics (e.g., naphthalene, anthracene, etc.). The rules for naming derivatives of benzene are presented in this section.

Figure 27-12 in Section 27-7 summarizes the classification of hydrocarbons that we studied in this chapter.

Many organic compounds involve some groups of atoms having bonds other than C–C or C–H single bonds. It is found that a given grouping of atoms generally undergoes similar reactions, no matter in what compound it occurs. Such a group of atoms is often viewed as a potential reaction site, called a **functional group.** The consistent behavior of each functional group enables us to organize large amounts of information about organic chemistry. The next major portion of Chapter 27 deals with the most common functional groups and with the resulting characteristic classes of compounds. It is suggested that you organize your study of each of these classes of compounds to emphasize the functional group common to all of the compounds in that class, the system for naming those compounds, and some of their common uses. We sometimes use group symbols to represent unspecified groups of atoms, such as alkyl (R) or aryl (Ar, not to be confused with argon). This is a commonly used compact notation.

Organic halides, Section 27-8, can be thought of as hydrocarbons in which one or more hydrogen atoms have been replaced by a halogen atom.

The group of atoms –O–H **(hydroxyl)** attached to a carbon atom in an aliphatic or aromatic group is characteristic of the two related classes of compounds known as the **alcohols** and the **phenols**. When the hydroxyl group is bonded to an aliphatic carbon atom, the compound is classed as an alcohol; when the carbon is aromatic, the compound is a phenol. Examples of structures and naming of several compounds of these two classes are presented in Section 27-9. Because of electronegativity differences within the molecule, alcohols are polar molecules. Alcohols can be considered as derivatives of water in which one hydrogen atom has been replaced by an organic group.

Similarly, **ethers** can be considered as water derivatives in which both H atoms have been replaced by organic groups. Thus, an ether is a compound in which an oxygen atom is bonded to two organic groups, R–O–R', where R and R' may each be either an alkyl or an aryl group. Section 27-10 describes these compounds. They are used for a variety of purposes such as anesthetics, artificial flavors, refrigerants, and solvents.

Two related groups of compounds, the **aldehydes** and the **ketones,** both contain the $-\overset{\overset{\textstyle O}{\|}}{C}-$ group, called the **carbonyl** group. Ketones have two alkyl or aryl groups bonded to the carbonyl group, while in aldehydes one of these is replaced by a hydrogen atom. The structures, nomenclature, and some uses of these two classes of compounds are presented in Section 27-11. The aldehydes are usually called by their common names, which indicate the number of carbon atoms in the compound. Ketones are named according to the two alkyl or aryl groups bonded to the carbonyl group. Many important natural substances are aldehydes and ketones. Their aliphatic or aromatic ends make them mix readily with organic compounds, while the polar carbonyl group makes them miscible with water. As a result, many ketones are useful as solvents, of which the most widely used is acetone.

The **amines** are basic compounds that can be considered as derivatives of ammonia, NH_3, in which one or more of the hydrogen atoms have been replaced by an alkyl or an aryl group. These compounds, their structures, and their nomenclature are the subject of Section 27-12.

Carboxylic acids (Section 27-13) contain the **carboxyl group,** $-\overset{\overset{\textstyle O}{\|}}{C}-OH$, often written –COOH. Most carboxylic acids are weak acids. You have seen acetic acid, CH_3COOH, many times in weak acid problems or examples. The carboxylic acids can be classed as either **aliphatic** or **aromatic,** depending on the nature of the organic group bonded to the –COOH group. Many carboxylic acids occur in common natural products. The naming and structures of some carboxylic acids are described. Be sure that you understand how we name the common groups that are derived from the carboxylic acids (the **acyl** groups) and how these are related to the parent acid.

Section 27-14 describes the preparation and properties of four important types of derivatives of carboxylic acids, known as the **acid anhydrides,** the **acyl halides,** the **organic esters,** and the **amides.** Each of these can be viewed as resulting from the replacement of the –OH group of the carboxylic acid by another group, such as a halogen atom in the acyl halides or the –OR' group in the esters. Many commonly recognizable flavors and aromas come from organic esters, and many common fats, oils, and waxes are organic esters. The Chemistry In Use essay on page 1083 relates some dietary aspects of fats and oils to their chemical structures. Amides find use as medicinal agents such as headache and fever remedies, as insect repellents, and as the starting materials for polyamides, of which nylon is an example.

The summary of the classification of compounds according to functional groups in Figure 27-21 will be a great aid in your study of this chapter.

In the rest of the chapter, we survey some important types of reactions of organic compounds. The next three sections deal with three fundamental classes of organic reactions. In a **substitution reaction**, Section 27-16, an atom or a group of atoms attached to a carbon atom is removed and another atom or group of atoms takes its place, so that *no change in the degree of saturation* occurs.. This is the typical reaction of the saturated hydrocarbons (alkanes and cycloalkanes). Some types of substitution reactions are known by particular names, which indicate the type of group that replaces another group; examples are halogenation and nitration. In an **addition reaction**, Section 27-17, the number of groups attached to carbon increases, so that the molecule becomes *more nearly saturated*. Again, subcategories are named according to what groups are added—halogenation, hydrogenation, and hydration. The third fundamental type of reaction, an **elimination reaction**, Section 27-18, involves a decrease in the number of groups attached to carbon, so that the *degree of unsaturation increases*. Dehydrohalogenation, and dehydration reactions are particular types of elimination reactions.

The final section of the chapter discusses **polymerization reactions**, which produce the **polymers** that seem to be everywhere in our surroundings. As you learn here, these are two basic types of these reactions. In **addition polymerization**, polymers are formed by addition reactions; polyethylene, polypropylene, and natural and synthetic rubbers are formed by this kind of reaction. **Condensation polymerization** is based on condensation reactions, in which molecules combine by splitting out or eliminating a small molecule; polyesters and polyamides (e.g., Nylon) are examples of synthetic condensation polymers. The section closes with a discussion of **polypeptides**, which are addition condensation polymers based on the **peptide bond**. Proteins and other biologically important polypeptides, will be discussed further in Chapter 28.

Study Goals

This is quite a long and diverse chapter. Be systematic in your study. Organize your study around the classes of compounds. Look at Figures 27-12 and 27-20 to help you organize the material.

1. Understand the basic ideas that lie behind the study of organic chemistry. (Chapter introduction; Exercises 1 through 4)
2. Know the structures, properties, some sources, and uses that are typical of each type of aliphatic hydrocarbons: (a) alkanes and cycloalkanes, (b) alkenes and cycloalkenes, (c) alkynes. (Sections 27-1 and 27-3 through 27-4; Exercises 5 through 12, 20, 21, 24 through 26, and 31)
3. Know how to name compounds of the classes in Study Goal 2. (Sections 27-2 through 27-4; Exercises 13 through 19, 22, and 27 through 30)
4. Understand the types of isomers that are possible for compounds of the classes in Study Goal 2. Be able to name these isomers uniquely. (Sections 27-1 through 27-4; Exercises 13, 14, 16, and 23)
5. Know the structures, properties, nomenclature, some sources, and some uses that are typical of aromatic hydrocarbons. (Sections 27-5 and 27-6; Exercises 32 through 38)
6. Know the structures, properties, nomenclature, some sources, and some uses that are typical of alkyl and aryl halides. (Section 27-8; Exercises 39 through 43)
7. Know the structures, properties, nomenclature, some sources, and some uses that are typical of alcohols and phenols. (Section 27-9; Exercises 44 through 54)
8. Know the structures, properties, nomenclature, some sources, and some uses that are typical of ethers. (Section 27-10; Exercises 55 through 59)
9. Know the structures, properties, nomenclature, some sources, and some uses that are typical of aldehydes and ketones. (Section 27-11; Exercises 64 through 67)
10. Know the structures, properties, nomenclature, some sources, and some uses that are typical of amines. (Section 27-12; Exercises 60 through 63)
11. Know the structures, properties, nomenclature, some sources, and some uses that are typical of carboxylic acids and their derivatives. (Sections 27-13 and 27-14; Exercises 68 through 78)
12. Know the characteristics of some important classes of organic reactions: (a) substitution, (b) addition, and (c) elimination. Know what classes of organic compounds can undergo each type of reaction. (Sections 27-16 through 27-18; Exercises 79 through 94)
13. Describe the basic features of polymerization reactions. Recognize various types of polymerization reactions and polymers. (Section 28-19; Exercises 95 through 112)

331

14. Apply the ideas of this chapter to unclassified compounds and reactions. Be able to (a) identify the functional groups present, (b) identify the class(es) represented, or (c) name the compounds. (Mixed Exercises, 113 through 120)
15. Answer conceptual questions based on this chapter. (Conceptual Exercises 121 and 122)
16. Apply your knowledge about chemical bonding, structure, electronegativity, and stoichiometry to the new material in this chapter. (Building Your Knowledge exercises 123 through 126)

Some Important Terms in This Chapter

Write the meanings *in your own words*. Check the Key Terms list and the chapter reading. Then rewrite your definitions, still in your own words, to improve them. Study other new terms, and review terms from preceding chapters if necessary.

catenation

alkanes

alkenes

alkynes

alkyl group

saturated hydrocarbons

unsaturated hydrocarbons

aliphatic hydrocarbons

aromatic hydrocarbons

constitutional isomers

structural isomers

geometrical isomers

alcohol

phenol

carboxylic acid

acid anhydride

acyl halide

ester

ether

amine

amide

aldehyde

ketone

substitution reaction

halogenation reaction

addition reaction

hydrogenation reaction

hydration reaction

elimination reaction

dehydrohalogenation

dehydration reaction

polymerization reaction

polymer

copolymer

addition polymerization

condensation polymerization

polyester

polyamide

peptide bond

Preliminary Test

Be sure to practice *many* of the additional textbook exercises.

True-False

_____ 1. Most hydrocarbons are polar.
_____ 2. The carbon atoms in saturated hydrocarbons are best described as sp^3 hybridized.
_____ 3. The straight-chain hydrocarbons are so called because the carbon atoms lie in a straight line.
_____ 4. In saturated hydrocarbons, each carbon atom is bonded to either two or three hydrogen atoms.
_____ 5. All alkanes up to C_8H_{18} are gases at room temperature and atmospheric pressure.
_____ 6. Generally, the more branched an alkane is (for the same number of carbon atoms), the lower are its melting and boiling points.
_____ 7. Simple alkanes can exhibit either constitutional isomerism or geometrical isomerism.
_____ 8. Simple alkenes can exhibit either constitutional isomerism or geometrical isomerism.
_____ 9. Simple alkynes can exhibit either constitutional isomerism or geometrical isomerism.
_____ 10. Petroleum consists mainly of hydrocarbons.
_____ 11. Petroleum from all sources is identical, that is, composed of a mixture of the same compounds in the same amounts.
_____ 12. All aromatic hydrocarbons have a pleasant smell.
_____ 13. Many of the common flavoring agents are simple amines.
_____ 14. Alcohols are polar molecules.

_____ 15. Many reactions of carboxylic acids involve displacement of the –OH group by another atom or group of atoms.

_____ 16. All functional groups contain some atom other than carbon or hydrogen.

_____ 17. A single organic compound can have only one functional group.

_____ 18. The common source of the carbon atoms for the industrial preparation of most organic chemical is carbon dioxide from the air.

_____ 19. The main type of chemical reaction that the alkanes undergo is substitution.

_____ 20. Organic reactions usually form only a single product, so separation is usually not a serious problem.

_____ 21. The reaction in which a nitro group, –NO_2, is substituted for a hydrogen atom is called nitration.

_____ 22. Nitration reactions result when a hydrocarbon is mixed with NO_2 gas at high temperature.

_____ 23. Compounds with carbon-carbon double or triple bond undergo addition reactions much more readily than they undergo substitution reactions.

_____ 24. The alkynes are usually more reactive than the alkenes with respect to addition reactions.

_____ 25. Benzene is an unsaturated hydrocarbon, so it readily undergoes addition reactions like those of the alkenes and the alkynes.

_____ 26. Hydrogenation reactions are substitution reactions in which H atoms are substituted for other groups such as halogen atoms or hydroxyl groups.

Short Answer

1. The existence of the large number of organic compounds is primarily due to the ability of carbon atoms to _____ other carbon atoms.

2. A compound in which there is one continuous chain of carbon atoms, with no branching, is referred to as a straight-chain or _____ compound.

3. Each member of the series of normal saturated hydrocarbons differs from the next by one _____ (name, formula); such a series of compounds is called a _____.

4. The melting points of the normal alkanes generally _____ with increasing carbon content, and their boiling points generally _____.

5. Atomic arrangements that differ only by rotation around single bonds are called different _____ of the same compound.

6. The hydrogen atoms of alkanes are readily replaced by halogen atoms at high temperatures or in the presence of _____.

7. As the chain length increases for the normal alkanes, the heat of combustion _____ on a molar basis and _____ on a mass basis.

8. The process of "cracking" of petroleum involves _____ _____ to produce more gasoline.

9. Hydrocarbons containing one carbon-carbon double bond per molecule are called _____.

10. The suffix applied to all alkenes, whether open chain or cyclic, is _____.

11. The suffix applied to all alkanes, whether open chain or cyclic, is _____.

12. The suffix applied to all alkynes, whether open chain or cyclic, is _____.

13. The number of isomers of xylene is ____; these are examples of _____ isomers.

14. The number of isomers of trimethylbenzene is ____.

For each of the following hydrocarbons, either name the hydrocarbon or supply its formula, as required. These formulas should be sufficient to specify the compound uniquely and may be full structural formulas. Write the condensed chemical formula (e.g., C_8H_{18}) for each compound. Classify each compound as an alkane, an alkene, an alkyne, an aromatic hydrocarbon, or derived from one of these by substitution.

Formula	Name	Condensed Formula	Type

15.

$$H-\overset{\overset{\displaystyle H}{|}}{\underset{\underset{\displaystyle H}{|}}{C}}-\overset{\overset{\displaystyle H}{|}}{\underset{\underset{\displaystyle H}{|}}{C}}-H$$

_____ _____ ____

16.

_____ _____ ____

17.

$$CH_3-\overset{\overset{\displaystyle CH_3}{|}}{\underset{\underset{\displaystyle CH_3}{|}}{C}}-CH_2CH_2CH_3$$

_____ _____ ____

18. _____   2,4-dimethylhexane   _____ ____

19. $H_2C = CHCH_2CH_3$

_____ _____ ____

20. _____   *cis*-2-butene   _____ ____

21. _____   *trans*-2-butene   _____ ____

22.

_____ _____ ____

23.

_____ _____ ____

24. _____   o-xylene   _____ ____

25. _____   *p*-xylene   _____ ____

26. _____   acetylene   _____ ____

27. _____ ethyne _____ ___

28. $CH_3C \equiv CCH_3$ _____ _____ ___

29. _____ anthracene _____ ___

30. $CH_3CH_2CHCH_2CH_2CH_2CH_3$
　　　　　　　$|$
　　　　　　CH_2
　　　　　　　$|$
　　　　　　CH_2
　　　　　　　$|$
　　　　　　CH_3 _____ _____ ___

31. _____ cyclopentene _____ ___

32. $CH_3CH_2NO_2$ _____ _____ ___

33.

　Br　　　Br
　 \　　 /
　　C = C
　 /　　 \
　H　　　H _____ _____ ___

34.

　Br　　　H
　 \　　 /
　　C = C
　 /　　 \
　H　　　Br _____ _____ ___

35. _____ *trans*-2,3-dibromo-2-butene _____ ___

36. _____ tetrafluoroethene _____ ___

37. _____ difluorodichloromethane _____ ___

38. _____ bromobenzene _____ ___

- -

39. A compound in which one hydrogen atom of water has been replaced by an organic group is a(n) _____, whereas a compound in which both hydrogen atoms of water have been replaced by organic groups is a(n) _____.

40. The simplest aliphatic alcohol is _____; the most common one is _____.

41. Alcohols that contain more than one –OH group per molecule are called _____ alcohols.

42. Aromatic alcohols as a group are referred to as _____; the simplest member of this group is the compound _____.

43. Alcohols in which the –OH group is bonded to a carbon atom that is in turn bonded to one, two, or three other organic groups are called _____ alcohols, _____ alcohols, and _____ alcohols, respectively.

338

44. The secondary alcohol with the smallest number of carbon atoms is _____ (name, formula).

45. The secondary amine with the smallest number of carbon atoms is _____ (name, formula).

46. The polyhydric alcohol containing two equivalent carbon atoms is _____ _____ (name, formula);   this substance is frequently used as _____.

47. Dilute solutions of phenols are frequently used as _____.

- -

In each of the next twenty questions, 48 through 67, name the compound or give a formula, as required. For each compound, tell the class of which it is an example.

Formula	Name	Class
48. $CH_3CH_2CH_2OH$	_____	_____
49. _____	phenol	_____
50.	_____	_____
51. _____	pentanoic acid	_____
52. _____	3-chloropentanoic acid	_____
53. CH_3–$\overset{\overset{\displaystyle O}{\|}}{C}$–Cl	_____	_____
54. _____	methyl butyrate	_____
55. _____	glycerine	_____
56. _____	acetone	_____
57. CH_3–$\overset{\overset{\displaystyle O}{\|}}{C}$–H	_____	_____
58. CH_3–$\underset{\underset{\displaystyle H}{\|}}{N}$–$CH_2CH_3$	_____	_____
59. _____	aniline	_____
60. _____	N-methylacetamide	_____
61. CH_3CH_2–O–CH_2CH_3	_____	_____

62. _____ _____

63. $(CH_3)_2NH_2^+ Cl^-$ _____ _____

64. _____ oxalic acid _____

65.

$CH_3-\overset{O}{\overset{\|}{C}}-O-\overset{O}{\overset{\|}{C}}-CH_3$ _____ _____

66. _____ _____

67. _____ _____

- -

68. The three fundamental classes of reactions that are involved in most organic transformations are _____, _____, and _____.

69. The most common kind of reaction of the aromatic ring is _____.

70. The reaction by which a hydrogen atom on the benzene ring is replaced by the group $-NO_2$ is called _____; the resulting compound is named _____.

- -

For each of the following questions, 71 through 78, classify the reaction according to one (or sometimes more) of the following types: aliphatic substitution, aromatic substitution, addition, halogenation, nitration, elimination, dehydration, polymerization.

71. $CH_4 + Cl_2 \rightarrow CH_3Cl + HCl$ _____

72. $CH_3CH=CH_2 + Cl_2 \rightarrow CH_3\underset{Cl}{CH}-\underset{Cl}{CH_2}$ _____

73. $CH_3CH=CH_2 + H_2 \rightarrow CH_3CH_2CH_3$ _____

74. $CH_2=CH_2 + HOCl \rightarrow \underset{OH}{CH_2}-\underset{Cl}{CH_2}$ _____

75. $CH_2=CH-\underset{Cl}{C}=CH_2 \rightarrow \left[-CH_2-CH=\underset{Cl}{C}-CH_2-\right]_n$ _____

340

76. $CH_3C{\equiv}CH + 3Br_2 \rightarrow CH_3CBr_2CHBr_2$ _____

77. $+ HNO_3 \xrightarrow[H_2SO_4]{}$ ... $+ H_2O$

_____

78. ... $+ Cl_2 \rightarrow$ ... $+ HCl$

_____

Multiple Choice

_____ 1. How many constitutional isomers are possible for the compound with the molecular formula C_3H_7Br? Remember that bromine can form only one covalent bond to another atom.
   (a) 1   (b) 2   (c) 3   (d) 4   (e) 5

_____ 2. What is the molecular weight of n-pentane?
   (a) 44   (b) 70   (c) 42
   (d) The same as the molecular weight of isopentane.
   (e) The same as the molecular weight of 2-pentane.

_____ 3. In all saturated hydrocarbon, the carbon atoms have which of the following types of hybrid orbitals?
   (a) sp (b) sp^2 (c) sp^3 (d) sp^3d (e) sp^3d^2

_____ 4. Substances having the same molecular formula but different structural formulas are called
   (a) isotopes.   (b) hydrocarbons.   (c) isomers.
   (d) fractions.   (e) free radicals.

_____ 5. Which one of the following is an isomer of cyclohexane?
   (a) n-hexane   (b) benzene   (c) 1-hexene
   (d) 2-methylpentane   (e) 2-methylhexane

_____ 6. The possibility of geometrical isomerism in some alkenes is due to
   (a) the fact that alkenes are saturated hydrocarbons.
   (b) the restriction on the orientation of the p orbitals that make up the π bond.
   (c) the free rotation about σ bonds.
   (d) the acidity of substituted groups.
   (e) resonance in these substances.

_____ 7. Which one of the following compounds may exist as either of two geometrical isomers?
   (a) 1,1-dibromoethene   (b) 1,1-dibromoethane   (c) 1,2-dibromoethene
   (d) 1,2-dibromoethane   (e) 2-methylpropene

_____ 8. Which one of the following compounds is an alkane?

   (a) $CH_3CH_2\underset{\underset{H}{|}}{C}{=}CH_2$ (b) $CH_3CH_2{-}\underset{\underset{O}{\|}}{C}{-}CH_3$ (c) CH_3OH

(d) $CH_3CH_2CH_2CH_2Br$ (e) $CH_3CH_2CCH_2CH_3$
$|$
CH_3

_____ 9. The formulas

 HH$$and H$_3$CHrepresent

(a) geometrical isomers. (b) structural isomers. (c) the same compound. (d) ethylene.
(e) None of the preceding answers is correct.

_____ 10. What is the number of constitutional isomers of dichlorobenzene?
(a) 1 $\qquad$ (b) 2 $\qquad$ (c) 3 $\qquad$ (d) 4 $\qquad$ (e) 5

_____ 11. One of the following pairs of compounds is a pair of *constitutional* isomers. Which pair?

(a) $CH_3CH_2CH_2CH_3$ and CH_2—CH_2
$||$
CH_2—CH_2

(b) $CH_3CH_2CH_2CH_3$ and $CH_3CH_2CHCH_3$
$|$
CH_3

(c) HH$$and H$_3$CH

C=C$$C=C

H$_3$CCH$_3$$HCH_3$

(d) $CH_3CH_2CH_2CH_2CH_3$ and CH_3
$|$
H_3C—C—CH_3
$|$
CH_3

_____ 12. Which answer in Question 11 represents a pair of *geometrical* isomers?

OHO
$|||$
_____ 13. The molecule CH_3—CH—CH—CH_2—C—H contains the functional groups that are
$|$
NH_2
characteristic of
(a) aldehydes, amides, and ethers. $\qquad$ (b) amines, ketones, and esters.
(c) aldehydes, halides, and alcohols. $\qquad$ (d) alcohols, aldehydes, and amines.
(e) acids, alcohols, and halides.

_____ 14. The group of atoms characteristic of aldehydes is
OOO
$||||||$
(a) –C–C–C– $\quad$ (b) –C–O–C– $\quad$ (c) –C–O– $\quad$ (d) –C–H $\quad$ (e) –C≡C–

342

_____ 15. The group of atoms that is characteristic of esters is

(a) $-C-NH_2$
(b) $-C-O-C-$
(c) $-\overset{\overset{O}{\|}}{C}-H$
(d) $-\overset{\overset{O}{\|}}{C}-O-$
(e) $-C-\overset{\overset{O}{\|}}{C}-C-$

_____ 16. The compound
$$H_2C \overset{CH_2-CH_2}{\underset{CH_2-CH_2}{\diagdown \diagup}} CHOH \text{ is}$$

(a) an alkane.   (b) a carboxylic acid.   (c) an alkyne.   (d) an alcohol.   (e) an amine.

_____ 17. Which of the following is a secondary alcohol?

(a) CH_3OH
(b) CH_3CH_2OH
(c) $CH_3-\underset{\underset{OH}{|}}{C}-CH_3$
(d) $CH_3-\underset{\underset{OH}{|}}{\overset{\overset{CH_3}{|}}{C}}-OH$
(e) $CH_3-\underset{\underset{OH}{|}}{\overset{\overset{CH_3}{|}}{C}}-CH_3$

_____ 18. The structures of malonic acid and fumaric acid are, respectively,

$$\overset{H}{\underset{HOOC}{\diagup}} C=C \overset{H}{\underset{COOH}{\diagdown}} \quad \text{and} \quad \overset{HOOC}{\underset{H}{\diagdown}} C=C \overset{H}{\underset{COOH}{\diagup}}$$

Which of the following terms describes the relationship between these two acids?
(a) a conjugate acid-base pair   (b) constitutional isomers   (c) geometrical isomers
(d) polymers   (e) dicarboxylic acids

_____ 19. The reaction of methane with chlorine in the presence of ultraviolet radiation can produce
(a) chloromethane.   (b) dichloromethane.
(c) trichloromethane.   (d) carbon tetrachloride.
(e) all of the products listed in the other four responses

_____ 20. The nitration of *n*-butane is an example of a(an) _____ reaction.
(a) aromatic   (b) addition   (c) elimination   (d) substitution   (e) saponification

_____ 21. The reaction by which tetrafluoroethylene is converted into teflon is called
(a) fluorination.   (b) vulcanization.   (c) hydration.   (d) addition.   (e) polymerization.

Answers to Preliminary Test

True-False

1. False
2. True
3. False. The carbon chains are not branched. In their most stable conformations, they lie along a zigzag arrangement.

4. False. In branched hydrocarbons, some carbon atoms may be bonded to only one hydrogen atom, or some even to no hydrogens. Can you write examples of each of these cases of saturated hydrocarbons?

5. False. Only the lightest four normal alkanes are gases at room temperature and atmospheric pressure. The liquefied propane or butane gas that is available in tanks at room temperature is under high pressure.

6. True. This is described in Section 27-1 for the three isomeric pentanes and the five isomeric hexanes.

7. False. They can exhibit only constitutional isomerism, in which the atoms are linked together in a different order.

8. True

9. False. They cannot exhibit geometrical isomerism, because the environment of the triple bond is linear (*sp* hybridization).

10. True

11. False. Each oil field produces oil with a characteristic composition.

12. False. Read the introduction to the portion of the chapter on aromatic hydrocarbons. Some have foul odors, others are odorless. The term is used to describe benzene, its derivatives, and other compounds that exhibit similar chemical properties. Can you describe the bonding feature that all aromatic compounds have in common? See Sections 27-5 and 27-6.

13. False. Most amines have quite unpleasant odors. Many common flavoring agents are organic esters.

14. True

15. True. Examples of classes of compounds so formed are esters, amides, acyl halides, and acid anhydrides.

16. False. Any potential reactive site on a molecule is termed a functional group. This can include carbon-carbon double or triple bonds.

17. False. Many organic compounds have several functional groups. You have seen several examples of this in Chapter 27.

18. False. Most organic chemicals are produces from petroleum. This may become a major problem due to decreasing petroleum supplies. Coal is another source of starting materials for organic chemicals.

19. True

20. False. Usually a mixture of products is formed, and often these are isomers of one another. A common method of separation is fractional distillation.

21. True

22. False. The source of the $-NO_2$ group is usually nitric acid vapor, HNO_3, at high temperature. Aromatic hydrocarbons may be nitrated by concentrated aqueous solutions of HNO_3.

23. True. Addition reactions are the characteristic reactions of the alkenes and the alkynes. See Section 27-17.

24. True

25. False. The π electrons of the aromatic hydrocarbons (one of which is benzene) are not available for addition reactions because of delocalization. The main reactions of aromatic hydrocarbons such as benzene are substitution reactions.

26. False. Hydrogenation reactions are addition reactions. See Section 27-17.

Short Answer

1. form covalent bonds to
2. normal
3. methylene group (CH_2); homologous series
4. increase; increase
5. conformations
6. sunlight (or an ultraviolet light source)
7. increases; decreases. See the comparisons in Section 27-1.
8. breaking of long-chain hydrocarbons into shorter-chain hydrocarbons
9. alkenes
10. -ene
11. -ane
12. -yne
13. 3; structural. Can you draw and name these three isomers?
14. 3. Can you draw and name them?
15. ethane; C_2H_6; alkane
16. benzene; C_6H_6; aromatic hydrocarbon
17. 2,2-dimethylpentane, C_7H_{16}, alkane
18. $CH_3CHCH_2CHCH_2CH_3$; C_8H_{18}; alkane
      | |
     CH_3 CH_3
19. 1-butene; C_4H_8; alkene
20. ; C_4H_8; alkene
21. ; C_4H_8; alkene
22. methylcyclopentane; C_6H_{12}; alkane (cycloalkane)
23. cyclohexane; C_6H_{12}; alkane (cycloalkane)
24. ; C_8H_{10}; aromatic hydrocarbon
25. ; C_8H_{10}; aromatic hydrocarbon
26. HC≡CH; C_2H_2; alkyne

345

27. HC≡CH; C_2H_2; alkyne. This is another name for acetylene. You should be aware of the common, or trivial, names for many of the simple compounds in this and the next chapter.

28. 2-butyne; C_4H_6; alkyne

29. ; C_6H_5Br; substituted aromatic hydrocarbon

For aromatic hydrocarbons, it is understood that there is one hydrogen atom at each corner. Sometimes they are shown, but it is not necessary to show them explicitly.

30. 4-ethyloctane, $C_{10}H_{22}$; alkane. Remember to name according to the longest chain of carbon atoms, and number so that the substituted group (ethyl, in this case) has the smallest position number possible.

31. ; C_5H_8; alkene (cycloalkene)

32. nitroethane; $C_2H_5NO_2$; substituted alkane

33. *cis*-1,2-dibromoethene; $C_2H_2Br_2$; substituted alkene

34. *trans*-1,2-dibromoethene; $C_2H_2Br_2$; substituted alkene

35.
$$H_3C \quad Br$$
$$C{=}C$$
$$Br \quad CH_3$$
; $C_4H_6Br_2$; substituted alkene

36.
$$F \quad F$$
$$C{=}C$$
$$F \quad F$$
; C_2F_4; substituted alkene

37.
$$F$$
$$|$$
$$Cl{-}C{-}Cl$$
$$|$$
$$F$$
; CF_2Cl_2; substituted alkane. There is only one isomer.

38. Br ; C_6H_5Br; substituted aromatic hydrocarbon

39. alcohol or phenol; ether

40. methyl alcohol; ethyl alcohol

41. polyhydric

42. phenols; phenol

43. primary; secondary; tertiary

44. 2-propanol (or isopropyl alcohol); CH_3CHCH_3
$$|$$
$$OH$$

45. dimethylamine; $CH_3-\overset{\underset{|}{H}}{N}-CH_3$

46. ethylene glycol; $HOCH_2-CH_2OH$; permanent antifreeze
47. disinfectants
48. 1-propanol or normal propyl alcohol; alcohol
49. ; phenol.

    ("Phenol" is the name of the class of compounds and also of the compound itself.)
50. benzoic acid; carboxylic acid
51. $CH_3CH_2CH_2CH_2\overset{\overset{O}{\|}}{C}-OH$; carboxylic acid
52. $CH_3CH_2\underset{\underset{Cl}{|}}{CH}CH_2\overset{\overset{O}{\|}}{C}-OH$; carboxylic acid. Note that this is not an acyl chloride.

53. acetyl chloride; acyl halide (or acid halide)
54. $CH_3CH_2CH_2\overset{\overset{O}{\|}}{C}-OCH_3$; ester
55. $\underset{\underset{HO}{|}}{H_2C}-\underset{\underset{OH}{|}}{C}-\underset{\underset{OH}{|}}{CH_2}$; alcohol (a polyhydric alcohol)

56. $CH_3\overset{\overset{O}{\|}}{C}CH_3$; ketone
57. acetaldehyde (or ethanal); aldehyde
58. methylethylamine; amine
59. ; amine

60. $CH_3-\overset{\overset{O}{\|}}{C}-\overset{\underset{|}{H}}{N}-CH_3$; amide
61. diethylether; ether
62. methylphenylether; ether
63. dimethylammonium chloride; an amine salt
64. $HO-\overset{\overset{O}{\|}}{C}-\overset{\overset{O}{\|}}{C}-OH$; carboxylic acid (actually a dicarboxylic acid)
65. acetic anhydride; acid anhydride
66. benzaldehyde; aldehyde
67. diphenylketone; ketone
68. substitution, addition, and elimination. You should be able to recognize each of these.
69. substitution. Can you describe such a reaction and write an example of it?
70. nitration; nitrobenzene
71. aliphatic substitution, halogenation
72. addition, halogenation

73. addition, hydrogenation
74. addition
75. polymerization
76. addition (actually two steps of addition, as is common for alkynes), halogenation
77. aromatic substitution, nitration
78. aliphatic substitution, halogenation

Multiple Choice

1. (b). Can you convince yourself by drawing them?
2. (d). There is no such compound as "2-pentane."
3. (c). Carbon never forms compounds using the hybrid orbitals indicated in answers (d) or (e) because it does not have any d orbitals in its valence shell.
4. (c)
5. (c). You may need to write out structural formulas for each of the answers and figure out the chemical formula (i.e., C_xH_y). Remember that isomers have the same chemical formulas.
6. (b)
7. (c)
8. (e)
9. (c)
10. (c)
11. (d)
12. (c)
13. (d)
14. (d)
15. (d)
16. (d)
17. (c). This is 2-propanol, or isopropyl alcohol. It is the substance familiarly known as "rubbing alcohol."
18. (c). Each one is a dicarboxylic acid, but this does not describe the *relationship* between the two compounds.
19. (e). See the discussion of halogenation reactions in Section 27-16.
20. (d)
21. (e)

28 Organic Chemistry II: Shapes, Selected Reactions, and Biopolymers

Chapter Summary

This chapter deals with three important aspects of organic chemistry. In Chapter 27, we learned about some isomeric organic compounds. In the first portion of this chapter, we undertake a more systematic study of the geometrical aspects of organic chemistry, a topic known as **stereochemistry**. As you study this subject, pay very close attention to the definitions that are presented on the blue screens. **Isomers** are substances that have the same number and kind of atoms, but with the atoms arranged differently. We introduce many kind of isomerism by first discussing examples of the simplest classes of organic compounds, the hydrocarbons; all classes of organic compounds can exhibit isomerism.

Section 28-1 discusses **constitutional isomers**, which were introduced in Sections 27-1 through 27-7. These differ in the *order* in which their atoms are bonded together. **Stereoisomers** (Section 28-2) contain atoms linked together in the same order, but with their arrangements in space are different. Geometrical isomers differ in the orientation of groups about a plane or direction. The *cis* and *trans* isomers of alkenes (Section 27-3) are examples of geometrical isomers. Similarly, groups can be in either a *cis* or *trans* arrangement across a ring. A second kind of stereoisomerism, optical isomerism, depends on the existence of nonsuperimposable mirror image forms (**enantiomers**) of some molecules. As we saw when we studied optical isomerism in coordination compounds (Section 25-7), the optical isomers are characterized by their ability to rotate the plane of polarized light. Section 28-3 discusses conformations of a compound, which differ from one another in the extent of rotation about one or more single bonds.

The next few sections discuss some other important organic reactions. We have already seen many examples of Brønsted-Lowry acid-base behavior (Section 4-9.1, Chapter 10). Many organic compounds can act as weak Brønsted-Lowry acids or bases (Section 28-4). The more important organic acids contain carboxyl groups, abbreviated as —COOH, but alcohols and phenols are also weakly acidic compounds. Cations of weak organic bases act as weak acids, in the same way that ammonium salts give acidic solutions (Chapter 18); substituted amines are usually weaker bases, so their cations are usually stronger acids. Likewise, the amines and the anions derived from weak organic acids are common organic bases.

We have encountered many oxidation-reduction (redox) reactions in our prior study; we have learned to recognize such reactions by tabulating changes in oxidation numbers. Section 28-5 discusses oxidation reduction reactions in organic chemistry. In organic chemistry, it is more convenient to talk about oxidation as increase in oxygen content or as decrease in hydrogen content, and reduction as decrease in oxygen content or increase in hydrogen content. Keep these conventions in mind as you study the oxidation and reduction transformations presented in this chapter. The latter part of this section discusses some important aspects of combustion reactions, the most extreme type of oxidation reaction of organic compounds.

In Section 27-14 we learned about several classes of compounds that are structurally related to the carboxylic acids by replacement of the —OH portion of the carboxylic acid functional group. In Section 28-6, we learn about some reactions that produce these derivatives -- acyl halides, esters, and amides. Section 28-7 describes a reverse reaction, the hydrolysis of an ester to produce the parent alcohol and a salt of the parent acid of the ester. One such reaction, saponification, produces soaps.

The final three sections of Chapter 28 present brief descriptions of three important classes of **biopolymers** (*bio*logical *polymers*). **Carbohydrates**, discussed in Section 28-8, are composed of **monosaccharides**, or simple sugars. A monosaccharide can contain either five or six carbon atoms, and can contain an aldehyde group (an **aldose**) or a ketone group (a **ketose**). Most naturally occurring carbohydrates contain more than one monosaccharide unit, linked by **glycosidic bonds**. Examples are amylose (a starch) and cellulose, which differ in the stereochemistry of the glycosidic link.

Polypeptides and **proteins** (Section 28-9) are polymers of amino acids linked by **peptide bonds**. Twenty different amino acids are found widely in nature, most often as the L optical isomer. Proteins are large polypeptides with molecular weights usually greater than 5000 g/mol. **Enzymes** are proteins that catalyze specific biochemical reactions.

Nucleic acids (Section 28-10) are more complex polymers that consist of three kinds of monomers—a phosphate group, a simple carbohydrate group, and an organic base. The two types of nucleic acids differ in the carbohydrate groups that they contain; **ribonucleic acid (RNA)** contains ribose, whereas **deoxyribonucleic acid (DNA)** contains 2-deoxyribose. The purine and pyrimidine bases in nucleic acids form pairwise highly specific patterns of hydrogen bonds. The double helix of DNA contains genetic information about the organism; this information is transmitted when living cells of the organism replicate.

Study Goals

1. Understand and give examples of various kinds of isomerism in organic chemistry: (a) structural isomers and (b) stereoisomerism (geometrical isomerism and optical isomerism) (Sections 28-1 and 28-2; Exercises 1 through 4 and 6 through 9)
2. Describe different conformations of organic compounds. (Section 28-3; Exercises 5, 11, and 12)
3. Know the characteristics of some important types of organic reactions: (a) acid-base, (b) oxidation-reduction, (c) esterification, and (d) hydrolysis. Know what classes of organic compounds can undergo each type of reaction. (Sections 28-4 through 28-7; Exercises 13 through 35)
4. Know what classes of organic compounds can act as acids and what can act as bases. Understand ranges in strengths of acidic or basic behavior. (Section 28-4; Exercises 13 through 22)
5. Recognize oxidation and reduction reactions of organic compounds. (Section 28-5; Exercises 23 through 28)
6. Recognize reactions that form derivatives of carboxylic acids, and predict products of such reactions. (Section 28-6; Exercises 29 and 33)
7. Describe organic hydrolysis reactions and their products. (Section 28-7; Exercises 34 and 35)
8. Describe the main features of three classes of biopolymers: (a) carbohydrates; (b) polypeptides and proteins; and (c) nucleic acids. Describe the monomeric units and the types of reactions that form these biopolymers. (Sections 28-8 through 28-10; Exercises 36 through 50)

9. Apply the concepts of this chapter to answer questions of unidentified type. (Mixed Exercises, 51 through 54)
10. Answer conceptual questions based on this chapter. (Conceptual Exercises 57 and 58)
11. Apply concepts and skills from earlier chapters to the ideas of this chapter. (Building Your Knowledge exercises, 59 through 62)

Some Important Terms in This Chapter

Write the meanings *in your own words*. Check the Key Terms list and the chapter reading. Then rewrite your definitions, still in your own words, to improve them. There are many additional new terms in this chapter; study others listed in the Key Terms

stereochemistry

constitutional isomers

stereoisomers

geometrical isomers

cis

trans

optical activity

optical isomers

chiral

achiral

racemic mixture

conformation

Brønsted-Lowry acid

Brønsted-Lowry base

oxidation (of an organic compound)

reduction (of an organic compound)

combustion reaction

saponification reaction

hydrolysis

carbohydrate

monosaccharide

glycosidic bond

oligosaccharide

polysaccharide

peptide

polypeptide

protein

enzyme

nucleic acid

purine base

pyrimidine base

complementary base-pairing

deoxyribonucleic acid

ribonucleic acid

Preliminary Test

Be sure to practice *many* of the additional textbook exercises.

True-False

_____ 1. The different atomic arrangements obtained by rotation around a carbon-carbon single bond represent different compounds.

_____ 2. The different atomic arrangements obtained by rotation around a carbon-carbon double bond represent different compounds.

_____ 3. Saturated hydrocarbons cannot exhibit geometrical isomerism.

_____ 4. Reaction of an organic compound with O_2 is usually an exothermic reaction.

_____ 5. Alkoxides of low molecular weight are strong bases.

_____ 6. Ethers are very reactive compounds.

_____ 7. Some of the reactions of alcohols are similar to those of water.

_____ 8. Many reactions of carboxylic acids involve displacement of the –OH group by another atom or group of atoms.

_____ 9. Most carboxylic acids are weak acids.

_____ 10. The conversion of an alcohol to a ketone is an oxidation.

_____ 11. The conversion of an alcohol to an aldehyde is an oxidation.

Short Answer

1. Compounds that differ only in the order in which their atoms are bonded together are called _____.

2. Compounds in which the atoms are linked together in the same order but with different arrangements in space are called _____.

3. Compounds that differ only in the spatial orientation of atoms or groups about a plane or direction are called _____.

4. A compound that is *not* superimposable with its mirror image is said to be _____.

5. An equimolar solution of D-glucose and L-glucose is called a _____ mixture.

6. The eclipsed and staggered forms of ethane are different _____.

7. When a hydrocarbon undergoes complete combustion the two products are _____ and _____.

8. The reaction in which the net result is the addition of water to an alkene is a common method for the preparation of _____. (The answer is the name of a class of compounds, not a specific compound.)

9. In a process that is analogous to the reaction of active metals with water to form hydroxides, active metals react with alcohols to form _____; in both of these reactions, the gas _____ is liberated.

10. The reactions of alcohols with organic acids result in the formation of _____.

11. The hydrolysis of esters in the presence of strong soluble bases, a process which (with some esters) produces soaps, is called _____.

12. Aldehydes can be prepared by the oxidation of _____ alcohols, whereas ketones can be prepared by the oxidation of _____ alcohols.

- -

For each of the following questions, 13 through 20, classify the reaction according to one (or sometimes more) of the following types: neutralization, displacement, esterification, saponification, combustion, oxidation, reduction, hydrolysis.

13.
$$
\begin{array}{l}
CH_2OH \\
|\\
CHOH \;+\; 3HNO_3 \;\rightarrow\; \\
|\\
CH_2OH
\end{array}
\qquad
\begin{array}{l}
CH_2\!-\!ONO_2 \\
|\\
CH\!-\!ONO_2 \;+\; 3H_2O \\
|\\
CH_2\!-\!ONO_2
\end{array}
$$
_____

14. $CH_3\!-\!\overset{O}{\overset{||}{C}}\!-\!OCH_2CH_3 \;+\; NaOH \;\rightarrow\; CH_3\!-\!\overset{O}{\overset{||}{C}}\!-\!O^-, Na^+ \;+\; CH_3CH_2OH$

_____

15. $3C_2H_4 \;+\; KMnO_4 \;+\; 8H_2O \;\rightarrow\; 3CH_2\!-\!CH_2 \;+\; 2MnO_2 \;+\; 2KOH$
$$
\qquad\qquad\qquad\qquad\qquad\qquad\quad \underset{OH\;\;\;OH}{|\;\;\;\;|}
$$

_____

16. $CH_3CH_2OH \;+\; O_2 \;\xrightarrow{\;Cu\;}\; 2CH_3\overset{O}{\overset{||}{C}}H \;+\; 2H_2O$

_____

17. $CH_3\!-\!\underset{\underset{OH}{|}}{CH}\!-\!CH_3 \;\xrightarrow[\text{heat}]{\;Cu\;}\; CH_3\overset{O}{\overset{||}{C}}CH_3$

_____

18. $2CH_3CH_2CH_2\overset{O}{\overset{||}{C}}H \;+\; O_2 \;\rightarrow\; 2CH_3CH_2CH_2\overset{O}{\overset{||}{C}}\!-\!OH$

_____

19.

_____

20. $2C_8H_{18} \;+\; 25O_2 \;\rightarrow\; 16CO_2 \;+\; 18H_2O$

_____

355

Multiple Choice

_____ 1. Which one of the following *cannot* exist in isomeric forms?
(a) C_3H_8 (b) C_4H_{10} (c) C_5H_{12} (d) C_6H_{14} (e) C_7H_{16}

_____ 2. Vegetable oils can be converted to fats by _____ in the presence of a catalyst under high pressures and at high temperatures.
(a) substitution   (b) addition   (c) elimination   (d) vulcanization   (e) hydrogenation

_____ 3. Members of which of the following classes of compounds generally act as the weakest bases?
(a) alkoxides   (b) aliphatic amines   (c) phenoxides   (d) aromatic amines
(e) None of the other answers is correct, because organic compounds cannot be bases.

_____ 4. The reaction of propanoic acid with _____ produces propionyl chloride.
(a) sodium chloride   (b) hydrochloric acid   (c) hypochlorous acid
(d) chlorine   (e) phosphorus pentachloride

_____ 5. The reaction of 1-propanol with $K_2Cr_2O_7$ in acidic solution produces _____.
(a) propyne   (b) propene   (c) propanal   (d) propanone   (e) cyclopropanone

_____ 6. The reaction of methylamine with hydrochloric acid produces _____.
(a) ammonia   (b) ammonium methylate   (c) methylammonium chloride
(d) chloromethane   (e) chloromethylamine

> You will need much more practice in recognizing types of reactions and predicting their products. Answer many of the Exercises from the end of the textbook chapter.

Answers to Preliminary Test

True-False

1. False. These are just different conformations of the same compound. See Section 28-3.
2. True. These are geometrical isomers. See Sections 27-3 and 28-2.
3. False. Substituted cycloalkanes can exist as *cis* and *trans* isomers. Figures 28-1 and 28-2 show examples of this.
4. True. Remember that the term exothermic means that the process liberates heat. Another way of making the statement that appears in this question is "Most organic compounds burn." The production of heat energy is sometimes the reason we carry out these reactions.
5. True. They hydrolyze (react with water) to form the parent alcohol and a strong soluble base. This reaction is described in Section 28-7.
6. False. They burn very readily but otherwise are not particularly reactive.
7. True
8. True. Examples of classes of compounds so formed are esters, amides, acyl halides, and acid anhydrides. See Section 28-6.
9. True. See, for instance, the K_a values for common carboxylic acids, listed in Section 28-7.
10. True. See Section 28-5.
12. True. See Section 28-5.

Short Answer

1. structural isomers. See the definition at the beginning of Section 28-1.
2. stereoisomers. See the definition at the beginning of Section 28-2.
3. geometrical isomers.
4. chiral
5. racemic
6. conformations. These are *not* isomers. Be sure you understand the distinction. Study Section 28-3.
7. CO_2; H_2O
8. alcohols
9. alkoxides; hydrogen (H_2).
10. esters
11. saponification
12. primary; secondary
13. esterification (inorganic)
14. saponification, hydrolysis of an ester
15. oxidation, addition
16. oxidation, dehydrogenation
17. oxidation, dehydrogenation
18. oxidation
19. oxidation
20. oxidation (combustion, complete combustion)

Multiple Choice

1. (a)
2. (e)
3. (d). See the ranking of base strengths at the end of Section 28-4.
4. (e). This method of forming acyl chlorides is described at the beginning of Section 28-6.
5. (c). Oxidation of a primary alcohol under the conditions stated produces an aldehyde. See Section 28-5, under the heading "Oxidation of Alcohols."
6. (c). This is a Brønsted-Lowry acid-base reaction involving the organic base methylamine and the strong acid HCl. See Section 28-4.